A. Reza Hoshmand

EXPERIMENTAL RESEARCH DESIGN and ANALYSIS

A Practical Approach for Agricultural and Natural Sciences

CRC Press

Boca Raton   Ann Arbor   London   Tokyo

Library of Congress Cataloging-in-Publication Data

Hoshmand, A. Reza.
    Experimental research design and analysis : a practical approach for agricultural
and natural sciences / A. Reza Hoshmand.
      p.   cm.
    Includes bibliographical references and index.
    ISBN 0-8493-8635-7
    1. Agriculture—Experimentation—Statistical methods.
2. Experimental design. I. Title.
S540.S7H67   1993
630'.724—dc20   93-4657
                                                               CIP

No claim to original U.S. Government works
International Standard Book Number 0-8493-8635-7
Library of Congress Card Number 93-4657
Printed in the United States of America 1 2 3 4 5 6 7 8 9 0
Printed on acid-free paper

PREFACE AND ACKNOWLEDGMENTS

This book is for students and applied researchers of agricultural and natural sciences. It is intended to guide them in the design and analysis of their agricultural and natural science experiments. The book emphasizes the logical underpinnings of design and analysis, with the hope that the reader can extend those principles beyond the specifics discussed herein. In presenting the various designs, the reasons for selecting a particular design and its analysis are discussed. Practical examples from different areas of agriculture are given throughout, to show how practical issues of design and analysis are handled.

The reader is assumed to be familiar with statistical concepts such as the normal, binomial, chi-square, and t distributions. No previous exposure to analysis of variance is assumed. Since this is meant to be a basic experimental design book rather than a reference book of design and analysis, a number of advanced topics have not been addressed.

Regarding the book's organization: The first three chapters discuss the practical issues of design, assumptions of the analysis of variance, and the problems associated with the data in agricultural research; Chapters 4, 5, and 6 present the experimental design and analysis of 1-, 2-, and 3-factors; Chapter 7 discusses the concept of treatment means comparisons; Chapter 8 presents the design and analysis of experiments over time, and Chapter 9 covers regression and correlational analysis.

In the process of writing, I have attempted to provide a high degree of readability within the book. First, I use a step-by-step process to explain the concept of design and analysis. Second, I number the equations consecutively in each chapter as they are introduced. If the same equation is repeated at a later point within the chapter, the original equation number appears.

I wish to gratefully acknowledge the assistance provided by Wayne Yuhasz of CRC Press in the completion of this book. A special note of gratitude is due my wife Lisa, who not only provided support, but editorial assistance in the preparation of the manuscript. Finally, I wish to thank Susan Littell Smith for her clerical assistance.

A. Reza Hoshmand, Ph.D.
College of Agriculture
California State Polytechnic University
Pomona, CA

To my wife Lisa and
my children Anthony and Andrea

TABLE OF CONTENTS

Appendices

Chapter 1

THE NATURE OF AGRICULTURAL RESEARCH

1.1   FUNDAMENTAL CONCEPTS

Agriculturalists and other scientists in the biological fields who are involved in research constantly face problems associated with planning, designing, and conducting experiments. Basic familiarity and understanding of statistical methods that deal with issues of concern would be helpful in many ways. Practitioners (researchers) who collect data and then look for a statistical technique that would provide valid results will find that there may not be solutions to the problem and that the problem could have been avoided in the first place by a properly designed experiment. Obviously, it is important to keep in mind that we cannot draw valid conclusions from poorly planned experiments. Secondly, the time and cost involved in many experiments are enormous, and a poorly designed experiment increases such costs in time and resources. For example, an agronomist who carries out a fertilizer experiment knows the time limitations on the experiment. He/she knows that seeds must be planted in the spring and harvested in the fall. The experimental plot must include all the components of a complete design. Otherwise, what is omitted from the experiment will have to be carried out in subsequent trials in the next cropping season or the next year. This additional time and expenditure could be minimized by a properly planned experiment that will produce valid results as efficiently as possible.

Good experimental designs are products of the technical knowledge of one's field, an understanding of statistical techniques, and skill in designing experiments. Any research endeavor, be it correlational or experimental, may entail the following phases: conception, design, data collection, analysis, and dissemination. Statistical methodologies can be used to conduct better scientific experiments if they are incorporated into the entire scientific process, i.e., from the inception of the problem to experimental design, data analysis, and interpretation. The intent of this book is to provide practitioners with the necessary guidelines and techniques for designing experiments, as well as the statistical methodology to analyze the results of experiments in an efficient way.

Agricultural experiments usually entail comparisons of crop or animal varieties. When planning agricultural experiments, we must keep in mind that large uncontrolled variations are common occurrences. For example, a crop scientist who plants the same variety of a crop in a field may find variations in yield that are due to periodic variations across a field or to some other factor that the experimenter has no control over. Throughout this book we are concerned with the methodologies used in designing experiments that will separate, with confidence and accuracy, varietal differences of crops and animals from the uncontrolled variations.

It is essential that you become familiar with the terminology used in this book as we discuss the concept of experimentation. Agricultural experimentations are in response to questions raised by researchers who are interested either in comparing the effects of several conditions on some phenomena or in discovering an unknown effect of a particular process. An experiment facilitates the study of such phenomena under controlled conditions. Therefore, creation of controlled conditions is the most essential characteristic of experimentation. It has been said that wisdom consists not so much in knowing the right answers, as in knowing the right questions to ask (Gill, 1978). Hence, how we formulate our questions and hypotheses are critical to the experimental procedure that will follow.

Once we establish an hypothesis, we then look at a method to objectively test the validity of that hypothesis. Our final results and conclusions depend to a large extent on the manner in which the experiment was statistically designed and how the data were collected. Agricultural researchers know that it is difficult to avoid differences in yield of the same crop variety planted in two adjacent fields, or the weight gain in two animals fed the same ration, because of the uncontrolled variations. Such differences in yield resulting from experimental units treated alike are called *experimental* error. The intent of an optimal design is to provide a mechanism for estimation and control of experimental error in the field experiments conducted. Practitioners should keep in mind that it is very difficult to account for all the sources of natural variation even when an entire population of factors of interest are under study. The problem is further aggravated when we depend on a sample that is at best an approximation of the population or the characteristic of interest. Given this dilemma, what practitioners can hope for is a predictive model that minimizes experimental error. Such models are based on theoretic knowledge, empirical validity, and an understanding of experimental material. Before we discuss the various designs and any methodology for estimation and control of experimental error, a quick review of other commonly encountered terms is in order.

When discussing experimentation, you will encounter the term *experimental unit*. An experimental unit is an entity that receives a treatment. For example, for a horticulturist it may be a plot of land or a batch of seed; for an animal scientist it may be a group of pigs or sheep, and for an agricultural engineer it may be a manufactured item. Thus, an experimental unit may be looked upon as a small subdivision of the experimental material, which receives the treatment.

In choosing an experimental unit, the researcher must pay special attention to the size of the unit, the representative nature of the unit, and how independent the units are from one another. In terms of the size of the unit, technical and cost factors play a major role. The essential point to remember is how many units are needed to attain a specified precision in a most economical way.

So far as the representative nature of the unit is concerned, it is important that the conditions of the experiment be as representative as those to which the results are to be applied. For example, if an animal scientist is interested in a feeding experiment where the results are to apply to Holstein dairy cows, then ideally the sample of cows in the experiment should be selected from a population of Holstein dairy cows.

The independence of the experimental units from one another is also an important element of a properly conducted experiment. Independence of units implies that the researcher must ensure that the treatment applied to one unit has no effect on the observation obtained in another unit. Furthermore, occurrence of unusually high or low observation in one unit has absolutely no effect on what may be observed in another unit.

Replication refers to the repetition of the basic experiment or treatment. Thus, an animal scientist who treats 10 animals with a particular antibiotic has 10 replicates. Researchers replicate their experiments to control for experimental error. It is also used to estimate the error variance and to increase precision of estimates. It is important to keep in mind that increased replication of experimental units is desirable for increasing the accuracy of estimates of means and other functions of the measured variable (Gill, 1978; Das and Giri, 1986). However, the cost of increased replication plays an important role in the number of replicates a researcher is able to have.

As the experimenter makes a decision to use a particular design, the choice of the number of replicates must also be made. It has been suggested that present-day design practices utilize existing information to the fullest, and replication is sometimes not needed at all (Anderson and McLean, 1974). In some cases, such as the factorial fractional replication (discussed in Chapter 6, Section 6.4), only a part of all the treatment combinations are used in an experiment.

No matter how many replicates are used or not used, the important point to keep in mind is whether the experiment will provide valid results in terms of the estimates of the effects, and whether there are enough degrees of freedom for the error, to adequately test for the various effects.

Randomization is simply an objective method of random allocation of the experimental material or allocation of treatment to experimental units. We also use randomization to make certain that the order in which the individual trials of experiments are performed are determined randomly. The advantages of randomization are (1) it allows for protection against systematic error caused by subjective assignment of treatments, that is, each treatment will have an equal chance of being assigned to an experimental unit; (2) it helps in "averaging out" the effects of uncontrolled conditions or extraneous factors that persist over long or repeated experiments; and (3) it validates the statistical assumption that states that observations (or errors) are independently distributed random variables. In sum, we could say that randomization is the cornerstone of statistical theory in the design of experiments. Fisher (1956) stated that to ensure that the error estimate will be a good and valid one, we must randomize experimental treatments among experimental units. For further reading on this topic, Fisher (1956), Ogawa (1974), Gill (1978), Gomez and Gomez (1984), and Mead (1988) may be consulted.

We shall refer to the experimental variables as *factors*, or we may define a controllable condition in an experiment as a *factor*. For example, a fertilizer, a new feed ration, and a fungicide are all considered factors. Factors may be quantitative or qualitative and may take a finite number of values and types. Quantitative factors are those described by numerical values on some scale. Strength of a drug dosage (such as milligrams of a sulfa drug), the rate of application of a fertilizer, and temperature are examples of quantitative factors. Qualitative factors, such as type of protein in a diet, sex of an animal, age, or genetic makeup of a plant or animal are those factors that can be distinguished from each other, but not on a numerical scale.

Different factors are included in experimental designs for different reasons. The decision on including or excluding a factor, the levels of each factor in the experiment, and the criterion for selecting such factors are the responsibility of the experimenter who may be influenced by theoretical considerations. What is essential to remember is that factors should be related to one another in simple ways. The relation of one factor to another implies how the levels of that factor are combined with the levels of another factor. When choosing factors for any experiment, the researcher should ask the following questions:

1. What treatments in the experiment should be related directly to the objectives of the study?
2. Does the experimental technique adopted require the use of additional factors?
3. Can the experimental unit be divided naturally into groups such that the main treatment effects are different for the different groups?

4. What additional factors should one include in the experiment to interact with the main factors and shed light on the factors of direct interest?
5. How desirable is it to deliberately choose experimental units of different types?

A *treatment* in a broad sense refers to a controllable quantitative or qualitative factor imposed at certain levels by the experimenter. For an agronomist, several fertilizer concentrations applied to a particular crop or a variety of a crop is a treatment. Similarly, an animal scientist looks upon several concentrations of a drug given to an animal species as a treatment. In agribusiness, we may look upon the impact of an advertising strategy on sales as a treatment. To an agricultural engineer, different levels of irrigation may constitute a treatment. It should be kept in mind that a treatment is not a condition of the experiment unless the design consists only of treatment factors. It is because a condition, by definition, requires 1 level from every design factor for its specification (Lee, 1975). When there is but 1 criterion for classifying the imposed factor or condition, the levels of a factor may also be referred to as treatments.

Frequently, researchers include a treatment factor in their design not because they are interested in the treatment factor per se, but because the factor might prove to have importance, or the researcher wishes to control the effects of unavoidable variations in the experiment. For example, if time of day has an effect on taking blood samples of an animal, even though the experimenter had no real interest in studying such an effect, "time of the day" may be included in the design so that the effects would not enter in a haphazard way. Factors whose purpose is to systematically take into account the effects of extraneous variables are called *control factors*.

The *levels* of a factor refer to its presence or absence in an experiment. Another way of stating the level of a factor is the number of ways in which a factor is varied. So when a factor takes on different values, such as a high and a low dose of an antibiotic or 2 modes of an application such as drip or sprinkler irrigation, we are referring to these as the level of the factor (independent variable) of interest. Similarly, when a factor is varied in 3 ways, we refer to it as having 3 levels. In experiments where there are 2 or more factors, each with 2 or more levels, a treatment consists of a combination of 1 level of each factor. When all possible different combinations of 1 level from each factor, along with an equal number of observations for each treatment, are included in an experiment, we have a complete factorial experiment with equal replications.

Blocking refers to a methodology that forms the units into a homogeneous or pre-experimental subject-similarity groups. It is a method to reduce the effect of variations in the experimental material on the error of the treatment comparison (Dodge, 1985). For example, an animal scientist may decide to group animals based on age, sex, breed, or some other factor that he/she may believe has an influence on the characteristic being measured. Effective blocking removes a considerable measure of variation from the experimental error. A researcher must keep in mind that the selection of the source of variability to be used as the basis for blocking, as well as block shape and orientation, are crucial for effective blocking technique (Gomez and Gomez, 1984). Experimenters include a blocking factor in their design, to investigate observational differences among blocks and to increase the power of the design to detect treatment effects.

The last 3 concepts, namely *replication, randomization,* and *blocking,* are quintessential parts of experimental designs. Each of these concepts is explained fully as different designs are introduced in the subsequent chapters.

The importance of good design is inseparable from good research (results). The following examples point out the necessity for a good design that will yield good research. First, a nutrition specialist in a developing country is interested in determining whether mother's milk is better than powdered milk for children under age one. The nutritionist has compared the growth of children in village A, who are all on mother's milk, against the children in village B, who use powdered milk. Obviously, such a comparison ignores the health of the mothers, the sanitary conditions of the villages, and other factors that may have contributed to the differences observed without any connection to the advantages of mother's milk or the powdered milk on the children. A proper design would require that both mother's milk and the powdered milk be alternatively used in both villages, or some other methodology to make certain that the differences observed are attributable to the type of milk consumed and not to some uncontrollable factor. Second, a crop scientist who is comparing 2 varieties of maize, for instance, would not assign one variety to a location where such factors as sun, shade, unidirectional fertility gradient, and uneven distribution of water would either favor or handicap it over the other. If such a design were to be adopted, the researcher would have difficulty in determining whether the apparent difference in yields was due to varietal differences or resulted from such factors as sun, shade, soil fertility of the field, or the distribution of water. These 2 examples illustrate the type of poorly designed experiments that are to be avoided.

1.2 RESEARCH BY PRACTITIONERS

Researchers in the different disciplines within the food and fiber system are commonly concerned about problems and their solutions which will have implications for research, production, nutrition, handling, processing, marketing, and distribution, just to name a few. Many of the experiments in agriculture tend to be comparative and correlational in nature. That is, an agronomist or an animal scientist would compare a number of varieties of some crops or breeds of animals respectively under different treatment conditions to determine whether the yield/weight gain of one variety/breed is superior to the other and, if so, by how much. The determination of superiority of a crop variety or an animal breed is based on the statistical significance attached to the results. In Chapters 4, 5, 6, and 7, various designs are discussed in which tests are performed to determine the varietal differences. Correlational studies elaborate on the degree of association between variables, such as how the yield of a crop is affected by the application of fertilizer and the amount of rainfall. Correlational analysis (discussed in Chapter 9) allows the researcher to study the association between variables, alone or in combination.

The objectives of a study set in motion the plan to conduct an experiment, which in turn requires an experimental design that fits the objectives of the experiment. The experimenter has at his/her disposal a variety of experimental designs and techniques to obtain answers to research questions. The choice of the design determines the parameters of the experiment. This then allows the experimenter to make a number of technical decisions, such as on the size of the experiment, the manner in which treatments are allotted to the experimental units, and the grouping of the experimental units, etc., using in each case the fundamental concepts discussed in the previous section. Since there are many designs (from the very simple to the highly complex) available to researchers, the critical point to consider is which design will be suited for a particular experiment and how *efficient* the design is in providing answers to research questions, given limited resources.

Once a design is selected, the choice and the level of treatments in the experiment, the number of replications to be used, and how the analysis is to be performed are all critical planning elements. Improper attention to these factors will have major implications for how the experiment is conducted and analyzed. To illustrate the importance of the analysis in the design of experiments, consider an experiment in which an animal scientist is comparing a new sulfa drug (B) with an established drug (A) the performance of which has already been tested. The animal scientist argues that the emphasis should be placed on the new drug and feels it is not necessary to reexamine drug A as fully. Thus, in comparing drugs A and B, the investigator subjects 19 of the 20 pigs in the experiment to the new drug and only 1 to the established drug. In performing the analysis, the difference between the mean response to the drugs is tested, that is, $(\bar{x}_A - \bar{x}_B)$, and the standard error of the difference computed. The standard error of the difference for the samples of 1 and 19 is $\sigma\sqrt{(\frac{1}{1} + \frac{1}{19})} = 1.023\sigma$. To make the point clear, consider the situation where the experimenter appropriately designs the experiment, testing both drugs under an identical situation. This means that the sample of 20 pigs was equally divided between drugs A and B. The standard error for this design is $\sigma\sqrt{(\frac{1}{10} + \frac{1}{10})} = 0.447\sigma$. Notice the difference in the standard errors, under the 2 conditions. The smaller standard error of the second design, a highly desirable characteristic of any good experiment, improves the precision of the experiment by a factor of more than 2. Obviously, lack of consideration for statistical analysis can also lead to poorly designed experiments.

In this section a few other selected examples are given to illustrate the common types of comparisons made in agriculture and how practitioners design experiments to obtain answers to comparison questions.

Example 1.2.1

A crop scientist is interested in determining whether varieties of rice respond differently to a fertilizer treatment. In this particular instance the research question is simply about the varietal differences. For such a comparative study, often the researcher performs a field trial in which the experimental area is divided into plots and the different varieties are assigned, one to each plot. The yield is then measured or estimated for each plot, and varieties are thus compared. In making a comparison such as this, the researcher has to be cautious with respect to several factors that contribute to differences in yield. Yield differences due to fertility gradient of the soil, systematic trends, or difference in space and time may be of significant proportion. Research has shown that yields may vary from plot to plot by as much as 20 to 30% from their mean (Cox, 1954). Such a variation is considerable when a difference of a small percentage such as 5 to 10% may have practical implications for the adoption of a variety.

From a design point of view, an experiment such as this one would have to be arranged in such a way that the varietal differences are accurately separated from the uncontrolled variations mentioned above. As the aim of an experiment such as this one is to find out which variety is superior, rather than the absolute determination of the yield per acre, it would be economical to make a direct comparison of varieties, using separate experiments for each variety, and then compare the mean yield of each variety under representative conditions. However, if the researcher is interested in the absolute determination of the yield per acre for each variety, the design will be different than the one conducted by the crop scientist.

Example 1.2.2

In another experiment, a researcher is interested in determining whether hybrids or plant densities interact with the effect of 15-in. row spacing on grain yield and stalk breakage of corn grown under nonirrigated conditions. To get answers to his research questions, the researcher conducts a $2 \times 2 \times 4$ factorial experiment, replicated 3 times in a randomized complete block design arranged in a split-split-plot layout (this is discussed more fully in Chapter 6).

You will note that in this experiment there are 3 factors, namely, 2 corn hybrids (factor 1), 2 row spacings (factor 2), and 4 target plant densities (factor 3). As the emphasis in this experiment is on the interactions, the layout of the experiment would be such that hybrids are allocated to the main plots, row spacings to the subplots, and plant densities to the sub-subplots. In this experiment there are 3 levels of precision, the main plot factor receiving the lowest degree of precision, and the sub-subplot receiving the highest degree of precision. The design chosen for this experiment correctly places emphasis on the interactions rather than on the main effects. Stated differently, the present design sacrifices precision on the main effects in order to provide higher precision on the interactions.

Example 1.2.3

When plant breeders are interested in selecting qualitative traits, nursery tests can be conducted very easily. However, with complex traits such as yield, which is influenced by environment, the study should be conducted in several localities. Practitioners can use various techniques to evaluate genotype stability over a range of environmental conditions, using many crops. The most widely used technique is the regression analysis (Chapter 9 in this book), which was proposed by Yates and Cochran (1938), amplified by Finley and Wilkinson (1963), and further refined and adopted by other practitioners (Eberhart and Russell, 1966; Shukla, 1972; Ntare and Aken'Ova, 1985).

Other techniques such as genotype grouping, which does not use the regression analysis, have also been used in yield-stability studies (Francis and Kannenberg, 1978).

Example 1.2.4

A group of crop scientists (Peterman, Sears, and Kanemasu, 1985) was interested in determining genotypic differences in the rate and duration of spikelet initiation in 10 winter wheat cultivars. To study the differences among cultivars, the experiment was conducted on 10 cultivars with varying phenotypes. The cultivars were planted at 2 dates with 4 replications each. The scientists were interested in making 3 comparisons regarding these cultivars. Specifically, they compared the rate and duration of spikelet initiation among benchmark cultivars, between semidwarf and standard-height wheats, and between semidwarf pure lines and hybrids.

The experimental design for this study was a randomized, complete block (discussed in Chapter 4). Analysis of variance and least significant difference (Chapter 7, Section 7.2.1) was used to determine significant differences. The study also included a regression analysis (Chapter 9) to evaluate the correlation of interest.

Example 1.2.5 _____

In a study to evaluate soybean yield response to sprinkler irrigation (center pivot) management strategies, specifically those scheduled by growth stage vs. one scheduled by soil moisture depletion in different row widths, the scientists were interested in identifying the yield component mechanisms that respond to the sprinkler irrigation strategies and in evaluating the effect of sprinkler irrigation on reproductive organ abscission (Elmore, Eisenhauer, Specht, and Williams, 1988).

This experiment was conducted over a 3-year period in a research station in Nebraska. The scientists used a split-split-plot randomized complete block experimental design (see Chapter 6, Section 6.2) with 4 replications. The main plots were irrigation treatments and included a nonirrigated check, irrigation scheduled by soil moisture depletion, irrigation commenced no earlier than flowering, and irrigation commenced no earlier than pod elongation. The researchers limited the application of water to 1.5 in./week in order to simulate a minimum-capacity center pivot system. The subplots consisted of 10- and 30-in. row widths, and were 30 ft long. The sub-subplots were planted with the 3 cultivars of soybean. To further minimize error, the harvest areas within the subplots were limited to 20 ft long and consisted of the center 2 of the 4 rows that are 30 in. wide, and the center 6 of the 10 rows that are 10 in. wide.

Standard analyses of variance for a split-split plot design were used. The single degree of freedom comparisons (see Chapter 7, Section 7.3.1) were made for 1982 vs. 1983, 1984; and 1983 vs. 1984. The least significant difference (LSD) was used for mean comparisons.

Example 1.2.6 _____

In determining the yield difference due to the application of P and K on kale, a horticulturist conducts a randomized complete block design experiment with 4 replications and 4 treatments. She is particularly interested in the following questions: (1) Is there a difference in yield between the fields receiving P and those receiving K? (2) Are there yield differences between the fields receiving P and K separately and those receiving P and K together? (3) Is there a response to fertilization at all? In response to these questions, the horticulturist designs the following set of treatments: (1) control, (2) P added to the plots, (3) K added to the plots, and (4) both P and K added to the plots.

This experiment is exactly the same as the one given in Chapter 7, Section 7.3.1. The reader is referred to this section of the book for the approach used to obtain answers to the researcher's questions.

Example 1.2.7

Agronomists have hypothesized that the successful establishment of legumes in tall fescue (*Festuca arrundinacea* Schreb.) pasture is limited by the amount of light that can be transmitted through the grass canopy. To ascertain the proportion of incident light transmitted to the ground through canopies of various biomass, the agronomists conducted a study to quantify the relationship between leaf area index (LAI), canopy biomass, and the light penetration into the canopy of an established tall-fescue pasture (Trott, Moore, Lechtenberg, and Johnson, 1988).

Regression analysis was used in this particular study, as it was an objective of the study to determine the relationship between canopy biomass and leaf area.

Example 1.2.8

Animal researchers were interested in determining if the quantity of beef produced per acre could be increased by applying N fertilizer at 0, 100, and 200 lb/acre to tall fescue-ladino clover pastures or by feeding a supplemental grain-protein mix (creep feed) to calves from mid-November to mid-April (Morrow et al., 1988).

The experimental design used for this study was a randomized complete block with 2 replications, where treatments were 0, 100, and 200 lb of elemental nitrogen per acre and creep feed vs. no creep feed for the calves, for a total of 12 pastures (experimental units).

The researchers used 6 test cows and 1 or 2 substitute cows. The test cows were randomly assigned to each of the 12 pastures. Each cow remained on her assigned pasture system for the entire year and received no supplemental feed. Substitute cows were grazed with the test cows and were used for replacement if a test cow lost a calf or failed to breed back during the cycle. Self-feeders were used for creep feed treatments.

In performing the analysis of variance, the sources of variation tested were on pasture replications (2), sex of calf (2), creep feeding (2), years or environment (4), levels of nitrogen fertilization (3), and all the interactions. The researchers used the LSD test for mean comparisons.

The above examples are only a small sample of the variety of research questions that are addressed by researchers in the various fields of agriculture. In the subsequent chapters of this book, you will be introduced to a number of research designs that are often used by practitioners in agriculture.

REFERENCES AND SUGGESTED READINGS

Anderson, V. L. and McLean, R. A. 1974. *Design of Experiments: A Realistic Approach*. New York: Marcel Dekker.

Cochran, W. G. 1954. "The combination of estimates from different experiments." *Biometrics* 10:101-129.

Cochran, W. G. and Cox, G. M. 1957. *Experimental Designs*. 2nd ed. New York: John Wiley & Sons.

Cox, D. R. 1954. "The design of an experiment in which certain treatment arrangements are inadmissible." *Biomatrika* 41:287.

Cox, D. R. 1958. *Planning of Experiments*. New York: John Wiley & Sons.

Das, M. N. and Giri, N.C. 1986. *Design and Analysis of Experiments*. 2nd ed. New York: John Wiley & Sons.

Dodge, Y. 1985. *Analysis of Experiments with Missing Data*. New York: John Wiley & Sons.

Eberhart, S. A. and Russell, W.A. 1966. "Stability parameters for comparing varieties." *Crop Sci.* 6:36-40.

Elmore, R. W., Eisenhaure, D. E., Specht, J. E., and Williams, J. H. 1988. "Soybean yield and yield component response to limited-capacity sprinkler irrigation systems." *J. Prod. Agric.* July-September, 1(3):196-201.

Finley, K. W. and Wilkinson, G. N. 1963. "The analysis of adaptation in plant breeding programme." *Aust. J. Agric. Res.* 14:742-754.

Fisher, R. A. 1956. *Statistical Methods for Research Workers*. 12th ed. Edinburg: Oliver and Boyd.

Francis, T. R. and Kannenberg, L. W. 1978. "Yield stability studies in short season maize. I. A descriptive method of grouping genotypes." *Can. J. Sci.* 58:1029-1034.

Gill, J. L. 1978. *Design and Analysis of Experiments in the Animal and Medical Sciences*. Ames: Iowa State University Press.

Gomez, K. A. and Gomez, A. A. 1984. *Statistical Procedures for Agricultural Research*. 2nd ed. New York: John Wiley & Sons.

John, P. W. M. 1971. *Statistical Design and Analysis of Experiments*. New York: Macmillan.

Lee, W. 1975. *Experimental Design and Analysis*. San Francisco: W. H. Freeman.

Mead, R. 1988. *The Design of Experiments: Statistical Principles for Practical Applications*. Cambridge: Cambridge University Press, Chap. 9.

Morrow, R. E. et. al. 1988. "Cow-calf production on tall fescue-ladino clover pastures with and without nitrogen fertilization or creep feeding: fall calves." *J. Prod. Agric.* (2):145-148.

Ntare, B. R. and Aken'Ova, M. 1985. "Yield stability in segregating populations of cowpea." *Crop Sci.* March-April, 25(2):208-211.

Ogawa, J. 1974. *Statistical Theory of the Analysis of Experimental Designs*. Marcel Dekker.

Peterman, C. J., Sears, R. G., and Kanemasu, E. T. 1985. "Rate and duration of spikelet initiation in 10 winter wheat cultivars." *Crop Sci.* 25(2):221-225, March-April.

Shukla, G. K. 1972. "Some statistical aspects of partitioning genotype environment components of variability." *Heredity* 29:237-245.

Trott, J. O., Moore, K. J., Lechtenberg, V. L., and Johnson, K.D. 1988. "Light penetration through tall fescue in relation to canopy biomass." *J. Prod. Agric.* April-June, 1(2):137-140.

Yates, F. and Cochran, W. G. 1938. "The analysis of groups of experiments." *J. Agric. Sci.* 28:556-580.

Chapter 2

KEY ASSUMPTIONS OF
EXPERIMENTAL DESIGNS

2.1 INTRODUCTION

In designing an experiment, the researcher's main interest is in creating controlled conditions that easily measure those characteristics that are of interest to the researcher. That is, the experiment is designed in such a way as to uniformly maintain those factors that are not part of the treatment. In doing so the researcher must keep in mind that the design must satisfy the assumptions required for proper interpretation of data. Failure to meet the assumptions affects not only the significance level, but also the sensitivity of the F-test and the t-tests. The assumptions underlying most of the designs in this book are: (1) The effects of blocks, treatments, and error are additive. This implies no interaction. (2) The observations have a normal distribution. This means that experimental errors are normally distributed. (3) The observations are distributed independently. This assumption implies that experimental errors are independent. (4) The variance of the observations is constant, which means homogeneity of variance. This implies that the treatment effects are constant and that experimental errors have common variance. We must keep in mind that in certain conditions, not all of these assumptions are met. For example, when data are expressed as percentages (such as the percent of plants infected with a disease, or the percent of germinated seed in a plot), the observations have a binomial distribution, and hence the variance is not a constant. Similarly, when we are dealing with count data (such as the number of rare insects in a particular field, or the number of infested plants in a greenhouse), we have a poisson distribution where the variance is equal to the mean (more about the assumptions and their violations in the sections to follow).

15

2.2   ASSUMPTIONS OF THE ANALYSIS OF VARIANCE AND THEIR VIOLATIONS

Additivity: This assumption states that the effects of 2 factors are additive if the effect of one factor remains constant over all levels of the other factor. This means that each factor influences the dependent variable solely through its impact. For example, if treatment and blocks are the 2 factors of interest, this assumption implies that the treatment effect remains constant for all blocks, and the block effect remains constant for all treatments. One interpretation of additivity is that the blocks and the treatment effects do not interact; that is, all population interaction effects are zero.

We can also express the basic principle of additivity in the following manner. Denote the treatments as T_1, T_2,...T_n. It is then assumed that the observation obtained on any unit from a particular treatment such as T_1 differs from T_2 by a constant a_1 - a_2. As you would expect, there are constants a_1, a_2 ,...a_n for each treatment. The object of most of the experimental designs is to estimate the a_1 - a_2 differences. This difference is called the true *treatment effects*. If, for example, different "varieties" of a crop (*i*) are subjected to different "treatments" (*j*), then the true or expected yield of the *i*th variety when subjected to the *j*th treatment, under certain growing condition, is μ_{ij}. Thus, additivity implies that under the general experimental conditions of the test, the true mean yield of one variety is greater (or less) than the true mean yield of another variety by an amount, an additive constant, that is the same for each of the treatments concerned. Conversely, the true mean yield with one treatment is greater (or less) than the true mean yield with another treatment, by an amount that does not depend upon the variety concerned.

The mathematical model for the different experimental designs is called a linear additive model that may be written as

$$X_i = \mu + t_i + e_i \text{ (completely randomized design)} \tag{2-1}$$

$$X_{ij} = \mu + t_i + b_j + e_{ij} \text{ (randomized complete block design)} \tag{2-2}$$

In Equation 2-1, the value of the experimental unit (X_i) is made up of the general mean (μ) plus the treatment effect (t_i) plus an error term (e_i). Similarly, in Equation 2-2, the value of the experimental unit (X_{ij}) is made up of the general mean (μ) plus a treatment effect (t_i) plus a block effect (b_j) plus an error term (e_{ij}). This model implies that a treatment effect is the same for all blocks and that the block effect is the same for all treatments. What is crucial to remember is that in both models the terms are added, hence the term *additivity*.

There are instances in agricultural experimentation where the additivity assumption is not met. Let us take Equation 2-2 as an example of how the additivity assumption may be violated. The data may rarely be so obliging as to conform to Equation 2-2. More often than not, the variability observed in such an equation is also a function of the interaction between the treatment effect and the block effect. Such a condition requires a revision of the model to include an interaction term. This revised model may be written as

$$X_{ij} = \mu + t_i + b_j + (tb)_{ij} + e_{ij} \qquad (2\text{-}3)$$

where $(tb)_{ij}$ is the interaction effect of the ith treatment and the jth block. Another circumstance that leads to the violation of the additivity assumption is when the treatment effect increases yield by a percentage or proportion. This is referred to as the multiplicative treatment effect. This type of a condition is often observed when experiments are conducted to measure the number of insects per plot or the number of egg masses per unit area. When faced with a multiplicative model, it is appropriate to work with the logarithms of the original observations. Since $\log (xy) = \log x + \log y$, by transforming the original data into logarithms we are converting the multiplicative effect to the additive effect. The following hypothetical example is given to illustrate the concept of additivity and when this assumption is violated.

Example 2.2.1

In a randomized complete block design with 2 treatments and 2 blocks, a scientist found the set of data shown in Table 2.2.1. Do the data satisfy the additivity assumption?

TABLE 2.2.1.
Additive Effects of Treatment and Block Using Hypothetical Data

Treatment	Block		Block Effect
	I	II	(I - II)
A	160	120	40
B	140	100	40
Treatment effect (A - B)	20	20	

Solution

You will notice in this example that the treatment effect remains constant over all levels of the blocking factor. Similarly, the block effect remains constant for all treatments. Given this condition, we would say that the effects of treatment and block are additive.

On the other hand, when the treatment effect is not constant over blocks and block effect is not constant over treatments, the factors are said to have multiplicative rather than additive effect. Table 2.2.2 shows a set of hypothetical data with multiplicative effects of treatment and block. You will note that

TABLE 2.2.2.

Multiplicative Effects of Treatment and Block Using Hypothetical Data

Treatment	Block		Block Effect	
	I	II	(I - II)	(I - II)100/II
A	100	50	50	100
B	80	40	40	100
Treatment effect (A - B)	20	10		
(A - B)100/B	25	25		

the treatment effect for blocks I and II are 20 and 10, respectively, while the block effect is 50 for treatment A and 40 for treatment B. When the treatment and block effects are expressed in percentage terms, as shown in the last column of Table 2.2.2, you will observe that the treatment effect is 100% in both blocks, and the block effect is 25% for both treatments. What is important to remember is that a variable may interact with the treatments employed to the extent that the combined effects of treatments and error may be more multiplicative than additive. Therefore, it is desirable to detect nuisance variables in early stages of experimentation. This means making supplementary observation where appropriate and assigning the treatments to the experimental units in such a way as to detect variations.

The assumption of additivity is also violated when 1 or more aberrant observations (outliers) impact the data. It is crucial to determine whether the outlier is just an extreme value from the defined population or should be classified with another population. Outliers should not be excluded arbitrarily, as their exclusion may seriously bias estimates of the treatment effect and underestimate the experimental error. It is the responsibility of the experimenter to use common sense in dealing with outliers. The experimenter may choose from a variety of statistical methods devised by Dixon and Massey (1957), Anscombe and Tukey (1963), and Grubbs and Beck (1972) to determine whether an outlier is excluded (included) in the analysis.

Normality: This assumption states that errors are normally distributed. Certain types of data do not satisfy this assumption. The validity of such an assumption depends upon the measure chosen. For example, a measure such as count data where the number of rare insects found in a soil sample, the number of lesions per leaf, or the number of particular impurities in milk samples usually follow a poisson distribution. As the data in the form of counts are discrete, they do not follow the normal distribution that assumes a continuous variable. Percentage scores, such as seeds germinated in a greenhouse or percentage of plants infected with a disease, have binomial distribution and hence the variance, which depends on the unknown percentage, is not a constant.

The normality assumption is crucial in probability statements where decisions are based on tests of hypotheses as well as the reliability of estimates where confidence intervals are used. In most agricultural experiments, two statistical facts provide support to the normality assumption: first, the central limit theorem that states that as the sample size "n" becomes sufficiently large, the sampling distribution of the mean tends toward a normal distribution; second, it has been noted that the F-test of the hypothesis of treatment effects is known to be robust, i.e., the probabilities of errors of Type I and Type II are not affected severely by moderate departures from normality (Pearson, 1931; Donaldson, 1968; Tiku. 1971; Gill, 1978).

There are certain situations where a researcher, by looking at the data, may suspect departures from the normality assumption. For example, it is difficult to assume that when the mean values for the different treatments differ by a factor of 3 or more that there will not be greater variability. A researcher should also be cautious when there are substantial differences between blocks. It is difficult to assume that treatment differences can remain constant over blocks.

To evaluate the significance of departures of the obtained distribution from the assumed normal distribution, we can analyze the residuals, $e_{ij} = X_{ij} - \bar{X}_i$, of the sample. Depending on the size of the sample, we can use various tests. The test of goodness of fit that utilizes the χ^2 distribution provides a good approximation when sample size is large ($n = 100$). The same test could be used for smaller samples, however, with some loss of accuracy. Others have suggested that the Kolmogorov-Smirnov test is a better approximation for small samples (Siegel, 1956; Sokal and Rohlf, 1969).

In cases where nonnormality is suspected, one may transform the data to ensure near normality of distribution for the transformed data. The most common transformations used are square root, inverse square root, log, or reciprocal. These transformations are discussed later in Section 2.4. It should be recognized that while transformation improves the accuracy of the probability statements, interpretation of the results is greatly handicapped by the transformation.

Independent errors: This assumption states that normally distributed errors are independent of each other. In other words, the error of an observation is not correlated with that of another. The validity of this assumption is assured when proper randomization is applied in any designed experiment. There are instances, however, where the design requires a systematic assignment of treatments to the experimental units. In such a design (systematic design), the assumption of independence of error is violated. Figure 2.2.1 shows a systematic design and the location of the problem with respect to the violation of the assumption of independence of errors.

Replication I	A	B	C	D	E	F
Replication II	F	A	B	C	D	E
Replication III	E	F	A	B	C	D
Replication IV	D	E	F	A	B	C
Replication V	C	D	E	F	A	B
Replication VI	B	C	D	E	F	A

Figure 2.2.1.

Layout of a systematic design involving six treatments (A, B, C, D, E, and F) and six replications.

You will note in Figure 2.2.1 that treatment A and B are adjacent to each other in all replications, and B and E are always separated by 2 blocks. Furthermore, it should be noted that even when the design is not systematic, crop yields on neighboring plots tend to be positively correlated. Therefore, treatments that are close to each other tend to have similar errors more so than those that are further apart. Thus, in a systematic design such as Figure 2.2.1, the assumption of independence of errors is violated. As a measure of precaution, to ensure independence of experimental errors, it is highly desirable to properly randomize experimental layout. In chapters to follow we will discuss in detail the randomization and layout of experiments for each design.

Homogeneous variance: One of the most frequently occurring and serious violations of the assumptions of analysis of variance is the presence of heterogeneous variance among treatment groups. This means that the experimental error variance is not constant over all observations. Another way of stating the assumption of homogeneous variance is that the separate variance estimates provided by samples are estimates of the same population variance. When error variance is heterogeneous, the F- and t-tests tend to give too many significant results. This disturbance is usually only moderate if every treatment has the same number of replications (Scheffé, 1959). The validity of the homogeneity of variance depends upon (1) how careful the experimenter is in administering experimental treatments in a consistent manner to all subjects, (2) the type of measure taken, and (3) the treatment levels employed. Undoubtedly, the uniform application of treatments to subjects will tend to stabilize variances from group to group. However, despite appropriate experimental methodology, we can expect heterogeneity of variances. Heterogeneous variance is also associated with the data whose distribution is not normal. For example, when we have a Poisson distribution where the mean and the variance are equal ($\mu = \sigma^2$), the mean and the variance are either homogeneous or heterogeneous over treatments. Poisson distributions occur when the observations consist of counts. Once we recognize the nature of the functional relationship, data can be transformed on a scale on which the error variance is more nearly constant. Bartlett (1947) found several transformation methods such as the log, square root, and inverse sines extremely helpful. Such transformations are also useful in cases where treatment and environmental effects are not additive, as was mentioned earlier.

Several methods exist for detecting departures from homogeneity of variance. These will be discussed in the following section.

2.3   MEASURES TO DETECT FAILURES OF THE ASSUMPTIONS

Failure to take into account major departures of the data from the assumptions mentioned in the previous section have serious implications for the conclusions of any analysis. Minor departures, on the other hand, do not greatly disturb the conclusions reached from the standard analysis (Cochran and Cox, 1957; Montgomery, 1976; Snedecor and Cochran, 1980). Just as increased replications and proper randomization are essential to minimizing experimental error, so is the recognition of the departures of the data from the assumptions for the conclusions of an experiment. The reader is referred to Scheffé (1959) for an excellent review of the consequences of various types of failures of the assumptions.

In the previous section we discussed the assumptions and some of the theoretical conditions that lead to the violation of such assumptions. Now we will present specific examples of how to detect the failures of the assumptions and show procedures to transform the data to conform to the assumptions.

The assumption of *additivity*, which states that the block effects are approximately the same for all treatments, can be examined using the *Tukey's* test. The test requires that we compute the sum of squares for nonadditivity in the following manner:

$$SS \text{ nonadditivity} = \frac{N^2}{(\sum d_i^2)(\sum d_j^2)} \tag{2-4}$$

where

N = Each cell of the table of raw data multiplied by the corresponding treatment and block effect, and the sum of all the products or $N = \sum w_i\, d_i = \sum\sum X_{ij}\, d_i\, d_j$

$\Sigma\, d_i^2$ = Sum of the treatment effect squared

$\Sigma\, d_j^2$ = Sum of the block effect squared.

Equation 2-4 is basically the contribution of nonadditivity with 1 degree of freedom to the error sum of squares. This value is then tested against the remainder of the residual sum of squares to determine whether the hypothesis of additivity is tenable.

This test can be applied to any 2-way classification such as randomized complete block design experiments where the data are classified by treatments and blocks. Example 2.3.1 shows how this test is performed.

Example 2.3.1

An agricultural biologist attempting to determine insect infestation in a field has collected data over 4 periods in 4 traps as shown in Table 2.3.1. To find out whether the data meet the additivity assumption, we perform the Tukey's test.

TABLE 2.3.1.

Insects Caught over 4 Different Periods in 4 Traps in an Experiment

Period	Trap 1	Trap 2	Trap 3	Trap 4	Mean $\bar{X}_i$	Treatment Effect $d_i = \bar{X}_{i.} - \bar{X}_{..}$	$w_i = \Sigma\, X_{ij}\, d_j$
1	8	10	18	32	17.00	-5.75	400.00
2	12	16	26	46	25.00	2.25	560.00
3	16	20	26	42	26.00	3.25	423.00
4	10	22	20	40	23.00	0.25	144.50
Mean ($\bar{X}_{.j}$)	11.50	17.00	22.50	40.00			
$d_j = (\bar{X}_{.j} - \bar{X}_{..})$	-11.25	-5.75	-0.25	17.25		0.00	
$\bar{X}_{..}$					22.75		

Solution

Before we perform the Tukey's test, the data are analyzed using the method discussed in Chapter 4, Section 4.2.2. We need this analysis so that we can show the partitioning of the error sum of squares into its components. The result of the analysis is presented in Table 2.3.2.

TABLE 2.3.2.
ANOVA of the Data Presented in Table 2.3.1

Source of Variation	Degree of Freedom	Sum of Squares	Mean Square	F
Periods	3	195	65.0	7.39**
Traps	3	1,829	609.7	69.28**
Error	9	79	8.8	
Total	15			

** = significant at the 1% level.

Now we are ready to perform the Tukey's test. The steps in computing the intermediate values for performing the test are outlined below.

STEP 1: Compute the treatment effect (d_i) and the period effect (d_j) as shown in Table 2.3.1. Remember that both effects add to exactly 0.

STEP 2: Compute w_i for each period, using the following formula. The computed values are shown in Table 2.3.1.

$$w_i = \sum X_{ij}\, d_j \tag{2-5}$$

Thus, for our example we have:

$$w_1 = 8(-11.25) + 10(-5.75) + 18(-0.25) + 32(17.25) = 400.00$$
$$w_4 = 10(-11.25) + 22(-5.75) + 20(-0.25) + 40(17.25) = 144.50$$

STEP 3: Compute N, which is the sum of the product of w_i and d_i. Another way of stating N is that it is the sum of the product of each cell of Table 2.3.1 multiplied by the corresponding "treatment" and "period effect." Thus, we have:

$$N = \sum w_i\, d_i = \sum \sum X_{ij}\, d_i\, d_j \tag{2-6}$$

For our example, the value of N is:

$$N = 400(-5.75) + 560(2.25) + 423(3.25) + 144.5(0.25) = 370.88$$

STEP 4: Compute the sum of the value of the treatment and period effects as follows:

$$\sum d_i^2 = (-5.75)^2 + (2.25)^2 + (3.25)^2 + (0.25)^2 = 48.75$$

$$\sum d_j^2 = (-11.25)^2 + (-5.75)^2 + (-0.25)^2 + (17.25)^2 = 457.25$$

STEP 5: Perform the test using Equation 2-4 as follows:

$$SS \text{ nonadditivity} = \frac{(370.88)^2}{(48.75)(457.25)} = 6.17$$

This is the contribution of the nonadditivity to the error sum of squares with 1 degree of freedom. This value is tested against the residual as shown in Table 2.3.3.

TABLE 2.3.3.

ANOVA Table and the Test of Additivity

Source of Variation	Degree of Freedom	Sum of Squares	Mean Square	F
Error (Period × Trap)	9	79.00	8.8	
Nonadditivity	1	6.17	6.17	<1
Residual	8	72.83	9.10	

The test result indicates a strong evidence that the assumption of additivity is correct. Thus, the researcher is assured that the data gathered do meet at least the additivity assumption.

To test the assumption of *normality*, we have to look carefully at the error terms associated with each observation to determine whether they are randomly distributed or not. Recall from Equation 2-2 that any observation (X_{ij}) of a 2-way table in a randomized complete block design is simply made up of the general mean (μ), the treatment effect (t_i), the block effect (b_j), and the error term (e_{ij}). To find out the error term for any cell of a 2-way table, we can rewrite Equation 2-2 as follows:

$$e_{ij} = X_{ij} - \mu - t_i - b_j \tag{2-7}$$

Before we can determine the error term using Equation 2-7, we must find out the value of the treatment and block effects. Since the treatment and block effects are the difference between the treatment and block mean from the general mean, they can be written as:

$$\text{Treatment effect } (t_i) = \bar{X}_{i.} - \mu \tag{2-8}$$

$$\text{Block effect } (b_j) = \bar{X}_{.j} - \mu \tag{2-9}$$

Substituting Equations 2-8 and 2-9 into 2-7, we get

$$e_{ij} = X_{ij} - \mu - (\bar{X}_{i.} - \mu) - (\bar{X}_{.j} - \mu) \tag{2-10}$$

Using Equations 2-10 we are now able to determine the value of the error term for each cell of a 2-way table.

Let us use the data given in Table 2.3.1 to test the normality assumption for this example. Using Equation 2-10 we will remove the general mean, the block effect, and the treatment effect from each cell of the data. For example, to find the error term for e_{11}, we would have

$$e_{11} = 8 - 22.75 - (17 - 22.75) - (11.50 - 22.75)$$
$$= -14.75 - 17 + 22.75 - 11.50 + 22.75$$
$$= 2.25$$

The error terms for each of the cells are computed similarly and are presented in Table 2.3.4.

TABLE 2.3.4.
Error Components of the Insects Trapped Experiment

Period	Trap 1	2	3	4	Total
1	2.25	-1.25	1.25	-2.25	0.0
2	-1.75	-3.25	1.25	3.75	0.0
3	1.25	-0.25	0.25	-1.25	0.0
4	-1.75	4.75	-2.75	-0.25	0.0
Total	0.0	0.0	0.0	0.0	

Looking at the error terms, we note that the distribution does not appear to be normal, as there are 2 modal classes (see Figure 2.3.1). This test tells us that the normality assumption is untenable in the present case. To remedy this

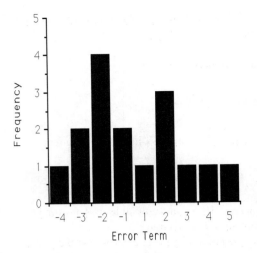

FIGURE 2.3.1.

Frequency distribution of the error terms.

problem, transformation of the data is required. In the subsequent section we will discuss different transformation techniques.

To test the presence of heterogeneous variances and whether the variances are functionally related to the mean, a simple procedure to use is a scatter diagram. The steps to follow are given below:

STEP 1: Compute the mean and the variance for each treatment across all replications.

STEP 2: Using the mean value and the variance, plot a scatter diagram. It should be noted that one could also use the mean and the range in plotting the scatter diagram.

STEP 3: Upon examining the scatter diagram, identify the relationship, if any, between the mean and the variance. Three possible outcomes could be observed: (1) homogeneous variance, (2) heterogeneous variance when there is a functional relationship between the variance and the mean, and (3) heterogeneous variance when there is no functional relationship between the variance and the mean. Figure 2.3.2 illustrates the presence and absence of homogeneity of variance.

In Figure 2.3.2 (a), we note that the relationship between the mean and variance is such that the variance of the treatments do not vary significantly from one another at different means and are thus considered homogeneous. However, in Figure 2.3.2 (b) and (c), the variances are heterogeneous as a result

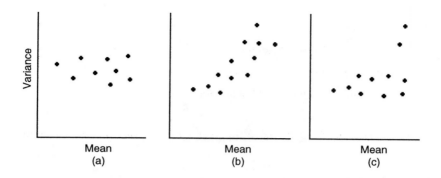

FIGURE 2.3.2.

Illustration of a homogeneous and heterogeneous variance: (a) homogeneous variance, (b) heterogeneous variance where variance is proportional to the mean, and (c) heterogeneous variance where variance and the mean are not functionally related.

of the proportionality of variance to the mean in the case of (b), and no functional relationship between variance and mean in the case of (c). The scatter diagram can be used as a signal for using the tests of the homogeneity of variance. There are several tests that could be applied to verify the homogeneity of variance. The reader may find the review of these methods given by Anderson and McLean (1974) helpful. In this text the following 2 tests are suggested for use by practitioners.

Bartlett (1937) introduced a homogeneity-of-variance test that involves computing a statistic whose sampling distribution is closely approximated by the chi-square distribution with $k - 1$ degrees of freedom. The test has become a well-established measure. However, it should be kept in mind that the test is a bit sensitive to nonnormality, especially if the tails of the distribution are too long. When this occurs, the test tends to show significance too often.

The test criterion, when there is $k > 2$ independent estimate of variance s_i^2 and all have the same number of degrees of freedom v using logarithms to base e is:

$$\chi^2 = \frac{M}{C} \tag{2-11}$$

where

$$M = v \left(k \ln \bar{s}^2 - \sum \ln s_i^2 \right) \tag{2-12}$$

$$C = 1 + \frac{k + 1}{3(k)(v)} \tag{2-13}$$

$$\hat{s}^2 = \frac{\sum s_i^2}{k} = \text{pooled variance} \tag{2-14}$$

$v = $ degrees of freedom per variance

$k = $ number of treatments or samples.

To see how the Bartlett's test is applied, let us use the following example.

Example 2.3.2

In a randomized experiment with 4 replications, the number of weeds found per plot after the fields were sprayed with different rates of a general herbicide are given in Table 2.3.5. Verify the homogeneity of variance for the data using Bartlett's test.

TABLE 2.3.5.

Number of Weeds per Plot Following Different Herbicide Treatments

Treatment	Rep. I	Rep. II	Rep. III	Rep. IV	Total	Mean	s_i^2	$\ln s_i^2$
A	15	12	17	20	64	16.00	11.33	2.425
B	12	5	7	4	28	7.00	12.66	2.538
C	11	7	8	6	32	8.00	4.66	1.390
D	18	10	11	9	48	12.00	16.66	2.813
E	16	18	20	14	68	17.00	6.66	1.896
Control	20	28	24	28	100	25.00	14.66	2.685
Total							66.60	13.747

Solution

STEP 1: Compute the variance for each of the treatments as shown in Table 2.3.5.

STEP 2: Compute the logarithm of each of the variances, the total for the variances, and the total for the logarithm of the variances as shown in Table 2.3.5.

STEP 3: Compute the estimate of the pooled variances where $k = 6$.

$$\hat{s}^2 = \frac{\sum s_i^2}{k} = \frac{66.60}{6} = 11.10$$

STEP 4: Compute M given that $v = 3$.

$$M = (3)[6(2.407) - 13.747]$$
$$= 2.085$$

STEP 5: Compute C, the correction factor, as:

$$C = 1 + \frac{(6+1)}{(3)(6)(3)} = 1.13$$

STEP 6: Compute the χ^2 as:

$$\chi^2 = \frac{2.085}{1.13} = 1.85$$

STEP 7: Compare the computed χ^2 with the table value at 5 degrees of freedom (one less than the number of treatments) given in Appendix A. You will note that $\chi^2_{.05, 5} = 11.07$. Thus, we cannot reject the null hypothesis and conclude that the variances are homogeneous.

When the samples are of unequal size, the test statistic is similar to Equation 2-11 except that computing M and C are as follows:

$$M = (\sum v_i) \ln \hat{s}^2 - \sum v_i \ln s_i^2 \tag{2-15}$$

$$C = 1 + \frac{1}{3(k-1)} \left(\sum \frac{1}{v_i} - \frac{1}{\sum v_i} \right) \tag{2-16}$$

where

$$\hat{s}^2 = \frac{\sum v_i s_i^2}{\sum v_i}$$
(2-17)

v = degrees of freedom per variance

k = number of treatments or samples.

Hartly (1950) proposed another measure for testing the homogeneity of variance. This test is based on the ratio of the largest to the smallest within-groups variance. This is known as the test of homogeneity of variance. If this ratio is nonsignificant, variances are said to be homogeneous. On the other hand, if the ratio is significant, the variances are said to be heterogeneous. The F-test for the homogeneity of variance is defined as:

$$F = \frac{s_1^2}{s_2^2}$$
(2-18)

where

s_1^2 = the larger sample variance

s_2^2 = the smaller sample variance

2.4   DATA TRANSFORMATION

Having discussed the various approaches that could be applied to determine whether the assumptions of the analysis of variance are violated, we are now ready to review data transformation as a means to remedy the violation of the various assumptions. Let us take the violation of the assumption of homogeneity of variance to show how data transformation improves the situation.

In Figure 2.3.2 (b) and (c), the functional relationship between the mean and the variance shows the presence of heterogeneous variances. Once the nature of the functional relationship between the mean and the variance is determined, the original data can be transformed to a scale on which the error variance is more nearly constant. In agricultural research, we often perform one of the following three commonly used transformations.

Square Root Transformation: The square root transformation is applied when the means and the variances are proportional for each treatment. That is,

$\sigma_j^2 = k\mu_j$. Transformation of the data allows the variance of the data to be nearly independent of the mean.

Researchers often encounter conditions where the mean and the variance are proportional to each other. This situation is not unusual when the data are in the form of frequency counts: for example, the number of weeds found in a plot, or when the dependent variable is the number of correct or "yes" responses. In such circumstances, the analysis of variance is performed on X' rather than X (the original data), where

$$X' = \sqrt{X} \tag{2-19}$$

Bartlett (1936) suggests that if most of the values of X are small, i.e., less than 10, and especially some with a value of 0, homogeneity of variance is likely to be produced by using the following transformation:

$$X' = \sqrt{X + 0.5} \tag{2-20}$$

Freeman and Tukey (1950) have suggested that for the Poisson distribution (counts of rare events), it is best to use transformation $\sqrt{X} + \sqrt{X + 1}$ for a better result.

We will use the data from Example 2.3.2 to show how it could be transformed before further analyses are performed on the data.

Solution

STEP 1: Transform the data, using the square root transformation. Observe that a large number of values in the data set are greater than 10 and there are no observations with a value of 0. Hence, the transformation $X' = \sqrt{X}$. The transformed data X' is shown in Table 2.4.1.

STEP 2: Observe that both the range and the estimates of the variance of the data as shown in Table 2.4.1 indicate that the transformed data are now on a scale on which the error variance is more nearly constant.

STEP 3: Perform the analysis of variance on the transformed data. Researchers sometimes prefer to report the results in the original scale as it is recognized that it is more easily understood on this scale. Hence, the means are reconverted to the original scale by squaring.

TABLE 2.4.1.

Square Root Transformation of the Data from Table 2.3.5

	Transformed Data							
Treatment	Rep. I	Rep. II	Rep. III	Rep. IV	Total	Mean	Range	s^2
A	3.87	3.46	4.12	4.47	15.92	3.98	1.01	0.18
B	3.46	2.23	2.65	2.00	10.34	2.59	1.46	0.41
C	3.31	2.65	2.00	2.45	10.41	2.60	1.31	0.29
D	4.24	3.16	3.31	3.00	13.71	3.43	1.24	0.30
E	4.00	4.24	4.47	3.74	16.45	4.11	0.73	0.09
Control	4.47	5.29	4.90	5.29	19.95	4.99	0.82	0.15

Logarithmic Transformation: Researchers often face situations where they suspect that the standard deviation (not variances) of the data is proportional to the mean or where the treatment and replication (block) effects are multiplicative. Under such conditions, logarithmic transformation of the data is appropriate. Once transformed, the data will not show any apparent relationship between the mean and the standard deviation. Furthermore, this transformation changes the multiplicative effect into additive, thus meeting the additivity assumption. With the use of some examples below, we will transform the data to illustrate the effectiveness of the logarithmic transformation.

Data that typically require transformation are whole numbers and may cover a wide range of values. The most common situation that gives rise to a logarithmic transformation is associated with the data on growth: for example, changes in the weight of animals caused by diets, the number of insects per plot, or the number of germinated seed per greenhouse. As we apply this transformation, we should keep in mind 2 data situations where logarithmic transformation is problematic. First, negative values cannot be transformed. Second, if the data set contains values of 0, its logarithm is minus infinity. If the number of 0s in a data set is large, it is best to use some other transformation. However, a small number of 0s in a data set can easily be handled by adding 1 to each data point before the data are transformed. You may use logarithms of any base, but it is easiest to use the common logarithm (to the base 10). To avoid negative logarithms, it is best to multiply the data set with a constant. Such manipulation of the data is legitimate as it has no consequence on subsequent analyses.

To show how logarithmic transformation changes the multiplicative effect into additive to meet the additivity assumption, let us use the hypothetical data given in Table 2.2.2. To transform the data, simply take the logarithm of each data point as shown in Table 2.4.2.

TABLE 2.4.2.

Logarithmic Transformation of the Multiplicative Effects of the Hypothetical Data from Table 2.2.2.

Treatment	Block		Block Effect
	I	II	
A	2.0000	1.6989	0.3010
B	1.9031	1.6020	0.3010
Treatment effect (A - B)	0.0969	0.0969	

You will note in Table 2.4.2 that the treatment effect is 0.0969 in both blocks and 0.3010 for both treatments. Thus, with the transformation, we have converted the multiplicative effect into additive effect.

Logarithmic transformation can also be applied to the data where the heterogeneity of variance is suspected. Using the data in Table 2.3.5, we want to determine the relationship that exists between the mean and the variance or the range. For this purpose, we have plotted the value of the mean against the range as shown in Figure 2.4.1.

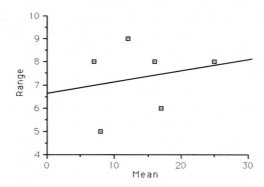

FIGURE 2.4.1

Relationship between treatment mean and the range of the data of Table 2.3.5.

Figure 2.4.1. shows that there is a linear relationship between the mean and the range, implying that the range increases proportionately to the mean. To remedy this problem, we perform a logarithmic transformation of the data as shown in Table 2.4.3. Since there are several data points with values of less than 10, it is appropriate to use the $\log (X + 1)$ rather than $\log X$ where X is the original data.

TABLE 2.4.3.

Transformed Data from Table 2.3.5 Using a Log $(X + 1)$ Scale Transformation

Treatment	Rep. I	Rep. II	Rep. III	Rep. IV	Total	Mean
	Transformed Data					
A	1.2041	1.1139	1.2552	1.3222	4.8954	1.2238
B	1.1139	0.7782	0.9031	0.6989	3.4941	0.8735
C	1.0792	0.9031	0.9542	0.8451	3.7816	0.9454
D	1.2788	1.0414	1.0792	1.0000	4.3994	1.0998
E	1.2304	1.2788	1.3222	1.1761	5.0075	1.2518
Control	1.3222	1.4624	1.3979	1.4624	5.6449	1.4112

Note in Table 2.4.3 that there seems to be a much more consistent pattern about the transformed data, though there is still some random variation. We now can perform the Bartlett's test of homogeneity of variance as before to see if the transformed data meet this assumption. As you recall from our earlier discussions, whenever we take the logarithms of any value less than 1, we have the problem of negative logarithms. To remedy this, we need to multiply each number by any constant we choose. This will not alter the Bartlett test we perform. In the present example, the value of all the variances computed are less than 1. Thus, they are multiplied by 10,000 and are recorded as the coded variance in Table 2.4.4.

TABLE 2.4.4.

Number of Weeds per Plot Following Different Herbicide Treatments

Treatment	Mean	s_i^2	Coded s_i^2	log Coded s_i^2
A	1.2238	0.0020	20.00	2.99
B	0.8735	0.0169	169.00	5.13
C	0.9454	0.0089	89.00	4.89
D	1.0998	0.0003	3.00	1.09
E	1.2518	0.0034	34.00	3.53
Control	1.4112	0.0192	192.00	5.26
Total	6.8055		507.00	22.89
Mean			84.50	
Log of Mean			4.44	

The Bartlett's test of homogeneity of variance, as given before, is:

$$\chi^2 = \frac{M}{C} \tag{2-11}$$

For the present example, the value of M and C are:

$$M = (3)[6(4.44) - 22.89]$$
$$= 11.25$$
$$C = 1 + \frac{(6+1)}{(3)(6)(3)} = 1.13$$

Thus, χ^2 is computed as follows:

$$\chi^2 = \frac{11.25}{1.13} = 9.96$$

Compare the computed χ^2 with the table value at 5 degrees of freedom (1 less than the number of treatments) given in Appendix A. From Appendix A you will note that $\chi^2_{.05, 5} = 11.07$. Thus, we cannot reject the null hypothesis, and therefore conclude that the variances are homogeneous.

We are now able to perform the analysis of variance on the transformed data. The results are shown in Table 2.4.5.

TABLE 2.4.5.
ANOVA of the Transformed Data

Source of Variation	Degree of Freedom	Sum of Squares	Mean Square	F
Replication	3	0.0554	0.1845	16.62**
Treatment	5	0.8136	0.1627	14.65**
Error	15	0.1668	0.0111	
Total	23			

** = significant at the 1% level.

The results indicate that both the replication and the treatment effects are significant at the 1% level.

Arc Sine or Angular Transformation: This transformation is most appropriate when researchers encounter data in the form of counts or the binomial proportions (p). In the binomial distribution, in contrast to the normal distribution, the mean and the variance are related in such a way that the variances tend to be small at the two ends of the range of values, and large in the middle. This means that small variances are associated with values close to 0 and 100%, while large variances are expected around the middle or 50%. Another way of stating this is that with binomial data (0 or 1 response), the variance is proportional to [mean (1 - mean)], and the appropriate transformation is arc sine $\sqrt{p}$. In the angular scale, the percentages near 0 or 1 are spread out so as to increase their variance. If all the error variance is binomial, the error variance in the angular scale is about $821/n$ (Snedecor and Cochran, 1980). It has been shown empirically that the variances are fairly stable between the range of $p = .30$ and $p = .70$. Thus, it is not necessary to transform the data for this range of values. Arc sine transformation is most appropriate for the data if the range of percentages is greater than 40.

In using the arc sine or angular transformation, it has been suggested by Bartlett (1947) that with n <50 (where n is the number upon which the percentage data are based), we should count a 0 proportion as $1/(4n)$ and a 100% proportion as $(n - 1/4)/n$ before transforming to angles. This will improve the equality of the variance in the angles. It should be kept in mind, however, that transformation will not improve the inequalities in variance if there are differing values of n. Under such conditions, it is appropriate to use a weighted analysis in the angular scale.

To illustrate the application of the arc sine transformation, let us assume that in a completely randomized design with 10 treatments and 4 replications, and $n = 25$ seeds, an ornamental horticulturist found the percentages of seeds germinated in a greenhouse that are shown in Table 2.4.6.

TABLE 2.4.6.

Percentages of Germinated Seeds in a Greenhouse

Treatment	Rep. I	Rep. II	Rep. III	Rep. IV	Total	Mean
A	42.2	36.4	43.2	52.5	174.3	43.6
B	24.5	29.7	26.2	29.0	109.4	27.4
C	0.0	0.0	0.0	0.0	0.0	0.0
D	44.0	70.1	81.2	69.3	264.6	66.2
E	96.0	98.0	95.4	100.0	389.4	97.4
F	20.0	28.1	24.9	28.4	101.4	25.3
G	13.5	14.8	20.1	12.1	60.5	15.1
H	0.0	0.0	0.0	0.0	0.0	0.0
I	55.6	62.8	64.6	58.9	241.9	60.5
J	33.2	39.7	42.1	37.5	152.5	38.1

You will note that the percentages of seeds that germinated range from 0 to 100%. Given this range of values, it is appropriate that we use the arc sine transformation. Remember from our earlier discussion that all 0 and 100% values need to be replaced by $1/(4n)$ and $(n - 1/4)/n$ respectively. Thus, for treatments C, E, and H we have:

Treatment C = [1/4(25)] = 0.01

Treatment E = [(25 - 1/4)/25] = 0.99

Treatment H = [1/4(25)] = 0.01

The transformed data using Appendix B are given in Table 2.4.7.

TABLE 2.4.7.
Transformed Data Using the Arc Sine Transformation

Treatment	Rep. I	Rep. II	Rep. III	Rep. IV	Total	Mean
A	40.5	37.1	41.1	46.4	165.1	41.3
B	29.7	33.0	30.8	32.6	126.1	31.5
C	0.6	0.6	0.6	0.6	2.4	0.6
D	41.6	56.9	64.3	56.4	219.2	54.8
E	78.5	81.9	77.6	5.7	243.7	60.9
F	26.6	32.0	29.9	32.2	120.7	30.2
G	21.6	22.6	26.6	20.4	91.2	22.8
H	0.6	0.6	0.6	0.6	2.4	0.6
I	48.2	52.4	53.5	50.1	204.2	51.1
J	35.2	39.1	40.5	37.8	152.6	38.2

We will now perform the analysis of variance of the angles following the procedures detailed in Chapter 4, Section 4.2.2. The results are shown in Table 2.4.8.

TABLE 2.4.8.
ANOVA of the Angles

Source of Variation	Degree of Freedom	Sum of Squares	Mean Square	F
Replication	3	420.77	140.26	0.93
Treatment	9	15,556.81	1,728.53	11.52**
Error	27	4,052.96	150.11	
Total	39			

Note: ** = significant at the 1% level.

Note that the error mean square in Table 2.4.8 is 150.11. This is greater than $821/n = 821/25 = 32.8$, implying that some variation in excess of the binomial may be present.

In summary, the intent of this chapter was to present the basic assumptions of experimental designs and to discuss the implications of the violation of these assumptions. Several approaches exist for detecting failures of the assump-

tions, and data transformation was suggested as a means to remedy the violation of the various assumptions.

REFERENCES AND SUGGESTED READINGS

Anderson, V. L. and McLean, R. A. 1974. *Design of Experiments: A Realistic Approach*. New York: Marcel Dekker.

Anscombe, F. J. and Tukey, J. W. 1963. "The examination and analysis of residuals." *Technometrics* 5:141-60.

Bartlett, M. S. 1936. "Square-root transformation in analysis of variance." *J. R. Stat. Soc. Supp.* 3:68-78.

Bartlett, M. S. 1937. "Some examples of statistical methods of research in agriculture and applied biology," *J. R. Stat. Soc. Supp.* 4:137-183.

Bartlett, M. S. 1947. "The use of transformations." *Biometrics* 3: 39-52.

Cochran, W. G. and Cox, G. M. 1957. *Experimental Designs*. 2nd ed. New York: John Wiley & Sons.

Dixon, W. J., and Massey, F. S. 1957. *Introduction to Statistical Analysis*. 2nd ed. New York: McGraw-Hill.

Donaldson, T. S. 1968. "Robustness of the F-test to errors of both kinds and the correlation between the numerator and denominator of the F-ratio." *J. Am. Stat. Assoc.* 63:660-67.

Freeman, M. F. and Tukey, J.W. 1950. "Transformations related to the angular and the square root." *Ann. Math. Stat.* 21:607-611.

Gill, J. L. 1978. *Design and Analysis of Experiments: In the Animal and Medical Sciences*, Vol. 1. Ames: Iowa State University Press, 154.

Grubbs, F. E. and Beck, G. 1972. "Extension of sample sizes and percentage points for significance tests of outlying observations." *Technometrics* 14:847-54.

Hartley, H. O. 1950. "The maximum F-ratio as a short-cut test for heterogeneity of variance." *Biometrika* 37:308-312.

Montgomery, D. C. 1976. *Design and Analysis of Experiments*. New York: John Wiley & Sons.

Pearson, E. S. 1931. "The analysis of variance in cases of non-normal variation." *Biometrika* 23:114-33.

Scheffé, H. 1959. *The Analysis of Variance*. New York: John Wiley & Sons.

Siegel, S. 1956. *Non-Parametric Statistics*. New York: McGraw-Hill.

Snedecor, G. and Cochran, W. G. 1980. *Statistical Methods*. 7th ed. Ames: Iowa State University Press.

Sokal, R. R. and Rohlf, F. J. 1969. *Biometry*. San Francisco: W. H. Freeman.

Tiku, M. L. 1971. "Power function of the F-test under non-normal situations." *J. Am. Stat. Assoc.* 66:913-16.

Tukey, J. W. 1949. "One degree of freedom for nonadditivity." *Biometrics* 5:232-242.

EXERCISES

1. Using the data given below, perform Tukey's test to determine if the data meet the assumption of additivity.

Treatment	Block			
	I	II	III	IV
A	20	32	22	24
B	18	20	23	25
C	19	21	28	27
D	24	24	26	28

2. To determine the impact of 2 newly developed fungicides on roses, a horticulturist used a randomized block experiment. Each block contained 6 plots, and each fungicide was used on 2 of the plots within each block. Two weeks after the spray, 50 leaves were examined on each plot, and the number of leaves showing rust problems were recorded as shown below.

Determine whether any transformation is required before an analysis could be performed.

| Block | Fungicide | | | | | |
	F1		F2		Control	
I	6	8	7	9	25	18
II	5	6	8	5	18	19
III	3	4	7	8	14	26

3. Using a randomized complete block design, an animal scientist performed an experiment in which the impact of linseed cake meal on weight gain (in pounds) of several groups of animals were observed. Perform a logarithmic transformation and analysis on the data.

| Treatment (Species) | Block | | | |
	I	II	III	IV
Pig- linseed cake	95	102	110	96
Pig- control	82	85	80	75
Sheep-linseed cake	112	130	114	129
Sheep-control	90	86	89	85
Chicken-linseed cake	3.0	2.9	2.6	2.8
Chicken-control	1.5	1.9	2.0	1.2

4. In a randomized experiment with 3 replications, the number of insects found per plot after the fields were sprayed with different rates of a general insecticide. Verify the homogeneity of variance for the data using Bartlett's test.

Treatment	Rep. I	Rep. II	Rep. III
A	10	12	7
B	9	15	10
C	5	16	12
D (Control)	15	22	19

Number of insects found per plot following different insecticide treatments

5. Slow germination limits establishment of perennial warm season forage
    grasses, and temperature is a major factor influencing it. A crop scientist
    conducted a completely randomized design experiment in which the fol-
    lowing percentages of seeds germinated were recorded.

Influence of temperature on germination percentage of seeds of warm-season
forage grasses

Species	Temperature,°C					
	9	12	15	20	25	30
Native big blue stem	6	32	36	30	27	20
Rountree big blue stem	36	58	68	64	59	49
Caucasian bluestem	0	1	8	10	12	14
Blackwell switch grass	4	14	18	8	9	12
Osage	30	52	56	54	44	32

(a) What type of transformation should be used?

(b) Perform an analysis of variance on the data.

Chapter **3**

DESIGNS FOR REDUCING ERROR

3.1 INTRODUCTION

Our main interest in this chapter lies in designs that reduce experimental error. The intent of an optimal design is to provide a mechanism for estimation and control of experimental error in many of the field experiments that are conducted. The success of reducing errors depends to a large extent on using general knowledge of the experimental material and appropriate groupings to detect true treatment effects.

Generally, there are 2 sources of experimental error to which a researcher must pay attention. First, the inherent variability in experimental units introduces errors into the experiment. Second, lack of uniformity or failure to standardize experimental technique contributes to experimental error. Lack of attention to either contributes to difficulties in assessing and interpreting results. Several strategies are available to reduce error and enhance the accuracy of the experiment. These approaches can be broadly grouped into (1) increasing the size of the experiment, (2) refining the experimental conditions, and (3) reducing variability in the experimental material. These strategies, along with their implications on experimental error, are discussed below.

3.1.1 Increasing the Size of the Experiment

As a means of increasing the accuracy of the experiment, the size of the experiment can be increased either by increasing the number of replications or by adding more treatments to the experiment.

As was mentioned in Chapter 1, *replication* refers to the repetition of the basic experiment or treatment. By having several replications in an experiment, the experimenter is attempting to reduce the error associated with the differences between the average results of the treatments concerned. It is also used to estimate the error variance that is a function of the difference among observations from an experimental unit receiving the same treatment. The error variance is steadily reduced in a randomized experiment as more replications are added. The rate of the reduction of such error is predictable from statistical theory. For example, if σ^2 is the error variance per unit and there are r replications, the error variance of the difference between the means for 2 treatments is $\sqrt{2\sigma^2/r}$. This result remains valid as long as increased replication does not translate into the use of less homogeneous experimental material and less careful technique. While more replication is helpful in improving the precision of the estimate, the cost associated with additional replication serves as a limiting factor in its use. Additionally, sensitivity of statistical methods depends on the number of replications. A discussion of how to determine the number of replications is given next.

Determination of Number of Replications: When comparing 2 treatments, it is essential that the experiment be large enough to ensure that if there is a true difference between the treatments, the experiment will obtain a significant result. Researchers have suggested that for crops and vegetable investigations, an experiment with 4 to 8 replications is sufficient to provide a reasonable degree of precision (Little and Hills, 1978). Similarly, Cochran and Cox (1964) have suggested a convenient method for estimating the number of replications required to detect a specified difference. We can also use a test of significance to determine the number of replications. It was mentioned earlier that the error variance is used as a measure of precision or accuracy of an experiment. The precision desired in an experiment could be set either by specifying the size of the true difference or by specifying the width of the confidence interval. In conducting an experiment, the researcher may be interested in knowing if 2 treatments differ in their effects by less than a certain quantity, say d. Should a greater difference than d be observed, the results are significant. The significance of the difference between 2 treatments (d) is tested by performing a t-test, as shown below:

$$ t = \frac{\bar{x}_i - \bar{x}_j}{\sqrt{2s^2/r}} \tag{3-1} $$

where

$$\bar{x}_i, \bar{x}_j \quad = \text{arithmetic means of treatments } (i) \text{ and } (j)$$

$$s^2 \quad = \text{variance}$$

$$r \quad = \text{replications.}$$

Given that d is the difference between $\bar{x}_i$ and $\bar{x}_j$, we can substitute the absolute value of d in Equation 3-1 and from it determine the number of replications needed for an experiment as follows:

$$t_o = \frac{|d|}{\sqrt{2s^2/r}} \tag{3-2}$$

where

$$t_o \; = \text{critical value of } t \text{ from the } t \text{ table}$$

$$s^2 \; = \text{variance}$$

Equation 3-2 serves as the basis for the calculation of the number of replications needed so that a significant difference between 2 treatments can be observed. Rewriting Equation 3-2, we get the following:

$$r = \frac{2t_o^2 s^2}{d^2} \tag{3-3}$$

Equation 3-3 uses the measure of error variance for a set of experimental units, along with the level of significance, to determine the number of replications needed for the results to be significant.

Another way to improve the precision of an experiment is to carefully select the treatments for the experiment. For example, if a horticulturist is interested in studying the effect of a fungicide on roses, it is more useful to the experimenter to determine how the experimental unit (roses) responds to the increasing doses of the selected fungicide, rather than to see if there is a significant difference between 2 succeeding doses. This allows the experimenter to conduct tests of significance that are more sensitive than merely looking at the differences between adjacent means in an array.

3.1.2 Refining the Experimental Conditions

Once the researcher has defined the problem and has selected a design that efficiently tests the researcher's hypothesis, it is essential to look at some of the critical elements of conducting an experiment. Some of these elements were touched upon in Chapter 1. Other refinements in the experimental conditions are the uniform application of the treatments, and selection of an appropriate measure for the treatment effects. Researchers can also improve precision by taking precautions that will prevent gross errors (unusually large or unusually small values that may have been mistakenly recorded) and control the external influences.

Poor attention to these elements will negate the superiority of the design and contribute to the likely increase in experimental error.

3.1.3   Reducing Variability in the Experimental Material

To avoid problems associated with the variability of the experimental material, uniform material is frequently prepared for the experiment. For example, in some agricultural experiments, researchers use inbred lines of plants and animals to ensure uniform experimental material. While appropriate in some conditions, the use of the inbred lines have been questioned by Biggers and Claringbold (1954). They have argued that inbred lines are not more homogeneous than randomly bred material, and have suggested the use of F_1 hybrids between the inbred lines as more appropriate than the inbred lines themselves.

Researchers can also use a sample chosen for its homogeneity from a large batch of experimental material. For example, in an animal experiment, the researcher may choose the initial weight as the criterion. Thus, 2 animals with the highest weight are grouped in a pair, and the 2 with the next highest weight in the second group, and this is continued until all animals (experimental material) have been paired in this fashion.

Undoubtedly, such pairing of animals and plants reduces the variability of the experimental material. However, the substantial increase in precision some-times may be at the cost of getting conclusions that are not representative of a wider class of experimental units.

3.2   APPROACHES TO ELIMINATE UNCONTROLLED VARIATION

Reducing the effect of uncontrolled variation on the error of the treatment comparisons is an essential element of a good design. Agricultural scientists in conducting experiments on varietal differences, for example, are concerned with arranging the experiment in such a way that they can, with confidence and accuracy, separate the uncontrolled variations from the varietal differences.

In the chapters to come you will be introduced fully to the different designs that allow the experimenter to separate the uncontrolled variation from the treatment effects under a variety of circumstances. Here we are interested in giving an overview of the different designs, from the simplest to the most complex, and how they can be used appropriately to remove the effect of the uncontrolled variation from the experiment.

In the most simple case of comparing 2 treatments, the researcher is inter-ested in evaluating the differences between treatments, rather than the treat-ments themselves. Thus, to be able to accurately evaluate the differences between treatments, the experimenter must select pairs of similar experimental units. For example, a horticulturist may choose a pair of roses that have the same genetic makeup, and an animal scientist may choose 2 male piglets from the same litter.

Another way of selecting a pair of observations is by means of *self-pairing*. This involves making observations on the same experimental unit on 2 different occasions. No matter which pairing method is used, the researcher is interested in having a number of pairs of experimental units; the 2 units in each pair are expected to provide, as nearly as possible, identical observations in the absence of treatment differences.

As was suggested in Section 3.1.3, there are many ways of pairing experimental units so that the experimenter is assured of uniform material. Proper randomization of the paired experimental units also provides added protection from the uncontrolled variations in an experiment.

When there are more than 2 treatments to be compared, each treatment must be applied to several different units. The reason for this is obvious. If each treatment is applied to only a single unit, the experimenter will have difficulty distinguishing between the difference that might be caused by the different treatments and the inherent difference between the units.

Researchers, when faced with t alternative treatments, can group the units into sets of t, the units in each set being expected to give, as nearly as possible, the same observation if the treatments are equivalent in their effect. Each set of the t units is called a *block*. An experiment in which the block differences are removed from the error, and the only recognizable difference between units is the treatment, is called a *randomized block* design. In agricultural research there are many ways of blocking the experimental units, and we shall return to this in Chapters 4, 5, and 6.

The statistical model for a randomized block design is

$$X_{ij} = \mu + \tau_i + \beta_j + \varepsilon_{ij} \left\{ \begin{array}{l} i = 1, 2, ..., a \\ j = 1, 2, ..., b \end{array} \right\} \tag{3-4}$$

where

$$\mu = \text{the overall mean}$$

$$\tau_i = \text{the } i\text{th treatment effect}$$

$$\beta_j = \text{the } j\text{th block effect}$$

$$\varepsilon_{ij} = \text{the random error term.}$$

In agricultural experiments the general principle for grouping plots into blocks is to make sure that the plots chosen minimize the uncontrolled variation from plot to plot within blocks. This is usually achieved by arranging the plots within a block, in a compact square area. Again, proper randomization within blocks will allow the experimenter to statistically assess uncontrolled variation between plots.

Even though blocking provides an added mechanism to remove uncontrolled variation from the results, experimenters should be cautious about the erratic variations that might result from the order in which plots are cultivated and harvested. For example, when harvesting takes more than a day, it is recommended that the plots in 1 block be harvested on the same day. This allows the constant differences between days to become identified with block differences and to not contribute to the error of the experiment. Furthermore, care must be taken in the collection of data and in the tests to be used in evaluating the data.

3.3 ERROR ELIMINATION BY SEVERAL GROUPINGS OF UNITS

So far we have discussed how the single grouping of units into blocks can be used to reduce error in an experiment. There are several other types of design that utilize the blocking principle. The Latin square, the combined Latin squares, and the Greco-Latin squares are such designs. Each of these designs allows for the removal of the known sources of variability from the experimental error. Hence, depending on the objective of the study, the researcher has at his/her disposal any one of these designs. In Section 3.3.1 a brief discussion of the use of the Latin square design is given simply to provide the reader with an overview of the contribution of this design in reducing error. For a detailed discussion of the Latin square, the reader is referred to Chapter 4 (Section 4.2.3). In Sections 3.3.2 and 3.3.3 we discuss the use of the combined and Greco-Latin squares, respectively.

3.3.1 Latin Square

This design is used to eliminate the effect of two known sources of uncontrolled variation. By using the row and column blocking, the design allows the experimenter to remove 2 sources of variability from the experimental error. A conditional requirement of the design is that the treatments be randomly assigned to the rows and columns, such that each treatment occurs only once in each row and column.

To satisfy the requirement that each treatment occurs once in each row and column means that the number of replications and treatments must be equal. The implication of this requirement is that when there are a large number of treatments in the experiment, there must be an equally large number of replications. Technically and economically, this places a limitation on the use of this design. The statistical model for a Latin square is

$$X_{ijk} = \mu + \alpha_i + \tau_j + \beta_k + \varepsilon_{ijk} \left\{ \begin{array}{l} i = 1, 2, \ldots, n \\ j = 1, 2, \ldots, n \\ k = 1, 2, \ldots, n \end{array} \right\} \tag{3-5}$$

where

X_{ijk} = the observation of the ith row and kth column for the jth treatment.

μ = the overall mean

α_i = the ith row effect

τ_j = the jth treatment effect

β_k = the kth column effect

ε_{ijk} = the random error.

The model is completely additive; that is, there is no interaction between rows, columns, and treatments.

Researchers should also recognize that when using a Latin square design, especially when there are only 3 or 4 treatments in an experiment, the degrees of freedom associated with error $(t-1)(t-2)$ are not adequate to estimate the σ^2. For example, when the number of treatments is 3, the degrees of freedom for error is only 2. This is not sufficient for an adequate estimate of σ^2. In such experimental conditions it is necessary to use more than 1 square in the experiment.

Although the Latin square design is a good design for many types of field experiments, it is suggested that it should not be used for experiments with fewer than 4 treatments or more than 8 (Kempthorn, 1952). The Latin square is most valuable to those experimenters who use only restrictions on randomization (not true factors of interest) for the rows and columns. Furthermore, the design is helpful in analyzing data where there are no interactions between factors. It should, however, be kept in mind that agricultural researchers have used the Latin square design to screen for major effects even when there exists a small amount of interaction.

In Appendix C a list of standard Latin squares is provided. Different Latin squares can be constructed using the randomization process. The randomization of a 3×3 standard Latin squares, for example, is accomplished by first randomizing the rows and then the columns. For a 4×4 Latin square, we first select at random 1 of the 4 tabulated squares from Appendix C and then randomize the rows and columns respectively. For the higher order squares, the rows, columns, and treatments are all randomized independently.

To illustrate the randomization process for a 4 × 4 Latin square, we can use a table of permutations, or another random selection procedure such as drawing pieces of paper from a bowl in which 4 pieces of paper with a number from 1 to 4 written on each are placed. Suppose that the first number drawn is 2. We draw a second piece of paper from the bowl. Suppose this number is 1. We continue this process until we have drawn all 4 numbers. This process is repeated 2 more times so that we have 3 sets of 4 numbers each. The first set of numbers is used to select a Latin square from Appendix C. The other 2 sets are used for randomizing the rows and columns of the selected 4 × 4 Latin square. Suppose the 3 sets of numbers drawn are: 2, 1, 4, 3; 2, 3, 1, 4; and 1, 4, 3, 2. Using the first set (2,1, 4, 3) we start with the second 4 × 4 tabulated square from Appendix C. The selected Latin square is:

A	B	C	D
B	C	D	A
C	D	A	B
D	A	B	C

Now we use the second set of numbers (2, 3, 1, 4) to randomize the rows of the above 4 × 4 Latin square. This is shown below:

Original Row
  Number

2	B	C	D	A
3	C	D	A	B
1	A	B	C	D
4	D	A	B	C

The third set of numbers (1, 4, 3, 2) is used to randomize the columns as illustrated below:

Original Column
        Number

1	4	3	2
B	A	D	C
C	B	A	D
A	D	C	B
D	C	B	A

The same procedure could be applied easily to larger squares. For a discussion of the theoretical basis of the randomization, the reader is referred to Yates (1933).

In the next section we will discuss the use of the multiple squares, which is used when there is not sufficient treatment for an adequate estimation of the σ^2. The multiple squares is an extension of the Latin square.

3.3.2 Combined Latin Square

The problem of inadequate degrees of freedom associated with error arises when the number of treatments in the experiment is not enough. The researcher can deal with this problem either by extending 1 of the sides of the square in space or time or by using a separate multiple Latin square.

Table 3.3.2.1 (*a*) shows the extension of the rows of a 4×4 Latin square. This form of a multiple Latin square implies that the 2 squares (columns 1 - 4,

TABLE 3.3.2.1.

Layout of a Multiple Latin Squares with Common Row Effects (*a*), and Completely Separate Squares Shown in (*b*).

a

Row	Column							
	1	2	3	4	5	6	7	8
1	A	B	C	D	A	B	C	D
2	B	C	D	A	B	C	D	A
3	C	D	A	B	C	D	A	B
4	D	A	B	C	D	A	B	C

b

Row	Column							
	1	2	3	4	5	6	7	8
1	A	B	C	D				
2	B	C	D	A				
3	C	D	A	B				
4	D	A	B	C				
5					A	B	C	D
6					B	C	D	A
7					C	D	A	B
8					D	A	B	C

and 5 - 8) have common row effects. When it is assumed that the row effects are consistent over several sets of columns, a less stringent requirement for the occurrence of each treatment in rows is allowed. This means that each treatment appears once in each column and twice or more in each row. This is not the case for separate multiple Latin squares.

As the rows (or columns) of a Latin square are extended, the resultant design can be looked at as a Latin rectangle rather than a square. This is not a problem, as the analysis of a Latin rectangle follows the model of a single Latin square design where the sum of squares for rows, columns, treatment, and error are computed for the analysis.

The completely separate multiple square, as shown in Table 3.3.2.1 (b), indicates that there is no relation between the rows (or columns) in the different squares. In either form of the Latin square, the number of treatments increases, thus allowing the researcher to estimate the σ^2 with adequate precision.

The difference in practice between the designs is best observed by considering what types of systematic variation are eliminated from the error by the 2 designs. What makes the 2 forms of the design different from one another is whether the row differences (or the column differences) might be expected to be similar for the different squares. For example, a horticulturist uses a leaf as an experimental unit, and plants and the leaf position are chosen as the 2 blocking criteria. If 2 groups of plants are used for the 2 squares, it would seem reasonable to assume that the differences in leaf position should be similar for plants in both groups. This example shows how 1 of the blocking criteria is consistent over squares.

Based on the assumption that the row effects are the same for both the original squares and for all columns, the form of the analysis differs. For example, the analysis for a design such as the one depicted in Table 3.3.2.1(a), where the width of the rows of a Latin square is extended, is presented in Table 3.3.2.2.

In Table 3.3.2.2 the notation used for the various sums of squares are:

$$R_1 \ldots R_4 = \text{row totals.}$$

$$G = \text{grand total.}$$

$$C_1 \ldots C_8 = \text{column totals.}$$

$$T_1 \ldots T_4 = \text{treatment totals.}$$

$$x_{ijkl} = \text{a particular observation.}$$

The form of the analysis of variance for 2 independent squares such as the one shown in Table 3.3.2.1 (b) differs slightly. For this design, the sum of squares for rows and columns is calculated separately for each square. Additionally, we calculate a sum of squares for the overall difference between squares. Table 3.3.2.3 shows the ANOVA table for this design.

TABLE 3.3.2.2.

The ANOVA Table for a Latin Rectangle

Source of Variation	Degree of Freedom	Sum of Squares
Rows	1	$\dfrac{R_1^2 + R_2^2 + R_3^2 + R_4^2}{8} - \dfrac{G^2}{32}$
Columns	7	$\dfrac{C_1^2 + C_2^2 + \ldots + C_8^2}{4} - \dfrac{G^2}{32}$
Treatments	3	$\dfrac{T_1^2 + T_2^2 + T_3^2 + T_4^2}{4} - \dfrac{G^2}{32}$
Error	20	by subtraction
Total	31	$\sum (x_{ijkl}^2) - \dfrac{G^2}{32}$

TABLE 3.3.2.3.

Analysis of Variance for 2 Independent Latin Squares

Source of Variation	Degree of Freedom	Sum of Squares
Between squares	1	$\dfrac{G_1^2 + G_2^2}{16} - \dfrac{G^2}{32}$
Rows	9	(row $SS)_1$ + (row $SS)_2$
Columns	9	(column $SS)_1$ + (column $SS)_2$
Treatments	3	$\dfrac{T_1^2 + T_2^2 + T_3^2 + T_4^2}{4} - \dfrac{G^2}{32}$
Error	9	by subtraction
Total	31	$\sum (x_{ijk}^2) - \dfrac{G^2}{32}$

The notations for the various sums of squares given in Table 3.3.2.3 are similar to those given in Table 3.3.2.2, except that G_1 and G_2 represent the totals for square 1 and 2, and the subscripts 1 and 2 for the rows and columns refer to the rows and columns of each square, respectively.

3.3.3. Greco-Latin Square

This is an associated design of the Latin square which allows for 1 more restriction on randomization. Recall that in the case of the randomized block design, experimental units were grouped in 1 way. When Latin square design is used, the experimental units are simultaneously grouped in 2 ways. If it is desired to group the units in 3 or more ways, use is made of the Greco-Latin square.

In this design the treatments are grouped into replicates in 3 different ways such that the effects of the 3 sources of variation are equalized for all treatments. This means that we are able to systematically control 3 sources of variability by blocking in 3 directions. The additional grouping that is made possible by this design is usually represented by Greek letters. The sum of squares due to the Greek letter factor is computed directly from the Greek letter totals, and the experimental error is further reduced by this amount.

To see how this third grouping is incorporated into the design, consider a $n \times n$ Latin square, and superimpose on it a second $n \times n$ Latin square in which the treatments are denoted by Greek letters. When the second Latin square is superimposed on the first, the resultant Latin square is called a Greco-Latin square if each Greek letter appears once and only once with each Latin letter. Table 3.3.3.1 shows a Greco-Latin square with 4 treatments.

TABLE 3.3.3.1.

A 4 × 4 Greco-Latin Square

Row	Column 1	2	3	4
1	Aα	Bβ	Cγ	Dδ
2	Bδ	Aγ	Dβ	Cα
3	Cβ	Dα	Aδ	Bγ
4	Dγ	Cδ	Bα	Aβ

Experimenters use Greco-Latin square to investigate 4 factors. Factors could be assigned to the rows, columns, Latin letters, and Greek letters each at n levels with n^2 runs. The statistical model for a Greco-Latin square is given as:

$$X_{ijkl} = \mu + \theta_i + \tau_j + \omega_k + \rho_l + \varepsilon_{ijkl} \left\{ \begin{array}{l} i = 1, 2, \ldots, n \\ j = 1, 2, \ldots, n \\ k = 1, 2, \ldots, n \\ l = 1, 2, \ldots, n \end{array} \right\} \quad (3\text{-}6)$$

where

$$X_{ijkl}$$ = the observation in row i and column l for Latin letter j and Greek letter k

$$\theta_i$$ = the effect of the ith row

$$\tau_j$$ = the effect of the Latin letter treatment j

$$\rho_l$$ = the effect of column l

$$\varepsilon_{ijkl}$$ = the random error.

The statistical analysis of the results of the Greco-Latin square experiment is a straightforward extension of the Latin square. Table 3.3.3.2 is the ANOVA Table for the Greco-Latin square. The sum of squares for the various sources of variation is computed in the same manner as that of the Latin square given in Chapter 4, Section 4.2.3.

TABLE 3.3.3.2.
ANOVA Table for a Greco-Latin Square

Source of Variation	Sum of Squares	Degree of Freedom
Latin letter treatments	$\left(\sum\limits_{j=1}^{n} \dfrac{T_L^2}{n} \right) - C$	$n - 1$
Greek letter treatments	$\left(\sum\limits_{k=1}^{n} \dfrac{T_G^2}{n} \right) - C$	$n - 1$
Rows	$\left(\sum\limits_{i=1}^{n} \dfrac{R^2}{n} \right) - C$	$n - 1$
Columns	$\left(\sum\limits_{l=1}^{n} \dfrac{C_l^2}{n} \right) - C$	$n - 1$
Error	Error SS computed by subtraction	$(n - 3)(n - 1)$
Total	$\sum\limits_{i}\sum\limits_{j}\sum\limits_{k}\sum\limits_{l} X^2 - C$	$n^2 - 1$

In computing the various sum of squares in Table 3.3.3.2, the term C is the correction factor, computed as:

$$C = \frac{G^2}{N} \tag{3-7}$$

where

G = grand total; and

N = total number of observations.

In Appendix C several Greco-Latin squares are presented. The designs have been constructed for all numbers of treatments from 3 to 12, except 6 and 10 (Cochran and Cox, 1957). It is generally thought that no $n \times n$ Greco-Latin square exists whenever the size n leaves a remainder of 2 when divided by 4 (Cox, 1958).

The intent of this chapter was to give the reader an overview of the various approaches in reducing error. Error can be reduced by increasing the size of the experiment, refining the experimental technique, and reducing variability in the experimental material. The uncontrolled variation in an experiment can be reduced further by the design used in an experiment. The selection of the design is based on the objective of the experiment, and other factors such as technical and cost issues. In trying to minimize experimental error, several ways of grouping experimental units were discussed. The Latin square and its associated designs were reviewed to show how researchers can use these designs to reduce error. In the subsequent chapters of the book, each design is further detailed.

REFERENCES AND SUGGESTED READINGS

Biggers, J. D. and Claringbold, P. J. 1954. "Why use inbred lines?" *Nature* 174, 596.

Cochran, W. G. and Cox, G.M. 1964. *Experimental Designs.* New York: John Wiley & Sons.

Cox, D. R. 1958. *Planning of Experiments.* New York: John Wiley & Sons.

Kempthorn, O. 1952. *The Design and Analysis of Experiments,* New York: John Wiley & Sons.

Little, T. L. and Hills, F. J. 1978. *Agricultural Experimentation: Design and Analysis.* New York: John Wiley & Sons.

Yates, F. 1933. "The formation of latin squares for use in field experiments." *Emp. J. Exp. Agric.* 1:235-244.

Chapter **4**

SINGLE-FACTOR EXPERIMENTAL DESIGNS

4.1 INTRODUCTION

Many agricultural experiments involve the use of a single-factor experimental design. In this type of design, a single factor varies while all other factors are held constant. For example, when an agronomist is interested in finding whether one variety is superior to another, he/she will be using a single-factor experiment where the single variable factor is the variety, and the treatments or factor levels are different varieties. In such an experiment all cultural practices—water, fertilizer application, pest control, and other management activities—remain the same.

Single-factor experiments can be grouped under two distinct experimental designs. The first design involves a small number of treatments and is called the *complete block design*. As the name implies, complete block designs are characterized by blocks, each of which contains at least 1 complete set of treatments. The second grouping of designs is the *incomplete block designs*. These designs contain a large number of treatments and are also characterized by blocks. However, each block contains only a fraction of the number of treatments.

In this book you will be introduced to 3 types of the complete block designs (*completely randomized design, randomized complete block design,* and *Latin square design*) and 2 incomplete block designs (*lattice* and *partially balanced lattice designs*). As we discuss each of these designs, we will introduce you to the advantages of using the design, and the randomization procedures, plot layout, and analysis of variance for the particular design.

4.2 COMPLETE BLOCK DESIGNS

4.2.1 Completely Randomized Design (CRD)

This design is the simplest analysis of variance design from the standpoint of assignment of experimental units to treatment or treatment combinations. In this design, treatments are allotted to the experimental units entirely at random or by chance. More specifically, if a treatment is to be applied to 8 units, each unit has the same chance or probability of receiving the treatment, and consequently, there exists no systematic source of error. In a completely randomized design, the units are taken as a single group. Hence, as far as possible, the units forming the group should be homogeneous. This characteristic of the design suggests that it should be used where environmental effects are easily controlled. Agricultural field experiments do not lend themselves to such homogeneous conditions. Therefore, this design is recommended mostly for laboratory and greenhouse experiments where environmental effects are easily controlled. Additionally, a completely randomized design is a logical choice when performing a first-stage pilot experiment where experimental units and conditions are homogeneous.

The advantage of using the completely randomized design is in its simplicity. By using it, an experimenter may avoid making certain dubious assumptions. The inherent simplicity of the design extends to all aspects of the design, namely, the experimental layout, the model underlying the data analysis, and the computations involved in such analysis.

As far as the layout of the experiment is concerned, CRD only requires that treatments be assigned randomly to the experimental units. Any number of treatments and/or replicates may be used. In contrast, stratified designs require the division of treatments into strata or levels, on the basis of some additional measure, then random assignment to the experimental units or plots. Other designs, as will be discussed later, involve still other complications in laying out the experiment.

The model underlying the completely randomized design involves fewer assumptions than that of any other experimental design. As a consequence, the derivations of parameter estimates are simpler in this case than any other situation. Furthermore, fewer assumptions mean fewer violations of such assumptions. Hence, less can go wrong as far as the statistical inference.

Since the completely randomized design involves fewer terms, computation of such designs are simple. Furthermore, related analyses such as the analysis of covariance and the estimation of the missing data will also generally be simpler. Statistical analysis is generally easy to perform, whether the number of replicates are equal or not.

The major disadvantage in using the completely randomized design is its lack of accuracy, or its inefficiency. The error variance will usually be large in comparison to other designs. This outcome is due to the fact that randomization is not restricted in any way to ensure that the units that receive one treatment are similar to those that receive another treatment. This disadvantage is partly offset by the fact that no design yields as many degrees of freedom for the error variance as does the completely randomized design, assuming some fixed amount of data. The greater the degrees of freedom, the higher is the power of a test. When experiments require a large number of experimental units, completely randomized designs cannot ensure precision of the estimates of treatment effects.

Randomization: There are several methods for random allocation of treatments to the experimental units. You will be introduced to 2 such procedures as described below in a step-by-step fashion.

Example 4.2.1.1

An animal scientist has 5 treatments (A, B, C, D, and E), each replicated 3 times. Determine the random layout of the experiment for the scientist.

STEP 1: Determine the total number of experimental plots. This is simply the product of the number of treatments and replications:

$$n = t \times r$$

$$n = 5 \times 3$$

$$n = 15$$

STEP 2: Assign each experimental plot a number. In the present case, the plots will be numbered from 1 to 15 as shown in Figure 4.2.1.1.

STEP 3: Assign each treatment to an experimental plot, using either the random digits technique (method A), or an alternative to random allocation (method B).

1	2	3	4	5
6	7	8	9	10
11	12	13	14	15

FIGURE 4.2.1.1.

Layout of an experimental plot.

Method A—Random Digit Technique: To illustrate the use of a Table of Random Numbers, such as the one given in Appendix D, we present a partial Table of Random Numbers in Table 4.2.1.1. You may choose to randomly start at any point in the table you wish. However, for our example we start at the left-hand corner of the table. The first number is 85967. One may wish to select the first 3 digits, the middle 3 digits, or the last 3 digits of this number. Any of these choices are appropriate. We have elected to choose the first 3-digit number and read downward until 15 numbers have been selected.

TABLE 4.2.1.1.

A Partial Table of Random Numbers from Appendix D

85967	73152	14511
07483	51453	11649
96283	01898	61414
49174	12074	98551
97366	39941	21225
90474	41469	16812
28599	64109	09497
25254	16210	89717
28785	02760	24359
84725	86576	86944
41059	66456	47679
67434	41045	82830
72766	68816	37643
92079	46784	66125
29187	40350	62533
74220	17612	65522
03786	02407	06098
75085	55558	15520
09161	33015	19155

As you will note below, the 15 numbers selected from the first column of Table 4.2.1.1 are:

1.	859	6.	904	11.	410
2.	074	7.	285	12.	674
3.	962	8.	252	13.	727
4.	491	9.	287	14.	920
5.	973	10.	847	15.	291

Once the numbers are selected, rank the numbers in either ascending or descending order. In this example we have ranked the numbers in ascending order as shown below:

Random Number	Rank	Random Number	Rank	Random Number	Rank
859	11	904	12	410	6
074	1	285	3	674	8
962	14	252	2	727	9
491	7	287	4	920	13
973	15	847	10	291	5

Divide the 15 ranks into 5 groups, each consisting of 3 numbers as follows:

Group Number	Rank in the Group		
1	11	1	14
2	7	15	12
3	3	2	4
4	10	6	8
5	9	13	5

Now we are ready to assign the treatments to the experimental plots by using the group number as the treatment number and the corresponding ranks in each group as the plot number. In our example, the first group is assigned to treatment A and the last group (group 5) receives treatment E. Plots that will receive treatment A are numbers 11, 1, and 14. Similarly, treatment B is applied to plots 7, 15, and 12. The remaining treatments are applied in the same manner to the rest of the experimental units or plots. Finally, we have randomly assigned all treatments to the experimental plots as shown in Figure 4.2.1.2.

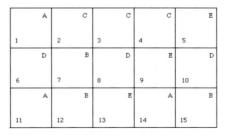

FIGURE 4.2.1.2.
A completely randomized layout for 5 treatments and 3 replications.

Method B—Alternative to Random Allocation: If a table of random numbers is not available, treatments can be allocated to the experimental units by drawing "lots" as shown below.

Let the number of *i*th treatment be written on r_j pieces of papers ($j = 1, 2, ..., k$). All the pieces of paper are then folded individually to protect the identity of the numbers written on them. Place them in a container and mix them thoroughly. Draw one piece at a time without replacement. The treatment that is drawn in the *j*th draw is allotted to the *j*th plot ($j = 1, 2, ..., k$). In the present example there should be 15 pieces of paper, 3 each with treatments A, B, C, D, and E written on them. Other procedures are also used to randomly assign treatments to experimental units.

Analysis of Variance: In a completely randomized design there are 2 sources of variation that one encounters in an experiment. The first is the treatment variation and the second is the variation due to experimental error. The treatment variation or treatment sum of squares (*SSTr*) is due to variability between sample results, while variability within sample results is measured by the error sum of squares (*SSE*). It is the relative size of these 2 sources of variation that determines whether differences observed among treatments are real or due to chance.

Table 4.2.1.2 gives the general format of a 1-factor analysis of variance table. As was mentioned earlier, computation of the analysis of variance for experiments with equal or unequal replication is simple. The following examples illustrate how the analysis of variance is applied to experiments with equal as well as unequal replications.

Equal Replications: In this type of experiment, the numbers of replications are equal. The methods of calculating each of the sum of squares and performing the analyses of variance are given in the following example.

TABLE 4.2.1.2.
General Format of a 1-Factor Analysis of Variance Table

Source of Variation	Degree of Freedom	Sum of Squares	Mean Square	F
Treatment	$t - 1$	SST_r	$MST_r = SST_r / t - 1$	MST_r / MSE
Error	$n - t$	SSE	$MSE = SSE/n - t$	
Total	$n - 1$	SST		

Example 4.2.1.2

An animal scientist in performing a study of vitamin supplementation randomly assigned 5 heifers to each of the following 8 treatment groups. The following weight gains (kg) were recorded from 5 replicates:

	Weight Gain, kg				
Treatment	Rep. I	Rep. II	Rep. III	Rep. IV	Rep. V
A	4.5	5.2	4.5	3.9	4.9
B	3.8	4.7	4.3	4.4	3.7
C	5.4	4.7	4.6	3.7	3.9
D	5.2	5.0	4.8	3.6	5.6
E	4.0	4.9	4.3	4.8	4.2
F	4.1	4.5	5.2	4.8	5.1
G	5.1	4.9	4.2	3.9	3.8
H (Control)	4.6	5.2	4.9	3.8	4.2

Perform the analysis of variance to determine if there are differences between the treatments.

Solution

STEP 1: Calculate the treatment totals (T), their respective means, and the grand total (G) as shown in Table 4.2.1.3.

TABLE 4.2.1.3.

Weight Gain by Animals from a Completely Randomized Design with 8 Treatments (t) and 5 Replications (r)

Treatment	Weight Gain, kg					Treatment Total (T)	Treatment Mean
A	4.5	5.2	6.2	3.9	4.9	24.7	4.94
B	5.6	4.7	4.3	4.4	6.1	25.1	5.02
C	6.4	6.7	6.8	6.1	6.9	32.9	6.58
D	5.2	5.0	6.8	3.6	5.6	26.2	5.24
E	4.0	4.9	4.3	4.8	4.2	22.2	4.44
F	7.1	6.5	6.2	6.8	6.1	32.7	6.54
G	6.1	4.9	4.2	3.9	6.8	25.9	5.18
H (Control)	4.6	4.0	4.9	3.8	4.2	21.5	4.30
Grand Total (G)						211.2	
Grand Mean							5.28

STEP 2: Determine the degree of freedom for each source of variation as shown below:

Total $d.f.$ $= (r)(t) - 1 = (5)(8) - 1 = 39$

Treatment $d.f.$ $= t - 1 = 8 - 1 = 7$

Error $d.f.$ $= t(r - 1) = 8(5 - 1) = 32$

STEP 3: Calculate the correction factor (C) and the various sums of squares using the following notations:

$$C = \frac{G^2}{N} \tag{4-1}$$

where

    C = correction factor

    G = grand total

    N = the total number of experimental plots $[(r)(t)]$

    X_i = an observation in the ith plot

    T_i = treatment total for the ith plot

    r_i = the number of replication of the ith treatment.

$$C = \frac{G^2}{N} = \frac{(211.2)^2}{40} = 1,115.14 \tag{4-2}$$

$$\text{Total } SS = \sum X_i^2 - C$$

$$\text{Total } SS = \left[(4.5)^2 + (5.2)^2 + \ldots + (3.8)^2 + (4.2)^2\right] - 1,115.14$$

$$= 1,160.26 - 1,115.14$$

$$= 45.12$$

$$\text{Treatment}\,SS = \frac{\sum T_i^2}{r} - C \tag{4-3}$$

$$= \frac{(24.7)^2 + (25.1)^2 + \ldots + (25.9)^2 + (21.5)^2}{5} - 1,115.14$$

$$= 1,140.83 - 1,115.14$$

$$= 25.69$$

$$\text{Error}\,SS = \text{Total}\,SS - \text{Treatment}\,SS \tag{4-4}$$

$$= 45.12 - 25.69$$

$$= 19.43$$

STEP 4: Calculate the mean square for the treatment and error variations by dividing each sum of squares by its respective degrees of freedom.

$$\text{Treatment}\,MS = \frac{\text{Treatment}\,SS}{t-1} \tag{4-5}$$

$$= \frac{25.69}{8-1}$$

$$= 3.67$$

$$\text{Error}\,MS = \frac{\text{Error}\,SS}{t\,(r-1)} \tag{4-6}$$

$$= \frac{19.43}{(8)(4)}$$

$$= 0.61$$

STEP 5: Calculate the F value using the treatment and error mean squares as:

$$F = \frac{\text{Treatment}\,MS}{\text{Error}\,MS} \tag{4-7}$$

$$= \frac{3.67}{0.61}$$

$$= 6.02$$

The ANOVA table for the present example is shown in Table 4.2.1.4.

TABLE 4.2.1.4.
Analysis of Variance of Animal Weight Gain Data in a Completely Randomized Design

Source of Variation	Degree of Freedom	Sum of Squares	Mean Square	F
Treatment	7	25.69	3.67	6.02^{**}
Error	32	19.43	0.61	
Total	39	45.12		

Note: Coefficient of Variation (cv) = 14.8%. Computation of the *cv* is presented in Step 7.
** = Highly significant or significant at 1% level.

STEP 6: Compare the computed F given in Table 4.2.1.4 with the tabular F value given in Appendix E. For our example, the tabular F (for 7 degrees of freedom for treatment and 32 degrees of freedom for the error source of variation) is 2.32 for the 5% level of significance and 3.25 for the 1% level of significance.

STEP 7: Make a decision on the significance of the differences among treatments by applying the following rules:

1. A treatment difference is said to be *highly significant* if the computed F value is greater than the tabular F at the 1% level of significance. Conventionally, 2 asterisks are placed on the computed F value in the analysis of variance.
2. A treatment difference is said to be *significant* if the computed F value is greater than the tabular F at the 5% level of significance but smaller than or equal to the tabular F value at the 1% level of significance. Conventionally, 1 asterisk is placed on the computed F value in the analysis of variance.
3. A treatment difference is said to be *nonsignificant* if the computed F value is smaller than or equal to the tabular F at the 5% level of significance. Conventionally, *ns* is placed on the computed F value in the analysis of variance.

What you observe in the present example is an F value that is highly significant. This means that the test indicates the existence of some differences among the treatments, but does not specify which treatments differ from one another.

You should keep in mind that if we had observed a nonsignificant test result, the outcome would not mean that all treatments are the same. What it would indicate is the failure of the experiment to detect any observable differences among the treatments. Failure to detect treatment differences could be a result of very small treatment differences or a very large experimental error, or both. When nonsignificant results are obtained from an experiment, it is helpful to take a look at the size of the experimental error and the numerical difference among treatment means. These will serve as a guide to the researcher in carrying out further trials. Should you note that both values are small, it implies that the treatment differences are so small that differences cannot be detected. Under such a condition, no additional trials will be helpful. On the other hand, if both values are large, additional trials may reduce the experimental error and possible treatment differences detected.

As a way of determining the reliability of the experiment, the coefficient of variation (cv) for the experiment should be calculated. The cv indicates the degree of precision with which the treatments are compared. In actuality, cv expresses the experimental error as a percentage of the mean. Thus, the higher the value of the cv, the lower is the reliability of the experiment. Conventionally, the cv value is reported below the ANOVA table for an experiment.

The coefficient of variation for the present example is calculated as shown below:

$$\text{Grand mean} = \frac{G}{n} \qquad\qquad (4\text{-}8)$$

$$= \frac{211.2}{40}$$

$$= 5.28$$

$$cv = \frac{\sqrt{\text{Error } MS}}{\text{Grand mean}} \times 100 \qquad\qquad (4\text{-}9)$$

$$= \frac{\sqrt{0.61}}{5.28} \times 100$$

$$= 14.8\%$$

As mentioned earlier, the higher the value of the coefficient, the less reliable the experiment. This cv indicates that there is reliability in the experiment. It should be noted that the coefficient of variation varies with the type of experiments and the characteristics measured. Gomez and Gomez (1984, p. 17) have reported that the acceptable range of cv is 6 to 8% for variety trials, 10 to 12% for fertilizer trials, and 13 to 15% for insecticide and herbicide trials. Furthermore, they point out that in a field experiment the cv for rice yield is about 10%, that for tiller number is about 20% and for plant height the cv is about 3%.

Unequal Replications: In the preceding example we performed the 1-factor analysis of variance when there were equal number of replications. Sometimes it is not possible to have an equal number of replications for all treatments. For instance, animal scientists may be performing experiments on different breeds of animals where the number of a particular breed may be limited to a few. In such situations we may not have equal number of replications.

In another situation, a researcher may start with an experiment that has equal replications. However, for previously unseen reasons, elements of the experimental unit may be destroyed or lost in the process. If such conditions prevail, the steps followed in performing the analysis of variance are similar to those for equal replications.

Example 4.2.1.3

In a recent study on maize hybrid response to nitrogen fertilization in the Northern Corn Belt, agronomists obtained the following data. Determine if there is a difference among the hybrids in yield.

	Yield, bu/acre		
Hybrid	Rep. I	Rep. II	Rep. III
P3747	133	154	175
P3732	125	151	161
Mo17 × A634		146	152
LH74 × LH 51		166	167
CP 18 × LH 54	122		159
Control	120	159	147

Solution

STEP 1: Calculate the treatment totals (T), their respective means, and the grand total (G) as shown in Table 4.2.1.5.

STEP 2: Determine the degree of freedom for each source of variation as shown below:

Total *d.f.* $= (n - 1) = 15 - 1 = 14$

Treatment *d.f.* $= (t - 1) = 6 - 1 = 5$

TABLE 4.2.1.5.

Effect of Hybrid on Grain Yield from a CRD Experiment with Unequal Number of Replications

Treatment	Yield, bu/acre			Treatment Total (T)	Treatment Mean
	Rep. I	Rep. II	Rep. III		
P3747	133	154	175	462	154.0
P3732	125	151	161	437	145.7
Mo17 × A634		146	152	298	149.0
LH74 × LH 51		166	167	333	166.5
CP 18 × LH 54	122		159	281	140.5
Control	120	159	147	426	142.0
Grand Total (G)				2237	
Grand Mean					149.6

Error $d.f.$ $= (n - 1)- (t - 1) = 14 - 5 = 9$

STEP 3: Calculate the correction factor (C) and the various sums of square using the following notations:

X_i = an observation in the ith plot.

T_i = treatment total for the ith plot.

G = grand total.

r_i = the number of replication of the ith treatment.

N = the total number of experimental plots $[(r)(t)]$.

$$C = \frac{G^2}{N} = \frac{(2,237)^2}{18} = 278,009.4 \qquad (4\text{-}1)$$

$$\text{Total } SS = \sum X_i^2 - C \qquad (4\text{-}2)$$

$$\text{Total } SS = [(133)^2 + (154)^2 + \ldots + (122)^2 + (159)^2] - 278,009.4$$
$$= 337,697 - 278,009.4$$
$$= 59,687.6$$

$$\text{Treatment}\,SS = \frac{\sum T_i^2}{r_i} - C \tag{4-3}$$

$$= \left[\frac{(462)^2}{3} + \frac{(437)^2}{3} + \frac{(298)^2}{2} + ... + \frac{(426)^2}{3} \right] - 278{,}009.4$$

$$= 292{,}996.5 - 278{,}009.4$$

$$= 14{,}987.1$$

$$\text{Error}\,SS = \text{Total}\,SS - \text{Treatment}\,SS \tag{4-4}$$

$$= 59{,}687.6 - 14{,}987.1$$

$$= 44{,}700.5$$

STEP 4: Calculate the mean square for the treatment and error variations by dividing each sum of squares by their respective degrees of freedom.

$$\text{Treatment}\,MS = \frac{\text{Treatment}\,SS}{t-1} \tag{4-5}$$

$$= \frac{14{,}987.1}{6-1}$$

$$= 2{,}997.2$$

$$\text{Error}\,MS = \frac{\text{Error}\,SS}{(n-1)-(t-1)} \tag{4-6}$$

$$= \frac{44{,}700.5}{9}$$

$$= 4{,}966.72$$

STEP 5: Calculate the F value, using the treatment and error mean squares as:

$$F = \frac{\text{Treatment}\,MS}{\text{Error}\,MS} \tag{4-7}$$

$$= \frac{2{,}997.42}{4{,}966.72}$$

$$= 0.604$$

The ANOVA table for the present example is shown in Table 4.2.1.6.

TABLE 4.2.1.6.

Analysis of Variance of Effect of Hybrid on Grain Yield from a CRD Experiment with Unequal Number of Replications

Source of Variation	Degree of Freedom	Sum of Squares	Mean Square	F
Treatment	5	14,987.1	2,997.42	0.604^{ns}
Error	9	44,700.5	4,966.72	
Total	14	59,687.6		

Note: $cv = 47.1\%$; ns = not significant.

STEP 6: Compare the computed F given in Table 4.2.1.6 with the tabular F value given in Appendix E. For our example, the tabular F (for 5 degrees of freedom for treatment and 9 degrees of freedom for the error source of variation) is 3.48 for the 5% level of significance and 6.06 for the 1% level of significance.

STEP 7: Make a statistical decision. Given that the computed value of 0.493 is less than the tabular value of 3.48 at the 5% level, the treatment difference is said to be nonsignificant.

The coefficient of variation for the present example is:

$$cv = \frac{\sqrt{\text{Error } MS}}{\text{Grand mean}} \times 100 \qquad (4\text{-}9)$$

$$= \frac{\sqrt{4,966.72}}{149.6} \times 100$$

$$= 47.1\%$$

This high cv indicates that the reliability of the experiment is questionable. The researcher may wish to reexamine the way in which the treatments are compared.

4.2.2 Randomized Complete Block Design (RCB)

In agricultural research the experimental unit, often being plots of land or animals, will by nature be different from place to place or animal to animal. Therefore, differences among treatment level means will reflect not only variations that are attributable to the different treatment but also variation that could directly be linked to individual differences prior to experimentation. It should be noted that some variation due to individual differences is inevitable. However, it is possible to isolate some of the variation so that it does not appear in the estimate of treatment effects and the experimental error. The randomized complete block design is one such experimental design that accomplishes this task. The design is used extensively in agricultural research where experimental material is divided into groups such that each group constitutes a single trial or replication. The design requires that each block must contain the same number of experimental units. Experimental units are assigned to k blocks so that the variability within each block is less than the variability among blocks. In this design the variation associated with an extraneous factor is isolated by means of a *blocking technique* whereby blocks are formed from experimental units that are homogeneous. The experimental units in the blocks are known as plots. Homogeneity within each block is achieved through matching units to each block, or using litter mates or subjects with similar heredities. Additionally, observation of each unit under all k treatment levels accomplishes homogeneity within each block. In agricultural experiments we often use neighboring plots of land, animals from the same litter, set of similar trees, etc., for such grouping.

The major advantages of this design are its accuracy of results, flexibility of design, and ease of statistical analysis. Accuracy of results in comparison to the completely randomized designs is due to further partitioning of the experimental units by blocking. In essence, blocking reduces the experimental error, by accounting for the differences among blocks rather than within blocks. This increased precision is due to the fact that error variance, which is a function of comparisons within blocks, is smaller because of homogeneous blocks. The fact that the design allows for any number of treatments and replicates to be used provides more flexibility to the experimenter. Statistical analysis performed on this design is not difficult or complicated. Even when the experimenter is faced with missing data, techniques such as the one developed by Yates (1936b) can be applied for analysis. It should be mentioned, however, that if the gaps in the data are numerous, this design is less convenient than the completely randomized design.

Before describing the randomization procedure for this design, it is relevant to discuss the blocking technique often used in agricultural experimentation.

Blocking technique: As mentioned earlier, the principal objective of blocking is to reduce experimental error. This is accomplished by accounting for a known source of variation among experimental units. To achieve appropriate and effective blocking technique, the experimenter must identify the experimental unit source of variability that can be used as the basis for blocking. In field experiments, soil characteristics, and slope of the land are examples of sources of variation that can be used for blocking purposes. Once such a source of variability is identified, it can be used to serve as a block. Then a decision must be made on the shape and orientation of the block. Following the suggestions of Gomez and Gomez (1984), the following guidelines can be used to maximize variability among blocks:

1. Use narrow and long blocks when the field gradient is unidirectional. Additionally, block orientation should be such that their lengths are perpendicular to the direction of the gradient, as shown in Figure 4.2.2.1.

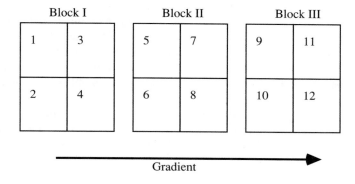

FIGURE 4.2.2.1

Layout of experimental plots when there is unidirectional field gradient.

2. Use the stronger gradient when the fertility gradient runs in 2 directions in a field. That is, when there is a weak and a strong fertility gradient, the weaker gradient should be ignored. The block orientation should be perpendicular to the direction of the strong gradient.
3. Choose 1 of the following alternatives when 2 equally strong gradients are encountered in a field:

   • Use should be made of blocks that are as square as possible.

   • Long and narrow blocks with their length perpendicular to the direction of one gradient should be used. Covariance technique should be applied for the other gradient.

   • Latin square design with 2-way blockings, 1 for each gradient, is quite helpful.

4. Use square blocks when the pattern of variability is not predictable in a field.

Randomization: The randomization process for the randomized complete block design entails numbering the treatments from 1 to k, in any order. Similarly, the units in each block are numbered from 1 to k. Then the k treatments are allotted at random to the k units in each block. The random allocation of treatments to the blocks may follow the procedures outlined in the completely randomized design discussed in Section 4.2.1. Essentially, the random allocation of each treatment to units in each block reduces the error variance, as now a portion of the variation is attributed to block differences. It should be kept in mind that if variation between blocks is not significantly large, grouping of the units as suggested in this design does not lead to any advantages; rather, some degree of freedom of the error variance is lost, without a decrease in the error variance. Hence, under such conditions, it is preferable to use the completely randomized designs. Another concern that a researcher must deal with in using this design is when the number of treatments is too large to allow for a homogeneous grouping of the units. For example, if the number of treatments is greater than 10, it is not suitable to use this design. In such cases, *incomplete block designs* that permit smaller groupings of homogeneous subjects may be preferable, or if it is possible to form homogeneous groups with larger number of units, this design can be used.

The layout for a randomized complete block design is shown in Figure 4.2.2.2, using a field experiment with 4 treatments (A, B, C, and D) and 3 replications. The steps followed in this layout are given below.

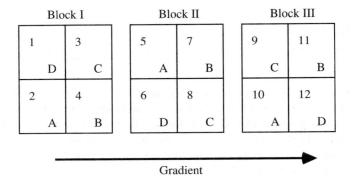

Gradient

FIGURE 4.2.2.2.

Randomized layout of an experimental area having 3 blocks and 4 treatments.

STEP 1: Following the blocking technique described earlier, divide the experimental area into *r* equal blocks. Each block represents a replication. In the present example, the experimental area is divided equally into 3 blocks or replications.

STEP 2: Subdivide each block into *t* experimental plot, where each *t* represents a treatment. Number the plots in ascending order from 1 to *t*, and then assign the treatments at random to the t plots. Random allocation may follow any of the earlier schemes described in section 4.2.1. In the present example, we have 4 treatments. Therefore, each block is subdivided into 4 equal-sized plots. The plots are usually numbered consecutively from top to bottom and from left to right as shown in Figure 4.2.2.1.

STEP 3: Using a Table of Random Numbers such as the one given in Appendix D, or Table 4.2.1.1, select four 3-digit random numbers. For the present example, we start at the intersection of the ninth row and seventh column of Table 4.2.1.1. The first number selected is 276. We read downward from 276 to select the remainder of the numbers as shown below:

Random Number	Sequence
276	1
657	2
645	3
104	4

STEP 4: Rank the numbers from smallest to the largest, as shown below:

Random Number	Sequence	Rank
276	1	2
657	2	4
645	3	3
104	4	1

STEP 5: Assign the 4 treatments to the 4 plots by using the sequence of occurrence of the random numbers as the treatment numbers and the ranks as the plot numbers. Hence treatment A will be assigned to plot 2, treatment B to plot 4, treatment C to plot 3, and treatment D to plot 1. The assignment of the treatments to the plots for blocks II and III follows the same steps as for block I. Figure 4.2.2.2 shows the randomized layout for the present example.

This design illustrates an important design principle. If a source of variability can be separated from treatment effects and experimental error, the sensitivity or power of the resulting experiment may be increased. Variations that cannot be estimated remain a part of the uncontrolled sources of variability in the experiment and thus are automatically a part of the experimental error.

Analysis of Variance: In this design the objective is to isolate the effect of not 1, but 2, variables of interest, namely the treatment and the block (replication). Data gathered from the t treatments and r blocks represent a 2-way classification with tr cells, each containing 1 observation. The data are orthogonal and hence the design is called an orthogonal design. The randomized complete block design accounts for 3 sources of variability. Variability due to treatment, block (replication), and error. The following example provides the procedural steps in performing an analysis of variance for this design.

Example 4.2.2.1 _____

In a study to determine whether method of placement of fertilizer (P, K) would have an impact on corn yield, the following yield data were obtained by an agronomist. Perform an analysis of variance.

Treatments	Grain Yield (bu/acre)		
	Rep I	Rep II	Rep III
Control	147.0	130.1	142.2
2 × 2-in band	159.4	167.3	150.5
Broadcast	158.9	166.2	159.1
Deep band	173.6	170.8	162.5
Disk applied	158.4	169.3	160.2
Sidedress	157.1	148.8	139.0

Solution _____

STEP 1: Calculate treatment totals (T), replication totals (R), and their respective means, and the grand total (G) as shown in Table 4.2.2.1.

STEP 2: Determine the degree of freedom for each source of variation as shown below:

Total $d.f.$ $= (r)(t) - 1 = (3)(6) - 1 = 17$

Treatment $d.f.$ $= t - 1 = 6 - 1 = 5$

Replication (Block) $d.f.$ $= r - 1 = 3 - 1 = 2$

Error $d.f.$ $= (t - 1)(r - 1) = (6 - 1)(3 - 1) = 10$

TABLE 4.2.2.1.

Corn Yield from 6 Different Application Methods with 3 Replications

Treatments	Grain Yield (bu/ac)			Treatment Total (T)	Treatment Mean
	Rep I	Rep II	Rep III		
Control	147.0	130.1	142.2	419.3	139.8
2 × 2-in band	159.4	167.3	150.5	477.2	159.1
Broadcast	158.9	166.2	159.1	484.2	161.4
Deep band	173.6	170.8	162.5	506.9	168.9
Disk applied	158.4	169.3	160.2	487.9	162.6
Sidedress	157.1	148.8	139.0	444.9	148.3
Rep. total (R)	954.4	952.5	913.5	2820.4	
Grand Mean (G)					156.7

STEP 3: Calculate the correction factor (C) and the various sums of squares using the following notations:

X_i = an observation in the ith plot

T_i = treatment total for the ith plot

G = grand total

r_i = the number of replication of the ith treatment

N = the total number of experimental plots $[(r)(t)]$.

$$C = \frac{G^2}{rt} = \frac{(2820.4)^2}{(3)(6)} = 441{,}925.3 \qquad (4\text{-}10)$$

$$\text{Total } SS = \sum_{j=1}^{i} \sum_{i=1}^{r} X_{ij}^2 - C \qquad (4\text{-}11)$$

$$= [(147.0)^2 + (159.4)^2 + \ldots + (139.0)^2] - 441{,}925.3$$

$$= 444{,}263.99 - 441{,}925.3$$

$$= 2{,}338.69$$

$$\text{Replication}\,SS = \frac{\sum R^2}{t} - C \qquad (4\text{-}12)$$

$$= \frac{(954.4)^2 + (952.5)^2 + (913.5)^2}{6} - 441{,}925.3$$

$$= 442{,}102.98 - 441{,}925.3$$

$$= 177.7$$

$$\text{Treatment}\,SS = \frac{\sum T_i^2}{r} - C \qquad (4\text{-}3)$$

$$= \frac{(419.3)^2 + (477.2)^2 + \ldots + (444.9)^2}{3} - 441{,}925.3$$

$$= 443{,}637.33 - 441{,}925.3$$

$$= 1{,}712.00$$

$$\text{Error}\,SS = \text{Total}\,SS - \text{Replication}\,SS - \text{Treatment}\,SS \qquad (4\text{-}13)$$

$$= 2{,}338.69 - 177.7 - 1{,}712.00$$

$$= 448.99$$

STEP 4: Calculate the mean square for each source of variations by dividing the treatment, replication, and error sum of squares by their respective degrees of freedom.

$$\text{Replication}\,MS = \frac{\text{Replication}\,SS}{r-1} \qquad (4\text{-}14)$$

$$= \frac{177.7}{2}$$

$$= 88.89$$

$$\text{Treatment}\,MS = \frac{\text{Treatment}\,SS}{t-1} \qquad (4\text{-}15)$$

$$= \frac{1{,}712}{5}$$

$$= 342.4$$

$$\text{Error}\, MS = \frac{\text{Error}\, SS}{(r-1)(t-1)} \tag{4-16}$$

$$= \frac{448.99}{10}$$

$$= 44.89$$

STEP 5: Calculate the F value using the treatment and error mean squares as:

$$F = \frac{\text{Treatment}\, MS}{\text{Error}\, MS} \tag{4-7}$$

$$= \frac{342.4}{44.89}$$

$$= 7.63$$

STEP 6: Compare the computed F with the tabular F value given in Appendix E. For this example, the tabular F (for 5 degrees of freedom for treatment and 10 degrees of freedom for the error source of variation) is 3.33 for the 5% level of significance and 5.64 for the 1% level of significance. Because the computed F value is greater than the tabular F at the 1% level of significance, it is therefore concluded that there is a significant difference among the 6 treatments.

To determine the degree of precision with which the treatments are compared, we can compute the coefficient of variation cv as follows:

$$cv = \frac{\sqrt{\text{Error}\, MS}}{\text{Grand mean}} \times 100 \tag{4-9}$$

$$= \frac{\sqrt{44.89}}{156.7} \times 100$$

$$= 4.28\%$$

The ANOVA table for the present example is shown in Table 4.2.2.2.

Efficiency of Randomized Complete Block Design: In our earlier discussion, we mentioned that an advantage of using block design over the completely randomized design was the ability to further partition the experimental units into blocks. In essence, such partitioning reduces the experimental error by accounting for the differences among blocks rather than within blocks. This increased precision is due to the fact that error variance, which is a function of comparisons within blocks, is smaller because of homogeneous blocks. To determine the magnitude of the reduction in experimental error due to blocking, the relative efficiency (*R.E.*) of a randomized complete

TABLE 4.2.2.2.

Analysis of Variance of Corn Yield Data in a Randomized Complete Block Design

Source of Variation	Degree of Freedom	Sum of Squares	Mean Square	F
Treatment	5	1,712.00	342.4	7.63[**]
Replication	2	177.7	88.85	
Error	10	448.99	44.89	
Total	17	2,338.69		

Note: cv = 4.28%; ** = significant at 1% level.

block design relative to that of the completely randomized design is computed as follows:

$$R.E. = \frac{(r-1)s_B^2 + r(t-1)s_e^2}{(rt-1)s_e^2} \qquad (5\text{-}17)$$

where

r = replications or blocks

t = treatments

s_B^2 = replication (block) mean square

s_e^2 = error mean square.

It should be noted that when the error degrees of freedom is less than 20, Fisher (1974) proposed an adjustment to account for the discrepancies in *d.f.* He suggests that the *R.E.* parameter should be multiplied by an adjustment factor as shown below:

$$R.E. = \frac{(r-1)s_B^2 + r(t-1)s_e^2}{(rt-1)s_e^2} \cdot \frac{[(r-1)(t-1)+1][t(r-1)+3]}{[(r-1)(t-1)+3][t(r-1)+1]} \qquad (4\text{-}18)$$

For the present example, the *R.E.* value is:

$$R.E. = \frac{(3-1)(88.85)+3(6-1)(44.89)}{(18-1)44.89} \cdot \frac{(11)(9)}{(13)(7)}$$

$$= (1.12)(1.08)$$

$$= 1.21$$

The result shows that by using the randomized complete block design over the completely randomized design, the efficiency of the experiment has improved by 21%. Thus, a block design has proved to be more efficient than a completely randomized design.

4.2.3 Latin Square Designs

In the previous section we have seen how effective single grouping of the units into blocks were in reducing the experimental error. Such reduction in error is a major improvement over the completely randomized designs. Error reduction was made possible by removing the effect of a single factor from the experiment. Latin square design extends this further by allowing the removal of 2 sources of variation from the experimental error. As the name implies, the design has its origins in an ancient puzzle in which Latin letters can be arranged in a square matrix such that each letter appears once, and only once, in each row and column. This design enables an experimenter to isolate 2 known sources of variation among experimental units. The levels of 1 source of variation are assigned to rows, and the levels of the second source of variation are assigned to columns, of a Latin square; the t treatment levels are assigned to the t^2 cells of the square such that each level appears once in each row and once in each column. Another way of expressing the basic criteria for this design is that the experimental material should be arranged as 2 independent (row and column) blocks such that the differences among row blocks and column blocks reflect major sources of variation.

Latin square designs are appropriate to use for those experiments that meet the general assumptions of the analysis of variance model as well as the following conditions:

1. The number of treatments must equal the number of categories for each of the 2 factors or sources of variation; that is, the number of replicates equals the number of treatments.
2. There should be an absence of interactions among the 2 sources of variation and the treatment. This means that there is no interaction between the treatment effect and either block effect.

3. Treatment levels are randomly assigned to the cells of the square, with the proviso that each treatment level must appear once in each column and once in each row.

The advantages of using the Latin square design are in requiring fewer subjects, reducing systematic treatment biases through counterbalancing, and removing error variance through 2-way blocking. The basis for fewer subjects stems from 1 of the design requirements which states that the number of replications must equal the number of treatments. For example, an animal scientist, in conducting an experiment to evaluate the effects of 4 different rations on weight gain, may choose a design that takes into consideration other factors that might explain the weight-gain variability, such as the initial weight and age of animals. Thus, the animal scientist might block the sample subjects into 4 initial weight and 4 age categories. In such a situation where the scientist wants replications of the 4 treatments with 2 blocking factors, each also having 4 categories, $4 \times 4 \times 4 \times n = 64n$ subjects are required. In factorial experiments with large number of factors and numerous categories of each factor, the number of subjects required will indeed be very large. The Latin square design permits us to reduce the number of subjects required, yet assess the relative effects of various treatments when a 2-directional blocking restriction is imposed on the experimental unit. In the case of the feed ration experiment, we will need only 16 instead of 64 subjects. The disadvantage of the design lies in the requirement that the number of treatments must equal the number of replications. This is especially problematic when there are large numbers of treatments, hence requiring large numbers of replications. This has the added problem that as the square gets larger, the experimental error per unit also increases. On the other hand, smaller Latin squares do not have large degrees of freedom for the estimation of error. For example, a 2×2 square has no degrees of freedom, a 3×3 has only 2 degrees of freedom, and a 4×4 has only 6 degrees of freedom, for the estimation of error. For this reason, it is generally recommended to use Latin square when the number of treatments is not less than 4 and no more than 8. Another limitation of the design is when there is reason to believe that there is interaction among the treatment, row, and column. In such circumstances, it is not appropriate to use this design.

In using the Latin square design, the experimenter must keep in mind that if 1 of the 2 factors under consideration does not have a substantial impact on the variate under study, elimination of its variance may not influence the experimental error. Thus, using the Latin square may not be an improvement over the randomized complete block design. In agricultural field experiments, for example, where there is a strong indication of a 2-directional fertility gradient, adoption of a Latin square is desirable. Similarly, in an insecticide field trial where there is a predictable direction to insect migration, which is perpendicular to the fertility gradient of the field, use of the Latin square design is appropriate. Horticulturists and agronomists who perform greenhouse experiments may find using this design helpful where experimental pots are arranged perpendicular to a source of light or shade such that the difference among rows of pots and the distance from the light or shade account for two sources of variability. In animal experiments the experimenter may use the animals themselves and time as the 2 sources of blocking, or age and weight may be considered as the 2 sources of variation that may have a major effect on the variate under study.

Randomization: The process of randomization in Latin square design requires that a square be selected at random from a number of Latin squares of a given order. Appendix C presents examples of selected Latin squares. The procedure for the randomization of rows and column is outlined below.

STEP 1: Let a $k \times k$ Latin square be first written by denoting treatments by Latin letters A, B, C, D, etc., or randomly select a readily arranged square given in Appendix C. Suppose we randomly select a 4×4 square from Appendix C, as shown below:

A	B	C	D
B	C	D	A
C	D	A	B
D	A	B	C

STEP 2: Randomly reorder all rows (R_i) and columns (C_j) except the first, by using a Table of Random Numbers such as the one given in Appendix D, or Table 4.2.1.1. Select 4 three-digit random numbers. For the present example, we randomly start at the intersection of the first row and the seventh column of Table 4.2.1.1. The first number selected is 315. Reading downward from 315, the following numbers are selected:

Random Number	Sequence
315	1
145	2
189	3
207	4

STEP 3: Rank the numbers from smallest to the largest, as shown below:

Random Number	Sequence	Rank
315	1	4
145	2	1
189	3	2
207	4	3

STEP 4: Use the rank to represent the existing row numbers of a selected square and the sequence of occurrence to represent the row numbers after reordering of rows of the selected square. For the present example, the fourth row of the selected square becomes the first row of the reordered plan. After reordering all rows, the new plan will be:

D	A	B	C
A	B	C	D
B	C	D	A
C	D	A	B

STEP 5: Reorder the columns obtained in Step 4, using the same procedures applied in the previous step. Use the rank to represent the column number of the plan obtained in Step 4, and sequence of occurrence of numbers to represent the column number of the final plan. The 4 newly selected numbers from Table 4.2.1.1 and the sequence of occurrence are given below:

Random Number	Sequence	Rank
168	1	2
094	2	1
897	3	4
243	4	3

STEP 6: The final randomized layout for the experiment, based on the new row and column arrangement, is:

	Column Number			
Row Number	I	II	III	IV
1	A	D	B	C
2	B	A	C	D
3	C	B	D	A
4	D	C	A	B

Analysis of Variance: As explained earlier, the intent of the design is to further partition the sources of variation in an experiment. This means that blocking by rows (R_i) and columns (C_j) allows for removal of variation due to such sources from the experimental error. Hence, the design makes it possible to account for 2 more sources of variation than a completely randomized design and 1 more source of variation than a randomized complete block design.

The analysis is conducted by following a similar procedure as described in section 4.2.2. The following example illustrates the computational steps in performing the analysis of variance.

Example 4.2.3.1

An animal scientist designed an experiment to determine the effect of 4 diets (A, B, C, and D) on liver cholesterol in sheep. Recognized sources of variation were body weight and age. The experimenter randomly selected a 4×4 Latin square, and randomly arranged the results of the experiment from the different treatments as presented below.

Weight Groups	Age Groups			
	I	II	III	IV
1	1.15A	1.75D	1.78B	2.10C
2	1.65B	1.20A	1.56C	2.05D
3	1.30C	1.79B	1.93D	1.25A
4	1.10D	1.50C	1.20A	2.01B

TABLE 4.2.3.1.

Row, Column, and Treatment Totals and Means for Cholesterol Levels in Sheep Experiment

Row	Column				Row Total	Treatment Total
	I	II	III	IV		
1	1.15A	1.75D	1.78B	2.10C	6.78	4.80A
2	1.65B	1.20A	1.56C	2.05D	6.46	7.23B
3	1.30C	1.79B	1.93D	1.25A	6.27	6.46C
4	1.10D	1.50C	1.20A	2.01B	5.81	6.83D
Column Total	5.20	6.24	6.47	7.41		
Grand Total						25.32
Grand Mean						1.58

Solution

STEP 1: Calculate treatment totals (T), row totals (R_i), column totals (C_j), and the grand total (G) as shown in Table 4.2.3.1.

STEP 2: Determine the degree of freedom for each source of variation, given that t represents treatments as shown below:

Total $d.f.$ $= t^2 - 1 = (4^2) - 1 = 15$

Treatment $d.f.$ $=$ Column $d.f.$ = Row $d.f.$ = 4 - 1 = 3

Error $d.f.$ $= (t - 1)(t - 2) = (4 - 1)(4 - 2) = 6$

STEP 3: Calculate the correction factor (C) and the various sums of squares using the following notations:

$$C = \frac{G^2}{t^2} = \frac{(25.32)^2}{16} = 40.07 \qquad (4\text{-}19)$$

$$\text{Total } SS = \sum X_{ijk}^2 - C \qquad (4\text{-}20)$$

$$= \left[(1.15)^2 + (1.75)^2 + (1.78)^2 + \dots + (2.01)^2 \right] - 40.07$$

$$= 41.88 - 40.07$$

$$= 1.81$$

$$\text{Treatment } SS = \frac{\sum T^2}{k} - C \tag{4-21}$$

$$= 1/4\left[(4.80)^2 + (7.23)^2 + (6.46)^2 + (6.83)^2\right] - 40.07$$

$$= 1/4\left[163.69\right] - 40.07$$

$$= 40.92 - 40.07$$

$$= 0.85$$

$$\text{Row } SS = \frac{\sum R_i^2}{r} - C \tag{4-22}$$

$$= 1/4\left[(6.78)^2 + (6.46)^2 + (6.27)^2 + (5.81)^2\right] - 40.07$$

$$= 1/4\left[160.77\right] - 40.07$$

$$= 40.19 - 40.07$$

$$= 0.12$$

$$\text{Col. } SS = \frac{\sum C_j^2}{c} - C \tag{4-23}$$

$$= 1/4\left[(5.20)^2 + (6.24)^2 + (6.47)^2 + (7.41)^2\right] - 40.07$$

$$= 1/4\left[162.75\right] - 40.07$$

$$= 40.69 - 40.07$$

$$= 0.62$$

$$\text{Error } SS = \text{Total } SS - \text{Treatment } SS - \text{Row } SS - \text{Col. } SS \tag{4-24}$$

$$= 1.81 - 0.85 - 0.12 - 0.62$$

$$= 0.22$$

STEP 4: Calculate the mean square for each source of variations by dividing the treatment, row, column, and error sum of squares by their respective degrees of freedom.

$$\text{Treatment}\, MS = \frac{\text{Treatment}\, SS}{t-1} \qquad (4\text{-}5)$$

$$= \frac{0.85}{3}$$

$$= 0.28$$

$$\text{Row}\, MS = \frac{\text{Row}\, SS}{t-1} \qquad (4\text{-}25)$$

$$= \frac{0.12}{3}$$

$$= 0.04$$

$$\text{Col.}\, MS = \frac{\text{Col.}\, SS}{t-1} \qquad (4\text{-}26)$$

$$= \frac{0.62}{3}$$

$$= 0.21$$

$$\text{Error}\, MS = \frac{\text{Error}\, SS}{(t-1)(t-2)} \qquad (4\text{-}27)$$

$$= \frac{0.22}{(3)(2)}$$

$$= 0.04$$

STEP 5: Calculate the F value, using the treatment and error mean squares as:

$$F = \frac{\text{Treatment}\, MS}{\text{Error}\, MS} \qquad (4\text{-}7)$$

$$= \frac{0.28}{0.04}$$

$$= 7.0$$

STEP 6: Compare the computed F value with the tabular F value given in Appendix E. For this example, the tabular F (for 3 degrees of freedom for treatment and 6 degrees of freedom for the error source of variation) is 4.76 for the 5% level of significance and 9.78 for the 1% level of significance. Because the computed F value is greater than the tabular F at the 5% level of significance, it is therefore concluded that there is a significant difference among the 4 treatments.

To determine the degree of precision with which the treatments are compared, we can compute the coefficient of variation cv as follows:

$$cv = \frac{\sqrt{\text{Error } MS}}{\text{Grand mean}} \times 100 \quad (4\text{-}9)$$
$$= \frac{\sqrt{0.04}}{1.58} \times 100$$
$$= 12.65\%$$

The ANOVA table for the present example is shown in Table 4.2.3.2.

TABLE 4.2.3.2.
ANOVA Table of 4 Different Diets Effect on Cholesterol Level of Sheep

Source of Variation	Degree of Freedom	Sum of Squares	Mean Square	F
Treatment	3	0.85	0.28	7.0*
Row	3	0.12	0.04	
Column	3	0.62	0.21	
Error	6	0.22	0.04	
Total	15	1.81		

Note: cv = 12.65 % ; * = significant at 5% level.

Efficiency of Latin square designs: To determine whether blocking by rows and columns was efficient in reducing the experimental error, we will compute the F value for the row and column as follows:

$$F \text{ (row)} = \frac{\text{Row } MS}{\text{Error } MS} \quad (4\text{-}28)$$
$$= \frac{0.04}{0.04}$$
$$= 1$$

$$F \text{ (column)} = \frac{\text{Column } MS}{\text{Error } MS} \quad (4\text{-}29)$$

$$= \frac{0.21}{0.04}$$

$$= 5.25$$

Since the computed F value for both row and column are equal to or greater than 1, we will compare them with the tabular F from Appendix E. The tabular F at 5% and 1% level of significance are 4.76 and 9.78, respectively. In comparing these results with the tabular F values, it appears that only column-blocking provides a significant difference. To determine the magnitude of the reduction in experimental error due to row- and column-blocking, the relative efficiency ($R.E.$) parameter of a Latin square design has to be computed. When computing the relative efficiency parameter for the Latin square we are making a comparison with (1) a completely randomized design (CRD); (2) a random-ized complete block design (RCB), in which the blocks are rows (first source of variation) of the Latin square; (3) a randomized block design, in which the blocks are columns (second source of variation) of the Latin square. The relative efficiency of the Latin square design as compared with each of the above 3 scenarios is computed below.

1. Relative Efficiency of a Latin square as compared to a CRD is:

$$R.E. = \frac{S_r^2 + S_c^2 + (t - 1)S_e^2}{(t + 1)S_e^2} \quad (4\text{-}30)$$

where

$\quad\quad\quad S_r^2$ = row mean square

$\quad\quad\quad S_c^2$ = column mean square

$\quad\quad\quad S_e^2$ = error mean square

$\quad\quad\quad t$ = number of treatments.

2. The efficiency of a Latin square design, relative to randomized complete block design with rows as blocks, may be called column efficiency and is given as:

$$R.E. \text{ (Col.)} = \frac{S_c^2 + (t - 1)\, S_e^2}{(t)(S_e^2)} \quad (4\text{-}31)$$

3. The efficiency of a Latin square design, relative to randomized complete block design in which columns are treated as blocks, is called row efficiency. Here the row variation is merged with the error variation, and is computed as:

$$R.E. \, (\text{Row}) = \frac{S_r^2 + (t-1)S_e^2}{(t)(S_e^2)} \qquad (4\text{-}32)$$

Thus, for the present example, we have:

$$R.E. = \frac{0.04 + 0.21 + (4-1)0.04}{(4+1)0.04}$$

$$= 1.85$$

The result shows that by using a Latin square design instead of a completely randomized design, the precision has increased by 85%.

Similarly, we compute the $R.E.$ of a Latin square design compared to the randomized complete block design as:

$$R.E. \, (\text{CRB, Col.}) = \frac{0.21 + (4-1)0.04}{(4)0.04}$$

$$= 2.06$$

$$R.E. \, (\text{CRB, Row}) = \frac{0.04 + (4-1)0.04}{(4)0.04}$$

$$= 1.00$$

It should be noted that the error degrees of freedom is less than 20, therefore the $R.E.$ values should be multiplied by the adjustment factor k as shown below:

$$k = \frac{\left[(t-1)(t-2)+1\right]\left[(t-1)^2+3\right]}{\left[(t-1)(t-2)+3\right]\left[(t-1)^2+1\right]} \qquad (4\text{-}33)$$

$$= \frac{\left[(4-1)(4-2)+1\right]\left[(4-1)^2+3\right]}{\left[(4-1)(4-2)+3\right]\left[(4-1)^2+1\right]}$$

$$= 0.93$$

Therefore, the adjusted *R.E.* values are:

$$R.E. \text{ (CRB, Col.)} = (2.06)(0.93)$$
$$= 1.92$$
$$R.E. \text{ (CRB, Row)} = (1.00)(0.93)$$
$$= 0.93$$

As the results show, the column-blocking of the Latin square design did increase the experimental precision by 92%. However, the additional row-blocking of the Latin square did not increase the precision over the RCB design in which the columns were treated as blocks.

Schematic partitioning of the total sum of squares for the completely randomized design, the randomized complete block design, and the Latin square is shown in Figure 4.2.3.1.

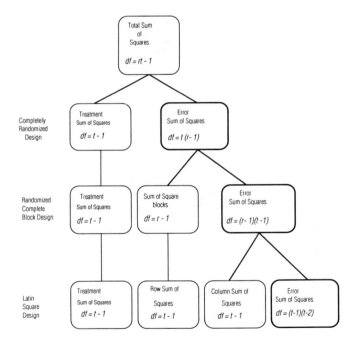

FIGURE 4.2.3.1.

Schematic partitioning of the total sum of squares for 3 designs. Squares with bold outline identify the experimental errors for testing treatment effects.

4.3 INCOMPLETE BLOCK DESIGNS

Many agricultural experiments involve varietal trials or other types of experiments in which the objective is to compare pairs of treatments for the purpose of selecting improved varieties of crops or breeds of animals. An agronomist, for example, may be interested in selecting a variety of a crop based on some genetic or economic characteristics. To determine which variety may be considered superior, an experiment would be to grow several varieties in a field, using an appropriate design. As mentioned in the previous sections of this chapter, when the number of varieties to be tested is small (less than 10), use of an ordinary randomized block design or a Latin square design may be appropriate. However, when the number of varieties tested is large, as is often the case with varietal trials, use of randomized block designs may not be appropriate because of the resultant increase in error variance due to larger block size. In factorial experiments, use is made of the confounding device to reduce the block size and hence ensure more precise estimation of the lower-order interactions that are confounded with blocks. In experiments other than varietal trials, where it is desired to make all comparisons among pairs of treatments with equal precision, use of complete block designs are also inefficient. That is, as the number of treatments increase, the block size increases, which contributes to the variability within the blocks. This loss of homogeneity contributes to increased experimental error and less precision. On the other hand, we have seen that the larger the number of replications of a treatment effect, the more the precision. Therefore, we need a set of designs to permit small groupings of subjects (fewer subjects per block than the number of treatments) for experiments that contain many treatments.

A set of designs for single-factor experiments were introduced by Yates in 1936(a) to accommodate large numbers of treatments with smaller blocks. These designs are called the *incomplete block designs*. In incomplete block designs, the blocks need not be of the same size and each treatment need not appear the same number of times. These characteristics of the designs are quite valuable, especially when there is a large number of treatment factors and resources are limited to permit large replications, or when there are natural or practical restrictions on block size.

Generally, incomplete block designs have the advantage of providing improved precision through reduced block size, without loss of information on any treatment effect. This is accomplished by partial confounding of the effects of all treatments with differences among incomplete blocks. That is, the variance between plots in the same block is smaller than the variance between plots in different blocks. This improved precision, while important, has drawbacks that should be carefully weighed before using these designs. For example, partial confounding leads to a more complex plan and analysis. Additionally, there is inflexibility in the number of treatments or replications or both. To ensure equal or nearly equal precision of comparisons of different pairs of treatments, the treatments are allocated to different blocks in such a way that each pair of treatments has the same or nearly the same number of replications, and each treatment has an equal number of replications (r).

An incomplete block design should be used whenever block size is large enough to not allow the experimental units within the same block to be relatively homogeneous. So, in cases where the experimental material is highly variable, but where it is possible to form small groups that are homogeneous, the design may be advantageous even with a small number of treatments. It should also be kept in mind that because of the complex nature of these designs, the data analysis requires access to computing facilities and services. In research experiments where missing or incomplete data (experimental units destroyed or injured during the course of the experiment) is a problem, the use of an incomplete design is not recommended. Such incomplete data present laborious computation of the block (or row and column) adjustments.

There are several types of incomplete block designs. The most commonly used in agricultural research is the lattice design. In the next 2 sections of this chapter, we will discuss 2 lattice designs: namely, the balanced lattice and the partially balanced lattice designs.

4.3.1   Balanced Lattice Designs

In 1937 Yates introduced quasifactorial designs called the lattice designs. The main characteristics of these designs are that the number of treatments must be perfect squares $(t = k^2)$, and that the block size k is the square root of the number of treatments $(k = \sqrt{t})$. And lastly, the number of replications is 1 more than the block size; that is, $r = (k + 1)$. The last criterion differentiates balanced lattice designs from the partially balanced designs that will be discussed later. To illustrate how to satisfy the first requirement of the design, the number of treatments may be 9, 16, 25, 36, 49, 81, etc. If the number of treatments that an experimenter wishes to work with is not a perfect square, it must be made a perfect square either by elimination or addition of treatments. Once the number of treatments is selected, the second and third requirement is easily met. For instance, if there are 49 treatments in the experiment, the block size is $k = \sqrt{t}$ or $k = 7$. The number of replications would be $r = (k + 1)$ or $r = 8$. The randomization and layout of the experimental plan, as well as the analysis, are discussed below.

Randomization: The randomization process is illustrated in a field experiment using a simple case with 9 treatments. The procedural steps for the layout of the design are:

STEP 1:   Determine the number of replications needed by dividing the experimental area into $r = (k + 1)$, such that each replication would have $t = k^2$ experimental plots. In the present case, $r = 4$, each containing 9 experimental plots, as shown in Figure 4.3.1.1.

	Replication I			Replication II			Replication III			Replication IV		
Block 1	1	2	3	10	11	12	19	20	21	28	29	30
Block 2	4	5	6	13	14	15	22	23	24	31	32	33
Block 3	7	8	9	16	17	18	25	26	27	34	36	37

FIGURE 4.3.1.1.

Division of the experimental area in a balanced design for nine treatments in blocks of three units.

STEP 2: Divide each replication into k incomplete blocks. For this example this would be 3 incomplete blocks each containing 3 experimental plots as shown in Figure 4.3.1.1.

STEP 3: Select from Appendix F a basic plan that corresponds with the number of treatments tested in the experiment. For the present case, the 3×3 balanced lattice design selected is shown in Table 4.3.1.1.

TABLE 4.3.1.1.

Basic Plan for a 3 × 3 Balanced Lattice Design for 9 Treatments in Blocks of 3 Units

Block Number	Rep. I			Rep. II			Rep. III			Rep. IV		
(1)	1	2	3	1	4	7	6	1	8	5	1	9
(2)	4	5	6	2	5	8	2	9	4	7	6	2
(3)	7	8	9	3	6	9	7	5	3	3	8	4

STEP 4: Use an appropriate randomization technique such as the one discussed in Section 4.2.1 to randomize the replication arrangement, the incomplete blocks within each replication, and the treatment arrangement within each block of the selected plan for field layout. To avoid being redundant, we have not illustrated the randomization technique for the present example. The reader is directed to Section 4.2.1 for the technique of randomly rearranging the replication, treatments, and incomplete blocks. Following the randomization technique of Section 4.2.1. the replications, blocks, and treatments are rearranged as shown in Table 4.3.1.2.

TABLE 4.3.1.2.

A Selected Basic Plan with Rearranged Replications, Blocks, and Treatments in a Balanced Lattice Design

Block	Rep. I			Block	Rep. II			Block	Rep. III			Block	Rep. IV		
(1)	8	6	1	(4)	8	7	9	(7)	8	5	2	(10)	3	4	8
(2)	3	5	7	(5)	3	2	1	(8)	4	7	1	(11)	6	2	7
(3)	4	2	9	(6)	6	5	4	(9)	3	6	9	(12)	9	5	1

STEP 5: Use the rearranged plan to layout the experiment in the field. This is shown in Figure 4.3.1.2. Note that every treatment appears in each of the 3 rows, and every pair of treatments appears once in the same

block. This feature allows each pair of treatments in a balanced lattice design to be compared with equal precision.

Analysis of Variance: The difference in the analysis between a balanced lattice design and a randomized complete block (RCB) is the added source of variation which is due to the incomplete blocks of the same replication. Computation and statistical analysis of the design is not difficult. However, the use of such designs requires additional care and resources. In comparison to the randomized complete block design, the potential source of gain in accuracy may come from the experimental material and the large number of treatments. The number of treatments in most incomplete block designs ranges from 6 to 200. This range increases to 1000, as in the case of the cubic lattice designs.

Example 4.3.1.1 _____

An animal scientist designed a 3×3 lattice design experiment to study the effect of 9 different diets on nonprotein nitrogen in the colostrum of rats. The plan compares 9 treatments in 4 replications. The results (mg/100 ml) are shown below:

Treatment No.	Rep. I	Rep. II	Rep. III	Rep. IV
1	69	89	90	80
2	92	72	132	72
3	74	81	127	98
4	68	74	98	56
5	80	83	92	63
6	83	72	87	76
7	86	55	102	84
8	66	76	85	81
9	78	71	96	95

Perform the analysis of variance on the data.

Solution _____

The steps involved in the analysis of variance are given below.

STEP 1: Select a basic plan from Appendix F (Basic Plans for Balanced and Partially Balanced Lattice Designs) and rearrange the replication, incomplete block, and treatment as shown:

	Replication I			Replication II			Replication III			Replication IV		
Block 1	T8	T6	T1	T8	T7	T9	T8	T5	T2	T3	T4	T8
Block 2	T3	T5	T7	T3	T2	T1	T4	T7	T1	T6	T2	T7
Block 3	T4	T2	T9	T6	T5	T4	T3	T6	T9	T9	T5	T1

FIGURE 4.3.1.2.

Allocation of nine treatments in a field layout with three blocks.

Block	Rep. I			Block	Rep. II			Block	Rep. III			Block	Rep. IV		
(1)	1	2	3	(4)	1	4	7	(7)	1	5	9	(10)	1	8	6
(2)	4	5	6	(5)	2	5	8	(8)	7	2	6	(11)	4	2	9
(3)	7	8	9	(6)	3	6	9	(9)	4	8	3	(12)	7	5	3

STEP 2: Using the randomization technique of section 4.2.1.1, rearrange the replications, incomplete blocks, and treatments. The final layout for the plan is given below:

Block	Rep. I			Block	Rep. II			Block	Rep. III			Block	Rep. IV		
(1)	8	6	1	(4)	8	7	9	(7)	8	5	2	(10)	3	4	8
(2)	3	5	7	(5)	3	2	1	(8)	4	7	1	(11)	6	2	7
(3)	4	2	9	(6)	6	5	4	(9)	3	6	9	(12)	9	5	1

STEP 3: Allocate the treatments to the rearranged plan as given below:

Block	Rep. I			Block	Rep. II			Block	Rep. III			Block	Rep. IV		
(1)	66	83	69	(4)	76	55	71	(7)	85	92	132	(10)	98	56	81
(2)	74	80	86	(5)	81	72	89	(8)	98	102	90	(11)	76	72	84
(3)	68	92	78	(6)	72	83	74	(9)	127	87	96	(12)	95	63	80

STEP 4: Calculate the replication totals (R), the block totals (B), the treatment totals (T), and the grand total (G) as shown in Table 4.3.1.3

STEP 5: Calculate the sum of block totals over all blocks (B_T) in which a particular treatment appears. For example, the (B_T) value for treatment 1 which appears in blocks 1, 5, 8, and 12 is $218 + 242 + 290 + 238 = 988$. Compute the (B_T) value for all 9 treatments as shown in Table 4.3.1.3.

STEP 6: Calculate the quantities (W) for each treatment whose sum must equal 0 using the following equation:

$$W = kT - (k + 1)B_T + G \qquad (4\text{-}34)$$

where

k = block size,

T, B_T, and G as defined above.

The W values for each of the 9 treatments are given in Table 4.3.1.3. Note that the sum of W values is equal to 0.

TABLE 4.3.1.3.

Milligrams per 100 ml of Nonprotein Nitrogen from 9 Diets in the Colostrum of Rats[a]

Block No.				Block Total (B)	Treatment Number	Treatment Total (T)	Block Total (B_T)	W
		Rep. I						
1	66(8)	83(6)	69(1)	218	1	328	988	15
2	74(3)	80(5)	86(7)	240	2	368	1,021	3
3	68(4)	92(2)	78(9)	238	3	380	1,027	15
	Rep. Total			696	4	296	992	-97
					5	318	1,016	-127
		Rep. II			6	318	989	-19
4	76(8)	55(7)	71(9)	202	7	327	964	108
5	81(3)	72(2)	89(1)	242	8	308	964	51
6	72(6)	83(5)	74(4)	229	9	340	988	51
	Rep. Total			673	Total (G)	2,983	8,949	0
		Rep. III						
7	85(8)	92(5)	132(2)	309				
8	98(4)	102(7)	90(1)	290				
9	127(3)	87(6)	96(9)	310				
	Rep. Total			909				
		Rep. IV						
10	98(3)	56(4)	81(8)	235				
11	76(6)	72(2)	84(7)	232				
12	95(9)	63(5)	80(1)	238				
	Rep. Total			705				

[a]Treatment numbers are shown in parentheses.

STEP 7: Determine the degrees of freedom associated with each source of variation. The source of variation and the total degrees of freedom depend on whether the treatments are adjusted or not. If the treatments are adjusted, the following sources of variation and the degrees of freedom are identified:

Replication $d.f.$ $\qquad = k = 3$

Treatment (unadj.) $d.f.$ $\qquad = k^2 - 1 = 8$

Block (adj.) $d.f.$ $\qquad = k^2 - 1 = 8$

Intrablock error $d.f.$ $\qquad = (k - 1)(k^2 - 1) = 16$

Treatment (adj.)$d.f.$ $\qquad = k^2 - 1 = 8$

Effective error $d.f.$ $\qquad = (k - 1)(k^2 - 1) = 16$

Total $d.f.$ $\qquad = 59$

If the treatments are not adjusted (the reasons for which will be explained below), then the sources of variation and the corresponding degrees of freedom are:

Replication $d.f.$ $\qquad = k = 3$

Treatment (unadj.) $d.f.$ $\qquad = k^2 - 1 = 8$

Block (adj.) $d.f.$ $\qquad = k^2 - 1 = 8$

Intrablock error $d.f.$ $\qquad = (k - 1)(k^2 - 1) = 16$

Total $d.f.$ $\qquad = 35$

STEP 8: Calculate the various sums of squares by first computing the correction factor (C) as:

$$C = \frac{G^2}{(k)^2 (k + 1)} \tag{4-35}$$

$$= \frac{(2{,}983)^2}{(9)(4)}$$

$$= 247{,}174.7$$

$$\text{Total } SS = \sum X_{ijk}^2 - C \tag{4-36}$$

$$= \left[(66)^2 + (83)^2 + (69)^2 + \dots + (80)^2 \right] - 247{,}174.7$$

$$= 255{,}597 - 247{,}174.7$$

$$= 8{,}422.3$$

$$\text{Replication} SS = \frac{\sum R^2}{k^2} - C \qquad (4\text{-}37)$$

$$= \frac{(696)^2 + (673)^2 + (909)^2 + (705)^2}{9} - 247,174.4$$

$$= 251,183.4 - 247,174.7$$

$$= 4,008.7$$

$$\text{Treatment (unadj.)} SS = \frac{\sum T^2}{(k+1)} - C \qquad (4\text{-}38)$$

$$= \frac{(328)^2 + (368)^2 + ... + (340)^2}{4} - 247,174.4$$

$$= 248,666.3 - 247,174.7$$

$$= 1,491.6$$

The sum of squares for blocks within replication adjusted for treatment is computed as:

$$\text{Block (adj.)} SS = \frac{\sum W^2}{(k^3)(k+1)} - C \qquad (4\text{-}39)$$

$$= \frac{(15)^2 + (3)^2 + (15)^2 + ... + (51)^2}{(27)(4)}$$

$$= 400.2$$

$$\text{Intrablock error } SS = \qquad\qquad\qquad\qquad (4\text{-}40)$$

$$\text{Total } SS - \text{Replication} SS - \text{Treatment (unadj.)} SS$$

$$- \text{Block (adj.)} SS$$

$$= 8,422.3 - 4,008.7 - 1,491.6 - 400.2$$

$$= 2,521.8$$

STEP 9: Compute the mean square for treatment, block (adj.), and the intrablock error as:

$$\text{Treatment (unadj.) } MS = \frac{\text{Treatment (unadj.) } SS}{k^2 - 1} \tag{4-41}$$

$$= \frac{1,491.6}{8}$$

$$= 186.5$$

$$\text{Block (adj.) } MS = \frac{\text{Block (adj.) } SS}{k^2 - 1} \tag{4-42}$$

$$= \frac{400.2}{8}$$

$$= 50.0$$

$$\text{Intrablock error } MS = \frac{\text{Intrablock error } SS}{(k - 1)(k^2 - 1)} \tag{4-43}$$

$$= \frac{2,521.8}{16}$$

$$= 157.6$$

STEP 10: Having obtained the values for the mean squares, we are now in a position to compute the adjusted treatment total $(\dot{T})$ as:

$$\dot{T} = T + \mu W \tag{4-44}$$

where

$$\mu = \frac{\text{Block (adj.) } MS - \text{Intrablock error } MS}{k^2 \left[\text{Block (adj.) } MS\right]} \tag{4-45}$$

This computation is necessary only if the intrablock error mean square is less than the block (adj.) mean square. In such conditions, the adjusted treatment totals ($\dot{T}$) for all treatments and the effective error mean square should be computed. They will in turn be used in performing the F test of significance. The equation for computing the adjusted mean square and the effective error mean square is given below.

$$\text{Treatment (adj.) } MS = \left[\frac{1}{(k + 1)(k^2 - 1)} \right] \left[\sum \dot{T}^2 - \frac{G^2}{k^2} \right] \qquad (4\text{-}46)$$

$$\text{Effective error } MS = (\text{Intrablock error } MS)(1 + k\mu) \qquad (4\text{-}47)$$

$$F = \frac{\text{Treatment (adj.) } MS}{\text{Effective error } MS} \qquad (4\text{-}48)$$

If the intrablock error mean square, on the other hand, is greater than the block (adj.) mean square, the value of μ is taken to be 0, and therefore there are no further adjustments necessary to the treatments. The F-test of significance is computed as the ratio of the treatment (unadj.) mean square to the intrablock error mean square. In the present example, the intrablock error mean square is greater than the block (adj.) mean square; therefore, μ is taken to be 0, and no adjustments are necessary.

STEP 11:   Compute the F value as:

$$F = \frac{\text{Treatment (unadj.) } MS}{\text{Effective error } MS} \qquad (4\text{-}49)$$
$$= \frac{186.5}{157.6}$$
$$= 1.18$$

STEP 12:   Compare the computed F with the tabular F value given in Appendix E. For this example, the tabular F (for 8 treatment degrees of freedom and 16 intrablock error degrees of freedom) is 2.59 and 3.89 for 5% and 1% levels of significance, respectively, and the computed F as shown in Table 4.3.1.4 is 1.18. The result suggests that there is no significant difference among the diets.

TABLE 4.3.1.4.

Analysis of Variance of Nonprotein Nitrogen from 9 Diets in the
Colostrum of Rats Using a 3×3 Balanced Lattice Design

Source of Variation	Degree of Freedom	Sum of Squares	Mean Square	F
Replication	3	4,008.7		
Block (adj.)	8	400.2	50.0	
Treatment (unadj.)	8	1,491.6	186.5	$1.18ns$
Intrablock error	16	2,521.8	157.6	
Total	35			

Note: cv = 15.1%. *ns* = Nonsignificant.

As in the previous sections, we can determine the degree of precision with which the treatments are compared by computing the coefficient of variation *cv* as:

$$cv = \frac{\sqrt{\text{Intrablock error} \, MS}}{\text{Grand Mean}} \times 100 \tag{4-50}$$

$$= \frac{\sqrt{157.6}}{83} \times 100 = 15.1\%$$

STEP 13:   Compute the relative efficiency (*R.E.*) to estimate the precision relative to randomized complete block designs as:

$$R.E. = \frac{\text{Block (adj.)} \, SS + \text{Intrablock error} \, SS}{(k-1)(k^2-1)(\text{Intrablock} \, MS)} \times 100 \tag{4-51}$$

$$= \frac{400.2 + 2,521.8}{(2)(8)(157.6)} \times 100$$

$$= 115.8\%$$

The result indicates that the experimental precision has increased by 15.8% over a randomized complete block design.

4.3.2   Partially Balanced Lattice Designs

These designs were developed by Bose and Nair (1939) to overcome the problems associated with the restrictive assumptions of the balanced lattice designs which require that the number of replications be determined by the number of treatments and the number of units per block. Partially balanced designs, while requiring treatments to be a perfect square, allow a flexible choice of the number of replications. Furthermore, all treatments need not be paired in the same block equally often, which implies that sometimes some treatments may never be paired in the same block. As a result, some treatments tested in the same incomplete block are compared with different levels of precision. This leads to a relatively more complicated data analysis. Such complications do not deter their use, especially when the number of replications required for a complete balance is too large for practical and economic reasons.

Partially balanced lattice designs are constructed similar to the balanced designs with the exception that the former contain fewer replications. A partially balanced lattice design with 2 replications is called a *simple lattice*, that with 3 replication a *triple lattice*, and that with 4 replications a *quadruple lattice*.

Randomization and Layout: With the exception of the modification in the number of replications, randomization and layout is similar to that of the balanced lattice designs. Modifications to a basic plan selected from Appendix F for designs with 2 or 3 replications follow the same procedures outlined in section 4.3.1. However, when the number of replications (r) exceeds three and is an even number, the basic plan to use for randomization and layout is:

- the first r replications of the basic plan (selected from Appendix F of the balanced lattice design) that has the same number of treatments, or
- the first r/p replications of the basic plan (selected from Appendix F of the balanced lattice design) that has the same number of treatments, repeated p times.

To apply the above criteria to a 5×5 quadruple balanced lattice design, for example, the basic plan to work with will be the first four replications of the selected plan, or the 5×5 simple lattice design plan repeated twice. Generally, it is preferable to work with the basic plans that do not involve repetition, as is the case with the first criterion. The advantage of using basic plans without repetition is the symmetry achieved by the balanced lattice designs in the arrangement of treatments over blocks. Examples are presented to illustrate the statistical analyses of designs with and without repetition.

Analysis of Variance for a Case without Repetition: We will use a 5×5 triple lattice experiment on grain yield of corn hybrids doublecropped with winter wheat to illustrate the steps in performing the analysis of variance. The treatments are 25 varieties of corn hybrids. Table 4.3.2.1 shows the yield data (bushels per acre) rearranged according to the basic plan of Appendix F. Treatment numbers are shown in parentheses next to the yield data.

TABLE 4.3.2.1.

Yield Data of 5 × 5 Triple Lattice Experiment on Hybrid Corn Doublecropped with Winter Wheat (bushels per acre, minus 20 bu.)

Block No.						Block Total (B)
			Rep. I			
1	39(1)	33(2)	40(3)	18(4)	22(5)	152
2	24(6)	40(7)	16(8)	35(9)	29(10)	144
3	26(11)	32(12)	18(13)	7(14)	15(15)	98
4	36(16)	12(17)	21(18)	24(19)	40(20)	133
5	15(21)	24(22)	39(23)	9(24)	18(25)	105
Rep. Total						632
			Rep. II			
1	33(1)	13(6)	18(11)	28(16)	26(21)	118
2	9(2)	20(7)	36(12)	35(17)	21(22)	121
3	36(3)	12(8)	19(13)	27(18)	25(23)	119
4	16(4)	19(9)	21(14)	22(19)	10(24)	88
5	19(5)	22(10)	9(15)	29(20)	38(25)	117
Rep. Total						563
			Rep. III			
1	13(1)	7(9)	30(11)	24(18)	12(22)	86
2	12(2)	25(10)	26(15)	45(20)	29(25)	137
3	21(3)	14(6)	20(13)	35(16)	45(24)	135
4	29(4)	30(8)	39(12)	34(19)	30(21)	162
5	8(5)	14(7)	33(14)	29(17)	16(23)	100
Rep. Total						620
Grand Total (G)						1815

Treatment No.	Treatment Total (T)	C_b
1	85	-108
2	54	-112
3	97	55
4	63	42
5	49	42
6	51	-81
7	74	
8	58	
9	61	
10	76	26
11	74	5
12	107	7
13	57	65
14	61	23
15	50	126
16	99	
17	76	
18	72	
19	80	
20	114	91
21	71	-32
22	57	-37
23	80	-107
24	64	40
25	85	-45

Solution _____

STEP 1: Compute the replication totals, the block totals (B), and the grand total as shown in Table 4.3.2.1.

STEP 2: Compute the treatment total (T) for all treatments. For example, the total for treatment number 1 is:

$$T_1 = 39 + 33 + 13 = 85 \qquad\qquad (4\text{-}52)$$

STEP 3: Calculate for each block, the C_b values where:

$$C_b = \text{total (over all replications) of all treatments in the block} - rB \qquad (4\text{-}53)$$

where

r = number of replications.

B = block total.

For example, the C_b value for block 1 in replication 1 is:

$$C_b = 85 + 54 + 97 + 63 + 49 - (3)(152) = -108$$

The C_b value for all the 15 blocks are computed and are shown in Table 4.3.2.1. The replicate totals (R_c) of the C_b values are -81, 126, and -45. Note that these replicate totals sum to 0.

STEP 4: Determine the degrees of freedom associated with each source of variation. The source of variation and the total degrees of freedom depend on whether the treatments are adjusted or not. If the treatments are adjusted, the following sources of variation and the degrees of freedom are identified.

Replication $d.f.$	$= r - 1 = 2$
Treatment (unadj.) $d.f.$	$= k^2 - 1 = 24$
Blocks within replications (adj.) $d.f.$	$= r\,(k - 1) = 12$
Intrablock error $d.f.$	$= (k - 1)(rk - k - 1) = 36$
Treatment (adj.)$d.f.$	$= k^2 - 1 = (24)$
Total $d.f.$	$= (rk^2 - 1) = 74$

where

r = number of replications.

k = block size.

STEP 5: Calculate the various sums of squares by first computing the correction factor (C) as:

$$C = \frac{G^2}{(r)(k)^2} \tag{4-54}$$

$$= \frac{(1{,}815)^2}{(3)(5)^2}$$

$$= 43{,}923$$

$$\text{Total } SS = \sum X_{ij}^2 - C \tag{4-55}$$

$$\text{Total } SS = \left[(39)^2 + (33)^2 + (40)^2 + \dots + (16)^2\right] - 43{,}923$$

$$= 51{,}288 - 43{,}923$$

$$= 7{,}365$$

$$\text{Replication} SS = \frac{\sum R^2}{k^2} - C \tag{4-37}$$

$$= \frac{(632)^2 + (563)^2 + (620)^2}{25} - 43{,}923$$

$$= 44{,}031.7 - 43{,}923$$

$$= 108.7$$

$$\text{Treatment} SS \text{ (unadj.)} = \frac{\sum T^2}{r} - C \tag{4-56}$$

$$= \frac{(85)^2 + (54)^2 + \dots + (85)^2}{3} - 43{,}923$$

$$= 46{,}473.6 - 43{,}923$$

$$= 2{,}550.7$$

$$\text{Block } SS \text{ (adj.)} = \frac{\sum C_b^2}{kr\,(r-1)} - \frac{\sum R_c^2}{k^2 r\,(r-1)} \tag{4-57}$$

$$= \frac{(-108)^2 + (-112)^2 + \dots + (40)^2}{(5)(3)(3-1)} - \frac{(-81)^2 + (126)^2 + (-45)^2}{(5)^2(3)(3-1)}$$

$$= 1{,}999.6 - 163.1$$

$$= 1{,}836.5$$

STEP 6: Compute the intrablock error SS as:

$$\text{Intrablock error } SS = TSS - \text{Rep. } SS - \text{Treatment (unadj.) } SS - \text{Block (adj.) } SS$$

$$= 7{,}365 - 108.7 - 2{,}550.7 - 1{,}836.5 \tag{4-40}$$

$$= 2{,}869.1$$

STEP 7: Calculate the mean square for the block (adj.), and the intrablock error as:

$$\text{Block } MS \text{ (adj.)} = \frac{\text{Block } SS \text{ (adj.)}}{r(k-1)} \tag{4-58}$$

$$= \frac{1{,}836.5}{12}$$

$$= 153.0$$

$$\text{Intrablock Error } MS = \frac{\text{Intrablock Error } SS}{(k-1)(rk-k-1)} \tag{4-59}$$

$$= \frac{2{,}869.1}{(5-1)[(3)(5)-5-1]}$$

$$= 79.70$$

STEP 8: Calculate μ, the weighting factor, to adjust the treatment totals. μ is computed as shown below:

$$\mu = \frac{\text{Block (adj.) } MS - \text{Intrablock error } MS}{k\,(r-1)\big[\text{Block (adj.) } MS\big]} \tag{4-60}$$

If the block (adj.) mean square is less than the intrablock error mean square, μ is assumed to be 0 and there is no need for adjustments for the block effects. Thus, the data is analyzed as if in randomized blocks. However, for the present example the block (adj.) mean square is greater than the intrablock error mean square, so the adjustment factor is calculated as:

$$\mu = \frac{153 - 79.70}{5(3-1)(153)}$$

$$= \frac{73.3}{1,530}$$

$$= 0.05$$

STEP 9: Calculate the adjusted treatment SS as:

$$\text{Treatment (adj.) } SS = \qquad\qquad\qquad (4\text{-}61)$$

$$\text{Treatment (unadj.) } SS - k\,(r-1)\,\mu\left\{\left[\frac{r}{(r-1)(1+k\mu)}\right]B_u - B_a\right\}$$

where

$\qquad\qquad B_u$ = unadjusted sum of squares for blocks within replications

$\qquad\qquad B_a$ = adjusted sum of squares for blocks within replications.

To compute equation 4-61, we first have to calculate B_u which is defined as:

$$B_u = \frac{\sum B^2}{k} - \frac{\sum R^2}{k^2} \qquad\qquad\qquad (4\text{-}62)$$

where

$\qquad\qquad B$ = block total

$\qquad\qquad R$ = replication total

$\qquad\qquad k$ = number of blocks.

For the present example, we have:

$$B_u = \frac{(152)^2 + (144)^2 + (98)^2 + ... + (100)^2}{5} - \frac{(632)^2 + (563)^2 + (620)^2}{(5)^2}$$

$$= 1,358.48$$

Substituting the value of B_u and B_a into the equation 4-61, we will have:

$$2,550.7 - 5(2)(0.05)\left\{\left[\frac{3}{(2)(1+(5)(0.05)}\right](1,358.48)-(1,836.5)\right\} = 2,653.9$$

STEP 10: Calculate the treatment adjusted mean square as:

$$\text{Treatment}\,MS\,(\text{adj.}) = \frac{\text{Treatment}\,SS\,(\text{adj.})}{k^2-1} \qquad [4\text{-}63]$$

$$= \frac{2,653.9}{24}$$

$$= 110.60$$

STEP 11: Compute the F value as:

$$F = \frac{\text{Treatment}\,MS\,(\text{adj.})}{\text{Intrablock error}\,MS} = \frac{110.60}{79.70} \qquad (4\text{-}64)$$

$$= 1.39$$

STEP 12: Compare the computed F shown in Table 4.3.2.2 with the tabular F value given in Appendix E. Because the computed F is less than the tabular F (for 24 treatment degrees of freedom and 36 intrablock error degrees of freedom), the F-test suggests that the experiment failed to show any significant difference among the treatments.

As in the previous sections, we can determine the degree of precision with which the treatments are compared by computing the coefficient of variation cv as:

$$cv = \frac{\sqrt{\text{Intrablock error}\,MS}}{\text{Grand Mean}} \times 100 \qquad (4\text{-}50)$$

$$= \frac{\sqrt{79.70}}{36} \times 100 = 24.79\%$$

STEP 13: Estimate the gain in accuracy over randomized blocks. This is accomplished by comparing the error mean square with the effective error variance. The error mean square in the randomized block is the pooled mean square for blocks and intrablock error as shown below:

TABLE 4.3.2.2.

Analysis of Variance of an Experiment on Hybrid Corn
Doublecropped with Winter Wheat Using a 5 × 5 Triple Lattice
Design

Source of Variation	Degree of Freedom	Sum of Squares	Mean Square	F
Replication	2	108.7		
Block (adj.)	12	1,836.5	153.00	
Treatment (unadj.)	(24)	2,550.7		
Treatment (adj.)	24	2,654.3	110.60	
Intrablock error	36	2,869.1	79.70	1.39^{ns}
Total	74	7,365.0		

Note: cv = 24.79%; ns = Nonsignificant.

$$\text{Randomized Block } MS = \frac{\text{Block } SS \text{ (adj.)} + \text{Intrablock error } SS}{(k^2 - 1)(r - 1)} \quad (4\text{-}65)$$

The effective error variance is computed as:

$$\text{Intrablock error } MS \left[1 + \frac{rk\mu}{(k+1)} \right] \quad (4\text{-}66)$$

For our example, the respective values are:

$$\text{Randomized block } MS = \frac{1,836.5 + 2,869.1}{48}$$

$$= 98.0$$

$$\text{Effective error variance} = 79.70 \left[1 + \frac{(3)(5)(0.05)}{(5+1)} \right]$$

$$= 87.9$$

Thus, the relative accuracy is:

$$\frac{98.0}{89.7} = 109.2\%$$

Analysis of Variance for a Case with Repetition: In this design, n replicates contained in the basic plan are repeated p times, so the total number of replications is $r = np$. To illustrate the case with repetition, we will apply the data from the triple lattice design used in the previous section and change it to a quadruple lattice for the present case. This means that for a case with repetition we take the first 2 replications of the 5×5 plan selected from Appendix F, and repeat each replication once to obtain the 4 replications needed for the experiment. Another way of stating this is in a 5×5 quadruple lattice design, the basic plan obtained is by repeating a simple lattice twice (where $n = 2$ for a simple lattice). That is, replications III and IV are a repeat of replications I, and II. The data are then rearranged according to the basic plan such that treatments (numbers in parenthesis) follow the same order within the corresponding replications as shown in Table 4.3.2.3. The steps in performing the analysis of variance are given below.

STEP 1: Compute the replication totals, the block totals (B), and the grand total (G) as shown in Table 4.3.2.3.

STEP 2: Compute the treatment total (T) for all treatments. For this example, the total for treatment number one is:

$$T_1 = 39 + 33 + 19 + 28 = 119$$

STEP 3: Determine the degrees of freedom associated with each source of variation. The source of variation and the total degrees of freedom depend on whether the treatments are adjusted or not. If the treatments are adjusted, the following sources of variation and the degrees of freedom are identified:

Replication *d.f.*	$= r - 1 = 3$
Treatment (unadj.) *d.f.*	$= k^2 - 1 = 24$
Blocks within replications (adj.) *d.f.*	$= r(k - 1) = 16$
Component (a) *d.f.*	$= n(p - 1)(k - 1) = 8$
Component (b) *d.f.*	$= n(k - 1) = 8$
Intrablock error *d.f.*	$= (k - 1)(rk - k - 1) = 56$
Total *d.f.*	$= (rk^2 - 1) = 99$

where

$r =$ total number of replications

TABLE 4.3.2.3.

Yields of 5 × 5 Simple Lattice Experiment on Hybrid Corn Doublecropped with Winter Wheat (Bushels per acre, minus 20 bu.)

Block No.						Block Total (B)	Treatment No.	Treatment Total (T)
		Rep. I						
1	39(1)	33(2)	40(3)	18(4)	22(5)	152	1	119
2	24(6)	40(7)	16(8)	35(9)	29(10)	144	2	72
3	26(11)	32(12)	18(13)	7(14)	15(15)	98	3	121
4	36(16)	12(17)	21(18)	24(19)	40(20)	133	4	66
5	15(21)	24(22)	39(23)	9(24)	18(25)	105	5	82
							6	71
Rep. Total						632	7	105
							8	70
		Rep. II					9	87
1	33(1)	13(6)	18(11)	28(16)	26(21)	118	10	92
2	9(2)	20(7)	36(12)	35(17)	21(22)	121	11	83
3	36(3)	12(8)	19(13)	27(18)	25(23)	119	12	112
4	16(4)	19(9)	21(14)	22(19)	10(24)	88	13	84
5	19(5)	22(10)	9(15)	29(20)	38(25)	117	14	60
							15	46
Rep. Total						563	16	112
							17	83
							18	84
		Rep. III					19	90
1	19(1)	20(2)	15(3)	18(4)	25(5)	97	20	121
2	16(6)	30(7)	20(8)	15(9)	21(10)	102	21	80
3	24(11)	18(12)	28(13)	10(14)	12(15)	92	22	87
4	38(16)	20(17)	12(18)	14(19)	28(20)	112	23	104
5	25(21)	20(22)	19(23)	10(24)	30(25)	104	24	49
							25	106
Rep. Total						507		
		Rep. IV						
1	28(1)	18(6)	15(11)	10(16)	14(21)	85		
2	10(2)	15(7)	26(12)	16(17)	22(22)	89		
3	30(3)	22(8)	19(13)	24(18)	21(23)	116		
4	14(4)	18(9)	22(14)	30(19)	20(24)	104		
5	16(5)	20(10)	10(15)	24(20)	20(25)	90		
Rep. Total						484		
Grand Total						2186		

$k =$ block size

$n =$ number of replication in the base design

$p =$ number of repetition.

STEP 4: Calculate the various sums of squares by first computing the correction factor (C) as:

$$C = \frac{G^2}{(n)(p)(k^2)} \qquad [4\text{-}67]$$

$$= \frac{(2,186)^2}{(2)(2)(25)}$$

$$= 47,785.96 \text{ or } 47,786$$

$$\text{Total } SS = \sum X^2 - C \qquad [4\text{-}68]$$

$$= \left[(39)^2 + (33)^2 + (40)^2 + \dots + (20)^2 \right] - 47,786$$

$$= 54,698 - 47,786$$

$$= 6,912$$

$$\text{Rep. } SS = \frac{\sum R^2}{k^2} - C \qquad [4\text{-}69]$$

$$= \frac{(632)^2 + (563)^2 + (507)^2 + (484)^2}{25} - 47,786$$

$$= 48,307.9 - 47,786$$

$$= 521.9$$

$$\text{Treatment } SS \text{ (unadj.)} = \frac{\sum T^2}{(n)(p)} - C \qquad [4\text{-}70]$$

$$= \frac{201,922}{(2)(2)} - 47,786$$

$$= 50,480.5 - 47,786$$

$$= 2,694.5$$

STEP 5: For each block in each repetition, calculate the B_T value, which is the sum of block totals over all replications in that repetition. For example, the B_T value for block 1 from replications I and III is:

$$B_T = 152 + 97 = 249 \qquad (4\text{-}71)$$

The B_T value for the remainder of the blocks in each repetition is computed and shown in Table 4.3.2.4.

STEP 6: Calculate for each block, the C_b values as:

$$C_b = \sum T - nB_T \qquad (4\text{-}72)$$

where

$T =$ Treatment total

$n =$ number of replications in the base design

$B_T =$ sum of block totals over all replications.

For example, the C_b value for block 1 in replication 1 which contains treatments 1, 2, 3, 4, and 5 is:

$$C_b = 119 + 72 + 121 + 66 + 82 - (2)(249) = -38$$

The C_b value for all the blocks are computed and are shown in Table 4.3.2.4.

STEP 7: Compute the block SS which contains 2 components, (a) and (b). Component (a), which is the difference between the totals of blocks that contain the same set of treatments, arises only when the design is repeated. Component (b), on the other hand, is present when there are no repetitions involved. To compute each component, we will use the information in Table 4.3.2.4 which has k rows and p columns. Thus, the sum of squares for component (a) is the sum of the rows × columns interactions over all repetitions. Each component is computed as follows:

$$\text{Component}(a)\ SS = \text{Total} - \text{Rows} - \text{Columns} \qquad (4\text{-}73)$$

TABLE 4.3.2.4.

B_T and C_b Values for Pairs of Blocks Containing the Same Set of Treatments

Block Number	Block Totals		Sum (B_T)	C_b
	Repetition 1			
	Rep. I	Rep. III		
1	152	97	249	-38
2	144	102	246	-67
3	98	92	190	5
4	133	112	245	0
5	105	104	209	8
Total	632	507	1,139	-92
	Repetition 2			
	Rep. II	Rep. IV		
1	118	85	203	59
2	121	89	210	39
3	119	116	235	-7
4	88	104	192	-32
5	117	90	207	33
Total	563	484	1,047	+92

where

$$\text{Total} = \frac{\sum B^2}{k} - \frac{\sum R^2}{pk^2} \qquad (4\text{-}74)$$

$$\text{Rows} = \frac{\sum B_T^2}{pk} - \frac{\sum R^2}{pk^2} \qquad (4\text{-}75)$$

$$\text{Columns} = \frac{\sum N^2}{k^2} - \frac{\sum R^2}{pk^2} \qquad (4\text{-}76)$$

The term B is the block total, R is the sum of B_T values for each repetition, N is the sum of block totals for each replications, and k and p are as defined earlier. For our example, the values are:

$$\text{Total} = \frac{(152)^2 + (144)^2 + ... + (104)^2 + (90)^2}{5} - \frac{(1,139)^2 + (1,047)^2}{50}$$

$$= 49,094.4 - 47,870.6$$

$$= 1,223.8$$

$$\text{Rows} = \frac{(249)^2 + (246)^2 + ... + (192)^2 + (207)^2}{10} - \frac{(1,139)^2 + (1,047)^2}{50}$$

$$= 48,257.0 - 47,870.6$$

$$= 386.4$$

$$\text{Columns} = \frac{(632)^2 + (507)^2 + (563)^2 + (484)^2}{25} - \frac{(1,139)^2 + (1,047)^2}{50}$$

$$= 48,307.9 - 47,870.6$$

$$= 437.3$$

$$\text{Component}(a)\ SS = 1,223.8 - 386.4 - 437.3$$

$$= 400.1$$

$$\text{Component}(b)\ SS = \frac{\sum C_b^2}{kr(n-1)} - \frac{\sum R_c^2}{k^2 r(n-1)} \qquad (4\text{-}77)$$

where

R_c = the total of C_b values over all blocks in a repetition.

For the present example, we have two R_c values which are -92 and +92. These replicate totals should add to 0.

$$\text{Component}(b)\ SS = \frac{(-38)^2 + (-67)^2 + ... + (-32)^2 + (33)^2}{20} - \frac{(-92)^2 + (92)^2}{100}$$

$$= 659.3 - 169.3$$

$$= 490$$

STEP 8: Calculate the adjusted sum of squares for blocks within replications as:

$$\text{Block } SS \text{ (adj.)} = \text{Component}(a)\, SS + \text{Component}(b)\, SS \qquad (4\text{-}78)$$
$$= 400.1 + 490$$
$$= 890.1$$

STEP 9: Calculate the intrablock error sum of squares as:

$$\text{Intrablock error } SS = \text{Total } SS + \text{Rep. } SS - \text{Treatment } SS \text{ (unadj.)} - \text{Block } SS \text{ (adj.)}$$
$$(4\text{-}79)$$
$$= 6{,}912 - 521.9 - 2{,}694.5 - 890.1$$
$$= 2{,}805.5$$

STEP 10: Calculate the mean square for block (adj.) and the intrablock error as:

$$\text{Block } MS \text{ (adj.)} = \frac{\text{Block } SS \text{ (adj.)}}{r\,(k-1)} \qquad (4\text{-}80)$$
$$= \frac{890.1}{16}$$
$$= 55.6$$

Page 122, Eq. 4-81:

$$\text{Intrablock Error } MS = \frac{\text{Intrablock Error } SS}{(k-1)\,(rk-k-1)} \qquad)$$
$$= \frac{2{,}805.5}{(5-1)\,[(4)\,(5)-5-1)]}$$
$$= 50.1$$

STEP 11: Calculate μ, the weighting factor, to adjust the treatment totals. If the block (adj.) mean square is the pooled mean square for the blocks, then the weighting factor is:

$$\mu = \frac{p\left[\text{Block (adj.)}MS - \text{Intrablock error }MS\right]}{k\left[(r-p)\text{ Block (adj.) }MS + (p-1)\text{ Intrablock error }MS\right]} \quad (4\text{-}82)$$

$$\mu = \frac{2(55.6 - 50.1)}{5\left[(4-2)55.6 + 50.1\right]}$$

$$= 0.0136$$

As before, p is the number of repetitions of the basic design, and is 2 for the present example. If block (adj.) mean square is less than the intrablock error mean square, μ is assumed to be 0 and there is no need for adjustments for the block effects. Since the opposite holds true for the present case, each treatment total is adjusted by applying a correction for each group of blocks in which the treatment appears.

STEP 12: Given that a correction is necessary, calculate the adjusted treatment sum of squares as:

$$\text{Treatment }SS \text{ (adj.)} = \text{Treatment }SS \text{ (unadj.)} - k(n-1)\mu\left\{\left[\frac{n}{(n-1)(1+k\mu)}\right]B_u - B_a\right\}$$

$$(4\text{-}83)$$

where

B_u = unadjusted sum of squares for component b of the blocks

B_a = adjusted sum of squares for component b of the block.

We have already computed B_u, which is 386.4 and was calculated as the rows sum of squares when computing component (a). B_a is the component (b) sum of squares and is 490. Thus, the adjusted treatment sum of squares is:

$$\text{Treatment } SS \text{ (adj.)} = 2{,}694.5 - 5(1)(0.0136)\left\{\left[\frac{2}{(1)(1+5)(0.0136)}\right]386.4 - 490\right\}$$

$$= 2{,}694.5 - 610.42$$

$$= 2{,}084.07$$

STEP 13: Calculate the adjusted treatment mean square as:

$$\text{Treatment } MS \text{ (adj.)} = \frac{\text{Treatment } SS \text{ (adj.)}}{(k^2 - 1)} \tag{4-84}$$

$$= \frac{2{,}084.07}{24}$$

$$= 86.84$$

STEP 14: Compute the F value as:

$$F = \frac{\text{Treatment (adj.) } MS}{\text{Effective error } MS} \tag{4-48}$$

$$= \frac{86.84}{50.1}$$

$$= 1.73$$

The corresponding cv value is computed as:

$$cv = \frac{\sqrt{\text{Intrablock error } MS}}{\text{Grand Mean}} \times 100 \tag{4-50}$$

$$= \frac{\sqrt{50.1}}{21.86} \times 100 = 32.3\%$$

STEP 15: Compare the computed F with the tabular F value given in Appendix E. For this example, the tabular F (for 24 treatment degrees of freedom and 56 intrablock error degrees of freedom) is less than the computed F value at the 1% level of significance. Thus, it is concluded that there is a highly significant difference among the treatments. Table 4.3.2.5 shows the summary of the results.

TABLE 4.3.2.5.
Analysis of Variance of Data for Hybrid Corn Using a 5 × 5 Quadruple Lattice Design

Source of Variation	Degree of Freedom	Sum of Squares	Mean Square	F
Replication	3	521.9		
Block (adj.)	16	890.1	55.6	
Component (a)	(8)	400.1		
Component (b)	(8)	490.0		
Treatment (unadj.)	(24)	2,694.5		
Treatment (adj.)	24	2,084.07	86.84	1.73**
Intrablock error	56	2,805.5	50.1	
Total	99	6,912.0		

Note: cv = 32.3%. ** = significant at 1% level.

STEP 16: Compute the error variance or the effective error mean square for the difference between 2 treatment means as shown below.

1. The error mean square for 2 treatments in the same block:

$$\text{Error } MS = MS \text{ error} \left[1 + (n-1)\mu \right] \tag{4-85}$$
$$= 50.1 \left[1 + (2-1)(0.0136) \right]$$
$$= 50.8$$

2. The error mean square for 2 treatments not the in the same block:

$$\text{Error } MS = MS \text{ error} \left(1 + n\mu \right) \tag{4-86}$$
$$= 50.1 \left[1 + 2(0.0136) \right]$$
$$= 51.5$$

3. The average effective error mean square is computed as:

$$\text{Average Error } MS = MS \text{ error}\left[1 + \frac{nk\mu}{k+1}\right] \tag{4-87}$$

$$= 50.1\left[1 + \frac{2(5)(0.0136)}{5+1}\right]$$

$$= 51.2$$

STEP 17: Calculate the relative efficiency (*R.E.*) to estimate the precision relative to randomized complete block designs as:

$$R.E. = \left[\frac{\text{Block (adj.) } SS + \text{Intrablock error } SS}{r(k-1) + (k-1)(rk-k-1)}\right]\left[\frac{100}{\text{Error } MS}\right] \tag{4-88}$$

We substitute the effective error mean square computed in Step 15 for each of the conditions to calculate the relative efficiency as shown below:

1. $R.E. = \left[\dfrac{890.1 + 2,805.5}{4(4) + (4)(20-5-1)}\right]\left[\dfrac{100}{50.8}\right]$

   $= 101.1\%$

2. $R.E. = \left[\dfrac{890.1 + 2,805.5}{4(4) + (4)(20-5-1)}\right]\left[\dfrac{100}{51.5}\right]$

   $= 99.6\%$

3. $R.E. = \left[\dfrac{890.1 + 2,805.5}{4(4) + (4)(20-5-1)}\right]\left[\dfrac{100}{51.2}\right]$

   $= 100.2\%$

In summary, this chapter has introduced the single factor experiments using the complete and incomplete block designs. Three different types of complete block designs —the completely randomized design (CRD), the randomized complete block design (RCBD), and the Latin square design—were discussed. For the incomplete block designs, the 2 most commonly used designs—the lattice and the partially balanced lattice designs—were detailed. Each design has its advantages and disadvantages. Using the coefficient of variation as a measure of degree of precision, the researcher is able to determine whether a certain design gives more precision than others.

REFERENCES AND SUGGESTED READINGS

Bose, R. C. and Nair, K. R. 1939. "On Construction of Balanced Incomplete Designs." *Ann. Eugenics* 9: 353-399.

Fisher, R. A. 1974. *The Design of Experiments*. London: Collier Macmillan.

Gomez, K. A. and Gomez, A. A. 1984. *Statistical Procedures for Agricultural Research*. New York: John Wiley & Sons.

Yates, F. 1936(a). "Incomplete randomized blocks," *Ann. Eugenics* 7: 121-140.

Yates, F. 1936(b). " A new method of arranging variety trials involving a large number of varieties," *J. Agric. Sci.* 26: 424-455.

Yates, F. 1937. "The design and analysis of factorial experiments," *Tech. Commun.*, 35, Rothamsted: Commonwealth Bureau of Soil Science.

EXERCISES

1. Field measurements were made to study the response of field-grown cassava (*Manihot esculenta* Crantz) to changes in the application of a fertilizer. The researcher conducted a completely randomized design to find out if there were any differences in the amount of the dry matter produced under 5 different fertilizer regimes. The following data were collected from the experiment with 4 replications:

Dry Matter Production (tons/ha) of Cassava as a Result of 5 Different Fertilizer Applications

Treatment	Rep. I	Rep. II	Rep. III	Rep. IV
Control	2.20	2.10	2.25	2.01
50 kg/ha	2.40	2.56	2.66	2.52
100 kg/ha	2.60	2.68	2.79	2.66
150 kg/ha	3.00	3.56	4.00	4.66
200 kg/ha	3.50	4.98	5.00	4.20

(a) Perform the analysis of variance.
(b) Are there differences between the treatments?
(c) Compute the coefficient of variation? What does this value mean?

2. Assume that the data collected in Exercise 1 has some observations missing, as shown below:

Dry Matter Production (tons/ha) of Cassava as a Result of 5 Different Fertilizer Applications.

Treatment	Rep. I	Rep. II	Rep. III	Rep. IV
Control	2.20	2.10	2.25	2.01
50 kg/ha	2.40	—	2.66	2.52
100 kg/ha	2.60	2.68	—	2.66
150 kg/ha	—	3.56	4.00	4.66
200 kg/ha	3.50	4.98	5.00	4.20

(a) Perform an analysis of variance on the data.
(b) What can you say about the results of this exercise as compared to Exercise 1?
(c) Compute the coefficient of variation.
(d) Is there any difference between the cv computed here and in exercise 1? What could you say about the differences?

3. Seed yield of early maturing high protein soybean lines adapted to the Mid-Atlantic area of the U.S. were evaluated in field test in a randomized complete block with 3 replications. Each plot contained four 20-ft rows, 30 in. apart. Each plot was evaluated for seed yield, and the following data were recorded:

Seed Yield (bu/ac) of Early Maturing Soybean Lines			
Line	Rep. I	Rep. II	Rep. III
CX797-115	32.2	33.5	33.3
CX797-21	35.8	36.2	36.8
CX804-3	34.5	35.4	36.6
K1085	37.2	36.4	38.3
K1091	39.8	36.2	40.2
Williams	37.8	38.2	41.1
Douglas	36.4	38.6	40.2

(a) Perform an analysis of variance
(b) Compute the coefficient of variation.
(c) Determine the relative efficiency of this experiment.

4. An ornamental horticulturist conducted a fertilizer experiment in a greenhouse where 5 fertilizer treatments (A, B, C, D, and E) were tested by arranging plants in a Latin-square design. Thus rows and columns in the table are rows and columns in the greenhouse. The data below shows the yield from the experiment.

A22	B23	C19	D12	E14
B20	C13	D16	E19	A18
C14	D10	E12	A26	B23
D19	E18	A20	B18	C14
E15	A24	B20	C17	D10

(a) Using a 1% level of significance, determine if the mean yields are not equal for the 5 fertilizers.
(b) Compute the coefficient of variation for the data.
(c) What can be said about the relative efficiency of the Latin-square design?

5. Plant breeders are interested in determining the spikelet initiation differences among 9 winter wheat cultivars. The number of spikelets per plant from a field experiment which followed a 3×3 balanced lattice design with 4 replications are given below. Each cultivar is given a treatment number and they are: Turkey (1), Pawnee (2), Scout (3), Larned (4), Newton (5), Hawk (6), Vona (7), HW 1010 (8), and Bounty 100 (9). The data collected from each replication are presented below.

Incomplete Block No.	Spikelet No. Rep. I			Incomplete Block No.	Spikelet No. Rep. II		
1	18.1(8)	18.4(6)	17.6(1)	4	18.2(8)	20.2(7)	16.5(9)
2	16.5(3)	18.7(5)	17.9(7)	5	15.2(3)	19.9(2)	17.8(1)
3	16.0(4)	18.0(2)	16.0(9)	6	17.8(6)	18.1(5)	16.4(4)
	Rep. III				Rep. IV		
7	17.1(8)	18.4(5)	18.6(2)	10	16.2(3)	15.9(4)	18.5(8)
8	16.2(4)	17.7(7)	16.9(1)	11	17.2(6)	18.9(2)	17.6(7)
9	16.5(3)	18.9(6)	16.2(9)	12	15.4(9)	18.9(5)	17.4(1)

(a) Perform the analysis of variance.
(b) From the analysis in (a), is it necessary to compute the adjusted treatment totals for all treatments?
(c) Compute the coefficient of variation.
(d) Compute the relative efficiency coefficient for this design. What do the results indicate?

6. An animal scientist is interested in evaluating the role of progestron in stimulating sexual receptivity in estrogen-treated gilts. She has used a 4×4 triple lattice design to conduct the experiment in which the 16 ovariectomized gilts were treated with estradil benzoate (EB). After EB treatment gilts were moved to an evaluation pen where bores were brought in. Gilts remained in the evaluation pen for 5 minutes, during which time the number of mounts attempted by the bore were recorded. The following data were collected from the experiment with 3 replications. The treatment numbers appear in parentheses.

Incomplete Block No.	Mounts, Number/5 min			
	Replication I			
1	7(01)	5(02)	4(03)	2(04)
2	5(05)	2(06)	1(07)	3(08)
3	4(09)	3(10)	2(11)	2(12)
4	1(13)	3(14)	1(15)	5(16)

Incomplete Block No.	Mounts, Number/5 min			
	Replication II			
1	6(01)	6(05)	6(09)	2(13)
2	4(02)	2(06)	3(10)	3(14)
3	3(03)	3(07)	1(11)	2(15)
4	1(04)	1(08)	3(12)	5(16)
	Replication III			
1	7(01)	5(06)	4(11)	6(16)
2	4(05)	4(02)	1(15)	3(12)
3	5(09)	2(14)	4(03)	2(08)
4	1(13)	3(10)	2(07)	2(04)

(a) Perform the analysis of variance.
(b) Estimate the gain in accuracy over randomized blocks.

7. Suppose the animal scientist in the previous exercise chose to conduct her experiment as a case with repetition. As you recall from your reading in this chapter, when the replicates contained in the basic plan are repeated p times so that the total number of replication is $r = np$, then we have a case of data analysis with repetition. For the present case, the basic plan of the first 2 replicates have been repeated. This changes a triple lattice to a quadruple lattice. The data collected from this experiment are shown below.

Incomplete Block No.	Mounts, Number/5 min				Incomplete Block No.	Mounts, Number/5 min			
	Rep. I					Rep. II			
1	7(01)	5(02)	4(03)	2(04)	5	6(01)	6(05)	5(09)	2(13)
2	5(05)	2(06)	1(07)	3(08)	6	4(02)	2(06)	3(10)	3(14)
3	4(09)	3(10)	2(11)	2(12)	7	3(03)	3(07)	1(11)	2(15)
4	1(13)	3(14)	1(15)	5(16)	8	1(04)	1(08)	3(12)	5(16)
	Rep. III					Rep. IV			
1	6(01)	4(02)	3(03)	1(04)	5	4(01)	5(05)	5(09)	1(13)
2	4(05)	1(06)	2(07)	4(08)	6	3(02)	3(06)	4(10)	4(14)
3	5(09)	2(10)	1(11)	2(12)	7	2(03)	4(07)	2(11)	3(15)
4	3(13)	5(14)	2(15)	4(16)	8	1(04)	3(08)	2(12)	4(16)

(a) Perform the analysis of variance.
(b) Estimate the gain in accuracy of this design over randomized blocks.

Chapter **5**

TWO-FACTOR EXPERIMENTAL DESIGNS

5.1 FACTORIAL EXPERIMENTS

In agricultural experiments where a number of independent variables or factors often have an impact on one another, it is not appropriate to use single-factor experiments discussed in the previous chapter. This chapter deals with the experimental designs that permit the experimenter to study a number of independent variables within the same experiment. Such experiments that consist of 2 or more combinations of different factors are referred to as *factorial experiments*. Another way of looking at factorial experiments is that they are ones in which all, or nearly all, factor combinations are of interest to the researcher.

Factorial experiments are different from varietal trials. In factorial experiments, comparisons are based on the main and interaction effects. Varietal trials, on the other hand, involve comparisons of different levels of only 1 factor, which is the treatment variable.

When a factor-by-factor experiment is conducted, we are unable to investigate the interaction effects. In such experiments the levels of 1 factor are changed 1 at a time while holding other factors constant. In agricultural experiments where factors are likely to interact with each other, factorial experiments are appropriate investigative tools. It should be kept in mind that even when interactions do not occur, the results of factorial experiments are more widely applicable, as the main treatment effects have been shown to hold over a wide range of conditions.

There are advantages and disadvantages in using factorial experiments. Advantages of performing factorial experiments lie in the purpose of the experiment. If the purpose of the experiment is to investigate the effect of each factor, rather than the combination of levels of the factors that produce a maximum response, then the researcher could either conduct separate experiments where each deals with a single factor, or include all factors simultaneously, performing a factorial experiment.

The main feature of factorial experiments is that they economize on experimental resources. For example, if factors are independent, all the simple effects are equal to the experiment's main effect. This implies that under such circumstances, the main effects are the only quantities needed to measure the consequence of variations in the other factor. Furthermore, factorial experiments allow us to estimate each main effect with the same precision as if we had performed a whole experiment on each factor alone. This means that if we have an experiment with n factors, all at 2 levels and all independent, the single-factor approach would require n times as much experimental resources as would a factorial experiment keeping the same level of precision. Thus, factorial experiments offer savings in time and material resources. An additional advantage of the factorial experiment is its ability to extend the range of validity of conclusions in a convenient way. For instance, in a fertilizer trial where the presence or absence of N, P, and K are of primary interest to us, we add 3 varieties of corn in the experiment. Here the object is not a comparison of the varieties of corn as much as the effects of fertilizers on varieties of corn. Should the experiment point to the fact that the fertilizer effects are essentially the same for the 3 varieties, we can apply the conclusions to a new variety, with a greater confidence than if the experiment had been confined to one variety.

It is also important to keep in mind that indiscriminate use of factorial experiments can lead to an increase in complexity, size, and cost of an experiment. Therefore, if the research is still in the exploratory stages and especially if the number of potentially relevant factors is large, the experimenter may wish to use only 2 levels of each factor (Gill, 1978). For practical reasons, others (Gomez and Gomez, 1984) have suggested that an experimenter may wish to use an initial single-factor experiment in order to avoid using large experimental areas and large expenditures.

Another disadvantage of factorial experiments lies in the less precise estimates that result from problems associated with heterogeneity of factors as the number of treatments increase in size. The larger the number of treatments, the more difficult it is to measure the influence of the primary trait of interest (even when good statistical design such as blocking is used to overcome the problems arising from natural variation). Further, it should be pointed out that in comparison to a single-factor experiment, a factorial experiment of comparable size would have a larger standard error. As the number of treatment combinations increases in an experiment, the standard error per unit also increases. This increase in the standard error can usually be kept small by forming blocks or by a device known as confounding.

The following examples illustrate the concepts related to factorial experiments where simple cases are involved.

Example 5.1.1

Consider an experiment where a horticulturist is interested in the application of a fertilizer trial with nitrogen (N), phosphorous(P), and potassium(K) as the 3 factors of interest, each at 2 levels. In the simplest case the horticulturist will have 8 treatment combinations where either N, P, or K is present or absent in a treatment. This is shown in Table 5.1.1.

TABLE 5.1.1.
The 8 Fertilizer Treatment Combinations

Treatment	N	P	K
1	No	No	No
2	No	No	Yes
3	No	Yes	No
4	No	Yes	Yes
5	Yes	No	No
6	Yes	No	Yes
7	Yes	Yes	No
8	Yes	Yes	Yes

Since this is an experiment with 3 factors, each at 2 levels, the experiment is a $2 \times 2 \times 2$ or a (2^3) experiment. In this situation, if we use all 8 combinations where the treatments include the selected levels of variable factors, we can refer to the experiment as a complete factorial experiment. On the other hand, if we use a fraction of all the combinations, we refer to this as an incomplete factorial experiment. This is shown in Example 5.1.2.

Example 5.1.2

Consider a situation similar to Example 5.1.1 where the 3 fertilizers (treatment factors) each being applied at 2 levels are considered by an agronomist in a fertilizer trial on corn. Suppose the agronomist considers the treatment factors to be a substitute for one another. In such a situation, the experiment is conducted with 4 treatments; a control and the above 3 factors. In this case, the experiment cannot be called a factorial as not all combinations of factors are included in the experiment.

In a factorial experiment, the total number of treatments is the product of the levels in each factor. For illustration, in Example 5.1.1. we had 3 factors, each at 2 levels. If we are to add an additional factor such as the application of a mineral, the number of treatments increases to $2 \times 2 \times 2 \times 2 = 16$ or (2^4) factorial. You will note that the number of treatments increases rapidly with an increase in either the number of factors or the level of each factor. For example, if we have a factorial experiment with 4 cultivars, 4 fertilizer application rates, 3 irrigation methods, and 3 weed control practices, the total number of treatments would be $4 \times 4 \times 3 \times 3 = 144$.

Example 5.1.3

In a comparison study to determine the main and interaction effects, an animal scientist has planned a dietary experiment in which 2 factors, (1) fiber at 3 levels, denoted as f_0, f_1, and f_2 and (2) protein at 2 levels, p_0 and p_1, are included. In this experiment the animal scientist forms the following 6 combinations taking 1 level from each factor $p_0f_0, p_0f_1, p_0f_2, p_1f_0, p_1f_1$, and p_1f_2. These combinations form treatments in factorial experiments. The researcher's interest lies in determining the main and interaction effects of this comparison. To determine the main effect of protein, for example, she wants to observe the difference of the totals between the first and the last 3 of the above 6 combinations. Hence, the total of the first 3 treatment combinations represents the effect of protein at its p_0 level, while the sum of the last 3 represents its effect at the p_1 level, given the same levels of fiber. Thus, the difference between the first and the last 3 treatment combinations provides a comparison between the responses from the 2 levels of protein. To determine the main effect of fiber, 2 independent comparisons among the totals are made, that is, $p_0f_0 + p_1f_0, p_0f_1 + p_1f_1$, and $p_0f_2 + p_1f_2$.

To determine the interaction effects, the animal scientist needs to make a comparison that would indicate to her whether factors act independently or the results are influenced by their interaction. To make such a comparison, she first determines the effect of protein when all levels of fiber are present. Thus, she writes the following 3 contrasts:

$$p_0f_0 - p_1f_0$$

$$p_0f_1 - p_1f_1$$

$$p_0f_2 - p_1f_2$$

The interaction effects are determined by taking the difference between the above contrasts in the following manner:

$$(p_0 f_0 - p_1 f_0) - (p_0 f_1 - p_1 f_1)$$

In the present case, there are a total of 2 interaction contrasts.

The above examples have given a brief view of the concepts of factorial experiments. Additional examples in section 5.2 show the detailed computation of the main and interaction effects.

The advantages of the factorial designs are in their efficiency (saving time and money) and, more importantly, their ability to study the joint effects of variables. Two-factor experiments allow the effect on the response due to increasing the level of each factor, to be estimated at each level of the other factor. This permits conclusions that are valid over a wider range of experimental conditions. Such is not the case with the single-factor experiments. Simultaneous investigation of 2 factors is necessary when factors interact with each other. Such interactions imply that a factor effect depends on the level of the other factor.

The major characteristic of the single-factor designs is that treatments consist solely of different levels of a single-variable factor. This means that only 1 factor varies while all other factors are held constant. If, in conducting an experiment, the experimenter recognizes that the variable of interest is affected by the different levels of other factors, then it is important to use a design that allows for 2 or more variable factors.

In conducting an experiment where we compare v treatments using $N = v$ b experimental units, the design is a 2-way or 2-factor classification if the N experimental units are divided into b homogeneous groups (e.g., parcels of land), and the v treatments are allocated to the experimental units, at random in each group and independently from group to group. In the 2-factor designs any observation is classified by the treatment it receives and the group to which it belongs. As mentioned in the previous chapter, groups are called blocks in agricultural field experiments, and the experimental plan is referred to as the block design. In such designs, treatment differences are what the experimenter is interested in, and block contributions, while accounted for, are eliminated from the experimental error. Thus, the experiment deals with 1 factor (treatments), and blocking is a restriction on the randomization.

When dealing with 2 or more factors in an experiment, one needs to take account of interactions between factors and measure the main and simple effects of each of the factors. A *replicated factorial design* is a 2-way arrangement in which the main effects and interactions of 2 factors are of equal interest. Let us consider a factorial experiment involving 2 factors A and B, with respective a and b levels to explain the different components of such designs. Table 5.1.2 shows an example of a hypothetical 2-factor experiment where rows of the table correspond to the level of factor A and the columns to

TABLE 5.1.2.

Data Matrix for a 2-Factor Design

	Levels	\multicolumn{5}{c}{Factor B}
		1	2	3	...	b
	1	X_{111}	X_{112}	X_{113}	.	X_{11b}
	2	X_{211}	X_{212}	X_{213}	.	X_{21b}
Factor A	3	X_{311}	X_{312}	X_{313}	.	X_{31b}
	.	.	.	.	.	.
	.	.	.	.	.	.
	a	X_{a11}	X_{a12}	X_{a13}	.	X_{a1b}

the levels of factor B, and X_{ijk} represents the observation taken under the ith level of factor A and the jth level of factor B in the kth replicate. The ab experimental groups of n observations may be considered as ab random samples, 1 from each of ab treatment populations. The treatment populations are assumed to have been drawn from the same infinitely large parent population; thus, any differences among the ab distributions are attributable solely to differences among the treatment effects. The underlying model of the 2-way classification is of the form

$$X_{ijk} = \mu + \alpha_i + \beta_j + (\alpha\beta)_{ij} + \varepsilon_{ijk} \qquad (5\text{-}1)$$

For $i = 1,...a$; $j = 1,...b$; $k = 1,...n$

where

X_{ijk} = kth observation at the ith level of A and jth level of B

μ = overall mean response; that is, the average of the mean responses for the ab treatments

α_i = effect of the ith level of the first factor (A), averaged over the b levels of the second factor (B). The ith level of the first factor adds α_i to the overall mean μ

β_j = effect of the jth level of the second factor

$(\alpha\beta)_{ij}$ = interaction between the ith level of the first factor and the jth level of the second factor; the population mean for the ijth treatment minus $\mu + \alpha_i + \beta_j$

ε_{ijk} = the random error component.

The effects, α_i and β_j, are usually referred to as *main effects* in contrast with the *interaction effect*, $(\alpha\beta)_{ij}$. Implicit in the statement of the model is the assumption that the random error component ε_{ijk} is independently and normally distributed with a mean of 0 and a variance σ_e^2 within each treatment population defined by a combination of levels of A and B. This is the only assumption we make for the design. The interaction effect and its relation to the main effect will be discussed further in Section 5.2.

5.2 MAIN EFFECTS AND INTERACTIONS IN A 2-FACTOR EXPERIMENT

As we discuss factorial experiments, it is important to elaborate on the concepts of simple effects, main effects, interactions, and how each is computed and represented in the analysis of variance of the results. To explain each of these concepts, we will use a 2×2 factorial experiment with factors A and B each at 2 levels. This factorial is called 2^2 as there are 4 treatment combinations in the experiment. The treatment combinations can be written as a_0b_0, $a_0b_1, a_1b_0,$ and a_1b_1. Suppose factor A is 2 cultivars of soybean (Hobbit and Mead) and factor B is 2 different levels of fertilizer application (no fertilizer application, and application of fertilizer at 100 lb per acre). Assume that there is no uncontrolled variation, and we obtain the observations shown in Table 5.2.1.

TABLE 5.2.1.

Yields of Soybean from 2 Cultivars and Their Means (Bu/Ac)

| Cultivar | Fertilizer Levels | | Row Ave. | Response due to b_1 |
	(b_0) 0 lb/ac	(b_1) 100 lb/ac		
Hobbit (a_0)	30	40	35	+10
Mead (a_1)	40	60	50	+20
Column Ave.	35	50	42.5	
Mead - Hobbit	10	20		

Looking at these results, we are able to explain the effect of the application of fertilizer on yield as well as the superiority of one cultivar over another. It appears that the yield of Hobbit and Mead cultivars increased by 10 and 20 bushels per acre, respectively, when fertilizer was applied. These increases are called the *simple effects* of fertilizer application. The simple effects of the cultivars could also be reported similarly. That is, Mead cultivar is superior to Hobbit; 10 bushels per acre, when no fertilizer is applied, and 20 bushels per acre increase in yield when fertilizer is applied.

When factors are *independent*, the *main effects* are the effect of the ith level of the first factor; averaged over the b levels of the second factor. Hence, the main effect for fertilizer application in the present example is the average of the 2 simple effects $(10 + 20)/2$, or 15 bushels per acre. The main effect can alternatively be derived as the difference between 2 column means 35 and 50, which is 15. Factors are said to be independent when, for example, the response to fertilizer is the same whether Hobbit or Mead cultivar is used, and when the difference between Hobbit and Mead cultivar is the same whether fertilizer is applied or not. To determine whether factors are independent or not, an experimenter may use his/her knowledge of the processes by which factors produce their effects, or simply test the assumption of independence using the information provided by the factorial experiment. For example, if using different cultivars affects the response to fertilizer, the difference between 20 and 10 bushels per acre, or 10, is an estimate of this effect. We can use a t-test for this purpose. If the difference proves significant, then we can say that the assumption of independence is rejected by the data. This would mean that factors are not independent in their effects.

Interaction between factors is defined as a condition in which observations obtained under the various levels of one treatment behave differently under the various levels of the other treatment. For the present example, we can measure interaction by taking the difference between 20 (superiority in yield when fertilizer is added) and 10 bushels per acre (superiority in yield of Mead over Hobbit when no fertilizer is added) and averaging it over the 2 factors. The difference of 10 bushels/acre is averaged over the 2 factors and the result of 5 bushels per acre is the interaction between fertilizer application and the cultivars used. The interaction effect for each factor is computed as:

$$A \times B = (1/2)(\text{simple effect of } A \text{ at } b_1 - \text{simple effect of } A \text{ at } b_0) \quad (5\text{-}2)$$

$$= (1/2)\left[\left(a_1 b_1 - a_0 b_1\right) - \left(a_1 b_0 - a_0 b_0\right)\right]$$

or,

$$A \times B = (1/2)(\text{simple effect of } B \text{ at } a_1 - \text{simple effect of } B \text{ at } a_0) \quad (5\text{-}3)$$

$$= (1/2)\left[\left(a_1 b_1 - a_1 b_0\right) - \left(a_0 b_1 - a_0 b_0\right)\right]$$

Thus, the interaction for our example using the above equation is:

$$A \times B = (1/2)\left[(60-40)-(40-30)\right]$$

$$= (1/2)(20-10)$$

$$= 5 \text{ bu/ac}$$

or

$$A \times B = (1/2)\left[(60-40)-(40-30)\right]$$

$$= (1/2)(20-10)$$

$$= 5 \text{ bu/ac}$$

If there is no interaction between factors as is shown by the hypothetical data in Table 5.2.2 and computed below, the results of an experiment are simply described in terms of the main effects and not the effects of all treatment combinations. The interaction effect between cultivar and fertilizer application for a data set with no interaction is:

$$A \times B = (1/2)\left[(60-50)-(40-30)\right]$$

$$= (1/2)(10-10)$$

$$= 0 \text{ bu/ac}$$

or

$$A \times B = (1/2)\left[(60-40)-(50-30)\right]$$

$$= (1/2)(10-10)$$

$$= 0 \text{ bu/ac}$$

When factors are independent in their effects, as in the case of data in Table 5.2.2, all the simple effects of a factor equal its main effect. This means that main effects are the only quantities we need to describe fully the consequences of all variations in the factor. This gain in describing the results is especially advantageous in experiments with more than 2 factors. The main effect of cultivar or factor A in the present example is computed as the average of the simple effects of cultivar over all levels of fertilizer or factor B as shown below:

$$
\begin{aligned}
\text{Main effect of cultivar} \ &= \ (1/2)(\text{simple effect of } A \text{ at } b_0 + \text{simple} \\
& \quad\quad \text{effect of } A \text{ at } b_1) \\
&= \ (1/2)[(a_1b_0 - a_0b_0) + (a_1b_1 - a_0b_1)] \\
&= \ (1/2)[(40 - 30) + (60 - 50)] \\
&= \ (1/2)(20) \\
&= \ 10 \text{ bu/ac}
\end{aligned}
$$

Similarly, the main effect of fertilizer (factor B) is computed as:

$$
\begin{aligned}
\text{Main effect of fertilizer} \ &= \ (1/2)(\text{simple effect of } B \text{ at } a_0 + \text{simple} \\
& \quad\quad \text{effect of } B \text{ at } a_1) \quad\quad\quad\quad\quad (5\text{-}4) \\
&= \ (1/2)[(a_0b_1 - a_0b_0) + (a_1b_1 - a_1b_0)] \\
&= \ (1/2)[(50 - 30) + (60 - 40)] \\
&= \ (1/2)(40) \\
&= \ 20 \text{ bu/ac}
\end{aligned}
$$

The data in the last column and row of Table 5.2.2 show the same results obtained here.

TABLE 5.2.2.
Yields of Soybean from 2 Cultivars and Their Means (Bu/Ac)

Cultivar	Fertilizer Levels (b_0) 0 lb/ac	(b_1) 100 lb/ac	Row Ave.	Response to b_1
Hobbit (a_0)	30	50	40	+20
Mead (a_1)	40	60	50	+20
Column Ave.	35	55	45	
Mead - Hobbit	10	10		

5.3 INTERPRETATION OF INTERACTIONS

Many of the agricultural experiments will show some type of interaction. To determine the presence of interaction among factors, we need to apply the appropriate significance test (Cochran and Cox, 1957; Goulden, 1952). Once the presence of interaction is established, we compute the interaction effect and account for it in the analysis of variance. The advantage of no interaction, as mentioned before, is the considerable economy in describing the results of an experiment. Furthermore, absence of interaction may shed light on the experimental condition which may have special physical significance.

The interpretation of a 2-factor interaction may best be clarified by graphical illustration. Figure 5.3.1 shows factor B to be represented along the x axis, and we have plotted each set of points joined by a light line to correspond to different levels of factor A.

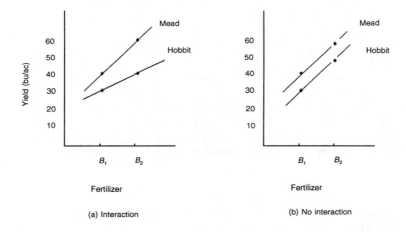

FIGURE 5.3.1.

Graphical representation of data in a 2-factor experiment (a) with interaction, (b) without interaction.

When factors are not independent, as is the case with Figure 5.3.1(a), the simple effect of a factor changes as the level of the other factor changes. In this situation we have interaction between factors. When there is no interaction between factors, as is the case in Figure 5.3.1.(b), the simple effect of a factor is the same for all levels of the other factors and equals the main effect. In this case the curves are parallel, indicating zero interaction among the factors.

It should be pointed out that while the separation of the treatment comparisons into main effects and interaction is a powerful tool in analyzing cases where interactions are small relative to main effects, much care in summarizing the results and detailed examination of the nature of interaction is required when interactions are large (Cochran and Cox, 1957).

In summary, interaction between factors can only be measured if 2 factors are tested together in the same experiment. Interaction between factors is absent when, for instance, the *difference* between the observations at 2 levels of *A* is the same for all levels of *B*. This implies that the simple effect of a factor equals the main effect. Under such circumstances the results from separate single-factor experiments are equivalent to factorial experiments with all factors tested together. Finally, when interaction is present we should examine the simple effects, and not the main effects, in reporting the results.

5.4 FACTORIALS IN COMPLETE BLOCK DESIGNS

Factorial experiments may be laid out in any convenient design such as complete block designs. In particular, if the total number of treatment combinations is not large, the designs discussed in Chapter 4 are frequently used. In Chapter 4 we discussed the procedures for the randomization and layout of such designs. The same procedures apply here with some modifications in factor composition. In factorial experiments we consider all factorial treatments to be unrelated to each other. In computing the treatment sum of squares for the analysis of variance, we take account of the factorial components corresponding to the main effects of individual factors and their interactions. Since the procedure for such partitioning is the same for all complete block designs, we will use an example for randomized complete block design to illustrate the step-by-step computation of the analysis of variance.

Example 5.4.1

Livestock breeders have continually changed animal types to meet perceived market demands and to adjust to changing economic pressures. To obtain data on measures of economic importance, such as average daily weight gain on 3 breeds of swine fed 3 different diets, an animal scientist used the following factorial treatment combination with 3 replications in conducting a performance test.

	Factorial Treatment Combinations		
Diet	Duroc (B_1)	Yorkshire (B_2)	Landrace (B_3)
D_1	B_1D_1	B_2D_1	B_3D_1
D_2	B_1D_2	B_2D_2	B_3D_2
D_3	B_1D_3	B_2D_3	B_3D_3

The randomized layout and the data recorded from the experiment are shown in Figure 5.4.1 and Table 5.4.1.

Before proceeding to the computation of the various components, let us denote the number of replications as (r), the level of factor A (breeds of swine) as a, and the level of factor B (diets) as b.

Rep. I

D_1B_1	D_2B_1	D_3B_2
D_2B_3	D_3B_3	D_1B_2
D_3B_1	D_2B_2	D_1B_3

Rep. II

D_3B_2	D_2B_2	D_3B_3
D_3B_1	D_2B_3	D_1B_1
D_2B_1	D_1B_3	D_1B_2

Rep. III

D_2B_2	D_2B_3	D_1B_1
D_1B_3	D_3B_1	D_1B_2
D_2B_1	D_3B_3	D_3B_2

FIGURE 5.4.1.

A sample layout of a 3×3 factorial experiment with 3 replications.

TABLE 5.4.1.
The Average Daily Weight Gain of 3 Breeds of Swine

Diet	Average Daily Weight Gain (lb)		
	Rep. I	Rep. II	Rep. III
	Duroc (B_1)		
D_1	1.80	2.00	2.20
D_2	1.50	1.74	1.85
D_3	2.05	2.08	2.25
	Yorkshire (B_2)		
D_1	2.00	1.80	1.95
D_2	1.85	1.65	1.58
D_3	1.90	2.22	2.15
	Landrace (B_3)		
D_1	2.15	2.12	2.18
D_2	1.75	1.86	1.95
D_3	2.17	2.05	2.28
Rep. Total (R)	17.17	17.52	18.39
Grand Total (G)			53.08

Solution

STEP 1: Calculate the replication totals (R) and the grand total (G) as shown in Table 5.4.1, and the treatment totals (T) as shown in Table 5.4.2.

TABLE 5.4.2.

Treatment and Diet Totals (lb) in a 3 × 3 Daily Weight Gain Experiment

Diet	B_1	B_2	B_3	Total
		Breed		
D_1	6.00	5.75	6.45	18.20
D_2	5.09	5.08	5.56	15.73
D_3	6.38	6.27	6.50	19.15
Total	17.47	17.10	18.51	53.08

STEP 2: Determine the degrees of freedom associated with each source of variation. The treatment component accounts for the variation due to breeds (factor A), diets (factor B) and the interaction between these 2 factors. The following sources of variation and the degrees of freedom are identified.

Replication d.f. $= r - 1 = 2$

Treatment d.f. $= ab - 1 = 8$

Breed (A) d.f. $= a - 1 = 2$

Diet (B) d.f. $= b - 1 = 2$

A × B d.f. $= (a - 1)(b - 1) = 4$

Error d.f. $= (r - 1)(ab - 1) = 16$

Total d.f. $= (rab - 1) = 26$

STEP 3: Calculate the various sums of squares by first computing the correction factor (C) as:

$$C = \frac{G^2}{rab} \tag{5-5}$$

$$= \frac{(53.08)^2}{3 \times 3 \times 3}$$

$$= \frac{2{,}817.49}{27}$$

$$= 104.35$$

$$\text{Total } SS = \sum X^2 - C \tag{5-6}$$

$$= \left[(1.80)^2 + (1.50)^2 + \dots + (2.28)^2 \right] - 104.35$$

$$= 105.52 - 104.35$$

$$= 1.17$$

$$\text{Replication } SS = \frac{\sum R^2}{ab} - C \tag{5-7}$$

$$= \frac{(17.17)^2 + (17.52)^2 + (18.39)^2}{9} - 104.35$$

$$= \frac{294.81 + 306.95 + 338.19}{9} - 104.35$$

$$= 104.44 - 104.35$$

$$= 0.09$$

$$\text{Treatment } SS = \frac{\sum T^2}{r} - C \tag{5-8}$$

$$= \frac{(6.00)^2 + (5.09)^2 + \dots + (6.50)^2}{3} - 104.35$$

$$= 105.19 - 104.35$$

$$= 0.84$$

$$\text{Error } SS = \text{Total } SS - \text{Rep. } SS - \text{Treatment } SS \tag{5-9}$$

$$= 1.17 - 0.09 - 0.84$$

$$= 0.24$$

A preliminary analysis of variance is shown in Table 5.4.3.

Comparing the results with the tabular F values given in Appendix E show that there exists a difference among the treatment means. We continue with our analysis to determine which component of the treatment is contributing to the difference.

TABLE 5.4.3.

Preliminary Analysis of Variance of the 3 Diets on 3 Breeds of Swine

Source of Variation	Degree of Freedom	Sum of Squares	Mean Square	F
Replication	2	0.09	0.045	3.0ns
Treatment	8	0.84	0.105	7.0**
Error	16	0.24	0.015	
Total	26	1.17		

Note: ns = not significant. * = significant at 1% level.

STEP 4: Calculate the 3 factorial components of the treatment sum of squares. In our example, factor A is the breeds of swine. Thus, to calculate the sum of squares we take the treatment totals shown in Table 5.4.2 and use them as follows:

$$SS_A = \frac{\sum A^2}{rb} - C \tag{5-10}$$

$$= \frac{(17.47)^2 + (17.10)^2 + (18.51)^2}{3 \times 3} - 104.35$$

$$= \frac{305.20 + 292.41 + 342.62}{9} - 104.35$$

$$= 104.47 - 104.35$$

$$= 0.12$$

To calculate the sum of squares for factor B, which represents the 3 diets, we use the treatment totals shown in Table 5.4.2 in the following manner:

$$SS_B = \frac{\sum B^2}{ra} - C \tag{5-11}$$

$$= \frac{(18.20)^2 + (15.73)^2 + (19.15)^2}{9} - 104.35$$

$$= \frac{331.24 + 247.43 + 366.72}{9} - 104.35$$

$$= 105.04 - 104.35$$

$$= 0.69$$

$$SS_{AB} = \text{Treatment} SS - SS_A - SS_B \tag{5-12}$$
$$= 0.84 - 0.12 - 0.69$$
$$= 0.03$$

STEP 5: Calculate the mean square for each component of the treatment sum of squares and the error mean square.

$$MS_A = \frac{SS_A}{a-1} = \frac{0.12}{3-1} = 0.060 \tag{5-13}$$

$$MS_B = \frac{SS_B}{b-1} = \frac{0.69}{2} = 0.345 \tag{5-14}$$

$$MS_{AB} = \frac{SS_{AB}}{(a-1)(b-1)} = \frac{0.03}{4} = 0.008 \tag{5-15}$$

$$\text{Error } MS = \frac{\text{Error } SS}{(r-1)(ab-1)} = \frac{0.24}{16} = 0.015 \tag{5-16}$$

STEP 6: Compute the F value for each source of variation as shown in Table 5.4.4.

TABLE 5.4.4.

Analysis of Variance of 3 Diets on 3 Breeds of Swine

Source of Variation	Degree of Freedom	Sum of Squares	Mean Square	F
Replication	2	0.09	0.045	3.00^{ns}
Treatment	8	0.84	0.105	7.00^{**}
Breed (A)	(2)	0.12	0.060	4.00^{*}
Diet (B)	(2)	0.69	0.345	23.00^{**}
Breed $\times$ Diet	(4)	0.03	0.008	0.53^{ns}
Error	16	0.24	0.015	
Total	26	1.17		

Note: Coefficient of Variation (cv) = 5.96%. ^{ns} = not significant, $*$ = significant at 5% level, $**$ = significant at 1% level.

STEP 7: Compare the computed F with the tabular F value given in Appendix E. It appears that both the breed main effect and the diet main effect are significant. Furthermore, the result suggests that there is no significant interaction between breeds of swine and the diets. This indicates that the yield difference among the breeds was not significantly affected by the diets.

As in the previous sections, we can determine the degree of precision with which the treatments are compared by computing the coefficient of variation cv as:

$$cv = \frac{\sqrt{\text{Error}\,MS}}{\text{Grand Mean}} \times 100 \qquad (5\text{-}17)$$

$$= \frac{\sqrt{0.015}}{2.056} \times 100 = 5.96\%$$

As mentioned at the beginning of this section, most types of experimental plans are suitable for factorial experiments. However, as the number of factors or the levels of a factor increase, so do the treatment combinations. This creates problems of maintaining homogeneous replications in randomized blocks which, in turn, leads to an increase in the experimental error. To overcome this problem, a number of designs has been developed. Most of these designs tend to reduce the block size, thus controlling increases in error. However, such reductions in block size are accompanied by a sacrifice in the accuracy of certain treatment comparisons. The split-plot experiments, discussed in Section 5.5, belong to this group of designs in which a factor or a group of factors and their interactions may be sacrificed.

5.5 SPLIT-PLOT OR NESTED DESIGN

In 2-factor experiments where the number of treatments are significantly large and cannot easily be accommodated by the complete block design, the split-plot design is used. In this design the *main plot* is being split into smaller *subplots* so as to accommodate a more precise measurement of the subplot factor and its interaction with the main plot factor. This means that the precision achieved by the subplot factor is at the expense of precision for the measurement of the main factor. These designs are also called *nested* designs because of the repeated sampling and sub-sampling that occur in the design.

To better understand how subplot factor is measured more precisely than the main factor, let us denote replications by r and the subplot treatments by a. In performing a test, each main plot is tested r times whereas each subplot treatment is tested ($r \times a$) times. For example, if we have 4 replications and 3 subplots within each replications, then the main plot gets tested 4 times while the subplot is tested 12 times. This increase in the number of times a subplot is tested contributes to the increased precision of the subplots. For this reason, it is extremely important to keep in mind which factor should be assigned to the main plot and which to the subplot. For example, if the experimenter is mainly interested in more precision in factor A than B, then factor A should be assigned to the subplot, and factor B to the main plot.

As a feature of the split-plot design, the subplot treatments are not randomized over the whole large block, but only over the main plots as shown in Figure 5.5.3. This means that randomization of the subtreatments is newly done in each main plot, and the main treatments are randomized in the large blocks. As a consequence of this difference in randomization, the experimental error for the comparison between the levels of one treatment is not the same as that for the comparison between the levels of the other treatment. Thus, the experimental error of the subtreatment is smaller than that for the main treatments.

Randomization and Layout: Given the nature of the design where main plots and subplots are the essential features, the randomization process requires that the main plot treatments be randomly assigned to the main plots first, followed by the assignment of the subplot treatments to each main plot. We may use any of the randomization schemes discussed in Chapter 4.

To provide an example of how such randomization and layout is carried out, let us denote the main plot treatments as a, the subplot treatments as b, and the number of replications as r. Assume that we are investigating the properties of 3 varieties of cotton bred for resistance to wilt as the subplot treatment and 4 dates of sowing as the main plot treatments, and the experiment is carried out in 4 replications. The steps in randomization and layout are given below.

STEP 1: Divide the experimental area into 4 blocks ($r = 4$). Further divide each block into 4 main plots as shown in Figure 5.5.1.

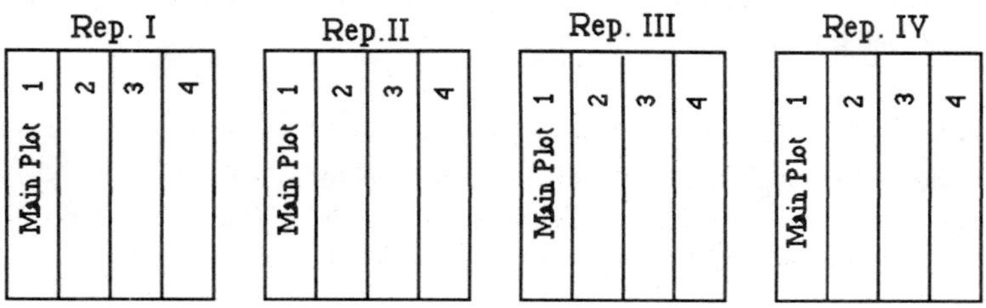

FIGURE 5.5.1.

Layout of an experimental area into 4 blocks and 4 main plots

STEP 2: Randomly assign the main plot treatments (4 dates of sowing: D_1, D_2, D_3, and D_4) to the main plots among the 4 blocks or replications as shown in Figure 5.5.2.

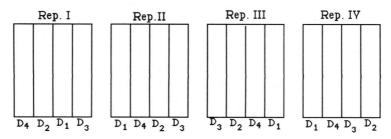

FIGURE 5.5.2.

Random assignment of the 4 main plot treatments to each of the 4 replications.

STEP 3: Divide each of the main plots into 3 subplots and randomly assign the cotton varieties (Factor B) to each subplot as shown in Figure 5.5.3.

Rep. I					Rep.II					Rep. III					Rep. IV			
V_1	V_3	V_2	V_2		V_3	V_1	V_2	V_1		V_3	V_1	V_2	V_3		V_2	V_3	V_3	V_1
V_3	V_2	V_1	V_3		V_1	V_2	V_1	V_3		V_2	V_3	V_3	V_2		V_1	V_2	V_1	V_2
V_2	V_1	V_3	V_1		V_2	V_3	V_3	V_2		V_1	V_2	V_1	V_1		V_3	V_1	V_2	V_3
D_4	D_2	D_1	D_3		D_1	D_4	D_2	D_3		D_3	D_2	D_4	D_1		D_1	D_4	D_3	D_2

FIGURE 5.5.3.

A possible layout of a split-plot experiment with 4 main plot treatments, 3 subplot treatments, and 4 replications.

Analysis of Variance: In performing the analysis of variance on an experiment where the split-plot design is used, the experimenter must consider a separate analysis for the main plot (factor A) and the subplot factor (factor B), respectively. As an example of a split-plot experiment, consider the following study.

Example 5.5.1

In a study carried out by agronomists to determine if major differences in yield response to N fertilization exist among widely grown hybrids in the northern Corn Belt, the subplot treatments were 5 hybrids of 105- to 110-day relative maturity (Pioneer 3747, 3732, Mo 17 × A634, A632 × LH 38, and LH 74 × LH 51), and the main plot treatments were N rates of 0, 70, 140, and 210 lb/acre broadcast applied before planting. The study was replicated 2 times, and the findings are as shown in Table 5.5.1.

TABLE 5.5.1.

Grain Yield Data of 5 Corn Hybrids Grown with 4 Levels of Nitrogen in a Split-Plot Experiment with 2 Replications

		N rate, lb/ac			
Replication	Hybrid	0	70	140	210
		Yield, bu/acre			
I.	P3747	130	150	170	165
	P3732	125	150	160	165
	Mo17 × A634	110	140	155	150
	A632 × LH38	115	140	160	140
	LH74 × LH51	115	170	160	170
II.	P3747	135	170	190	185
	P3732	150	160	180	200
	Mo17 × A634	135	155	165	175
	A632 × LH38	130	150	175	170
	LH74 × LH51	145	180	195	200

Solution

STEP 1: Calculate the replication totals (R), and the grand total (G) by first constructing a table for the nitrogen × replication totals shown in Table 5.5.2, and then a second table for the nitrogen × hybrid totals as shown in Table 5.5.3.

STEP 2: Determine the degrees of freedom associated with each source of variation. The following sources of variation and the degrees of freedom are identified:

Replication $d.f.$	$= r - 1 = 1$
Main plot factor $(A) d.f.$	$= a - 1 = 3$
Error (a) $d.f.$	$= (r - 1)(a - 1) = 3$
Subplot factor (B) $d.f.$	$= b - 1 = 4$

TABLE 5.5.2.

The Replication × Nitrogen (Factor A) Totals Computed from Data in Table 5.5.1

Replication	N rate, lb/ac 0	70	140	210	Rep. Total (R)
I.	595	750	805	790	2,940
II.	695	815	905	930	3,345
Nitrogen Total (A)	1,290	1,565	1,710	1,720	
Grand Total (G)					6,285

TABLE 5.5.3.

The Hybrid × Nitrogen (Factor B) Totals Computed from Data in Table 5.5.1

Hybrid	Yield Total (AB) N rate, lb/ac 0	70	140	210	Hybrid Total (B)
P3747	265	320	360	350	1,295
P3732	275	310	340	365	1,290
Mo17 × A634	245	295	320	325	1,185
A632 × LH38	245	290	335	310	1,180
LH74 × LH51	260	350	355	370	1,335

$A \times B$ d.f. $\qquad = (a-1)(b-1) = 12$

Error (b) d.f. $\qquad = a(r-1)(b-1) = 16$

Total d.f. $\qquad = rab - 1 = 39$

STEP 3: Compute the various sums of squares for the main-plot analysis by first computing the correction factor, using Equation 5-5 as follows:

$$C = \frac{G^2}{rab} \qquad (5\text{-}5)$$

$$= \frac{(6,285)^2}{2 \times 4 \times 5}$$

$$= 987,530.62$$

$$\text{Total } SS = \sum X^2 - C \qquad (5\text{-}6)$$

$$= \left[(130)^2 + (125)^2 + \dots + (200)^2 \right] - 987,530.62$$

$$= 1,007,775 - 987,530.62$$

$$= 20,244.38$$

$$\text{Replication } SS = \frac{\sum R^2}{ab} - C \qquad (5\text{-}7)$$

$$= \frac{(2,940)^2 + (3,345)^2}{20} - 987,530.62$$

$$= \frac{8,643,600 + 11,189,025}{20} - 987,530.62$$

$$= 991,631.25 - 987,530.62$$

$$= 4,100.63$$

$$SS_A \text{ (Nitrogen)} = \frac{\sum A^2}{rb} - C \qquad (5\text{-}18)$$

$$= \frac{(1,290)^2 + (1,565)^2 + (1,710)^2 + (1,720)^2}{2 \times 5} - 987,530.62$$

$$= 999,582.50 - 987,530.62$$

$$= 12,051.88$$

$$\text{Error } SS \ (a) = \frac{\sum RA^2}{b} - (C + Rep.\ SS + SS_A) \qquad (5\text{-}19)$$

$$= \frac{(595)^2 + (750)^2 + \dots + (930)^2}{5} - (987,530.62 + 4,100.63 + 12,051.88)$$

$$= 1,003,965 - 1,003,683.13$$

$$= 281.87$$

STEP 4: Compute the various sums of squares for the subplot (hybrid) analysis as shown below:

$$SS_B \text{ (Hybrid)} = \frac{\sum B^2}{ra} - C \qquad\qquad (5\text{-}20)$$

$$= \frac{(1,295)^2 + (1,290)^2 + (1,185)^2 + (1,180)^2 + (1,335)^2}{2 \times 4}$$

$$- 987,530.62$$

$$= \frac{7,919,975}{8} - 987,530.62$$

$$= 989,996.87 - 987,530.62$$

$$= 2,466.25$$

$$SS_{AB} \text{ (Nitrogen} \times \text{Hybrid)} = \frac{\sum (AB)^2}{r} - \left(C + SS_A + SS_B\right) \qquad (5\text{-}21)$$

$$= \frac{(265)^2 + (320)^2 + \ldots + (370)^2}{2}$$

$$- \left(987,530.62 + 12,051.88 + 2,466.25\right)$$

$$= \frac{2,005,825}{2} - 1,002,048.75$$

$$= 1,002,912.50 - 1,002,048.75$$

$$= 863.75$$

$$\text{Error } SS\ (b) = TSS - \text{All other sum of squares} \qquad\qquad (5\text{-}22)$$

$$= 20,244.38 - (4,100.63 + 12,051.88 + 281.87$$

$$+ 2,466.25 + 863.75)$$

$$= 480$$

STEP 5: Calculate the mean square for each variation as:

$$\text{Replication} MS = \frac{Rep.\ SS}{r - 1} = \frac{4,100.63}{1} = 4,100.63 \qquad\qquad (5\text{-}23)$$

$$MS_A = \frac{SS_A}{a-1} = \frac{12,051.88}{4-1} = 4,017.29 \qquad (5\text{-}24)$$

$$\text{Error}\,(a)\,MS = \frac{\text{Error }SS\,(a)}{(r-1)(a-1)} = \frac{281.87}{(1)(3)} = 93.96 \qquad (5\text{-}25)$$

$$MS_B = \frac{SS_B}{b-1} = \frac{2,466.25}{(5-1)} = 822.10 \qquad (5\text{-}26)$$

$$MS_{AB} = \frac{SS_{AB}}{(a-1)(b-1)} = \frac{863.75}{(3)(4)} = 71.98 \qquad (5\text{-}27)$$

$$\text{Error}\,(b)\,MS = \frac{\text{Error }SS\,(b)}{a(r-1)(b-1)} = \frac{479.97}{4(1)(4)} = 29.99 \qquad (5\text{-}28)$$

STEP 6: Compute the F value for each source of variation as shown in Table 5.5.4.

TABLE 5.5.4.
Analysis of Variance of Hybrid × N Response Experiment in a Split-Plot Design

Source of Variation	Degree of Freedom	Sum of Squares	Mean Square	F
Replication	1	4,100.63	4,100.63	
Nitrogen (A)	3	12,051.88	4,017.29	42.76**
Error (a)	3	281.87	93.96	
Hybrid (B)	4	2,466.25	822.10	27.41**
Nitrogen × Hybrid ($A \times B$)	12	863.75	71.98	2.40^{ns}
Error (b)	16	479.97	29.99	
Total	39	20,244.38		

Note: cv (a) = 6.17%, cv (b) = 3.49%. ^{ns} = not significant, ** = significant at 1% level.

STEP 7: Compare the computed F with the tabular F value given in Appendix E. For example, the tabular F for the effect of nitrogen is 9.28 at the 5% level of significance and 29.46 at the 1% level of significance. This indicates that grain yield was significantly affected by nitrogen. Similarly, the results show that in both replications grain yield was significantly affected by the hybrid. However, the interaction between N rate and hybrid were not significant.

Finally, we will compute the coefficient of variation (cv) for the mainplot and the subplot using Equation 5-17 as shown below:

$$cv = \frac{\sqrt{\text{Error}(a)\,MS}}{\text{Grand Mean}} \times 100 \tag{5-17}$$

$$= \frac{\sqrt{93.96}}{157.13} \times 100 = 6.17\%$$

$$cv = \frac{\sqrt{\text{Error}(b)\,MS}}{\text{Grand Mean}} \times 100$$

$$= \frac{\sqrt{29.99}}{157.13} \times 100 = 3.49\%$$

As before, the value of the coefficient of variation (cv) indicates the degree of precision with which the factors are compared and is a good measure of the reliability of the experiment. Once again, the coefficient of variation expresses the experimental error as a percentage of the mean. Thus, the smaller the value of cv, the greater the reliability of the experiment. As is apparent from our computations, the cv(b) is smaller than the cv(a), as would be expected. We had mentioned earlier that a feature of the subplot design is the sacrifice of precision on the mainplot in order to achieve greater precision in the measurement of the factor assigned to the subplot.

Missing Data: If for any reason (loss of animal during the experiment, crop destroyed by accident or a natural disaster, or errors made in recording the data), data on an observation is missing when an experiment is conducted, use is made of the following formula developed by Anderson (1946):

$$\widehat{M} = \frac{r\,U + b\,(A_iB_j) - (A_i)}{(r-1)(b-1)} \tag{5-29}$$

where

$\widehat{M}$ = the estimated value of the missing observation

r = Number of replications

U = total for unit containing the missing observation

b = the level of subplot factors

$A_i B_j$ = total of observed values of the treatment combination that contain the missing observation

A_i = total of all observations that receive the ith level of A.

To see how the above formula is applied, assume that in Example 5.5.1. the observation for hybrid P3732 receiving 140 lb of nitrogen fertilizer in replication I (which is 160 bushels) is missing. To find the estimated value for the missing observation, we have to first recompute the totals by subtracting the value of the missing observation as shown below:

$$U = 805 - 160 = 645$$

$$A_i = 1,710 - 160 = 1,550$$

$$A_i B_j = 340 - 160 = 180$$

$$r = 2 \text{ replications}$$

$$b = 5, \text{ the level of the sub-plot factor}$$

$$\widehat{M} = \frac{2(645) + 5(180) - 1,550}{(2-1)(5-1)}$$

$$= \frac{1,290 + 900 - 1,550}{4}$$

$$= 160$$

Interestingly, the estimated value is the same as the actual value. If several observations such as a, b, c, d, ...etc., are missing, we first guess the values of all the missing observations except a. We then use the above equation to find an approximation for a. With the estimated value for a, we use the estimating equation again to find the value for b, and so on. This iterative procedure is continued until we have found an estimated value for all the missing observations. Keep in mind that in performing the analysis of variance, 1 degree of freedom is subtracted from the total and error sum of squares for each missing observation. For a more elaborate explanation on how to deal with more than 1 missing observation, the reader is referred to Tocher (1952) and Bennett and Franklin (1954).

5.6   STRIP-PLOT DESIGN

It is sometimes desirable to get more precise information on the interaction between factors than on their main effect. The strip-plot is a 2-factor design that allows for greater precision in the measurement of the interaction effect while sacrificing the degree of precision on the main effects. In measuring the interaction effect between 2 factors, the experimental area is divided into 3 plots, namely the *vertical-strip plot*, the *horizontal-strip plot*, and the *intersection plot*. We allocate factor A and B, respectively, to the vertical and horizontal-strip plots, and allow the intersection plot to accommodate the interaction between these 2 factors. As in the split-plot design, the vertical and the horizontal plots are perpendicular to each other. However, in the strip-plot design the relationship between the vertical and the horizontal plot sizes is not as distinct as the main and subplots were in the split-plot design. The intersection plot, which is one of the characteristics of the design, is the smallest in size. In the next section, the randomization and layout of the strip-plot design is explained.

Randomization and Layout: Given the nature of the design where we have specifically divided the experimental area into 3 separate plots, the randomization procedure calls for the allocation of each factor to its respective plots separately. This means that the vertical and the horizontal factor each would have to be randomly assigned to their plots. To illustrate the steps in the randomization and layout, we have denoted factor A as the horizontal factor and B as the vertical factor. The levels of each factor and the numbers of replications are denoted by a , b, and r, respectively.

Assume we have a 2-factor experiment testing 4 cultivars of wheat (horizontal factor) and 4 nitrogen rates (vertical factor) in a strip-plot design involving 4 replications. The steps in randomization are:

STEP 1:   Divide the experimental area into 4 blocks or replications ($r = 4$) where each block is further divided into 4 horizontal strips. Randomly assign the 4 treatments ($a = 4$) to each of the strips, following any of the randomization schemes outlined in Chapter 4. Figure 5.6.1 shows the layout for this step.

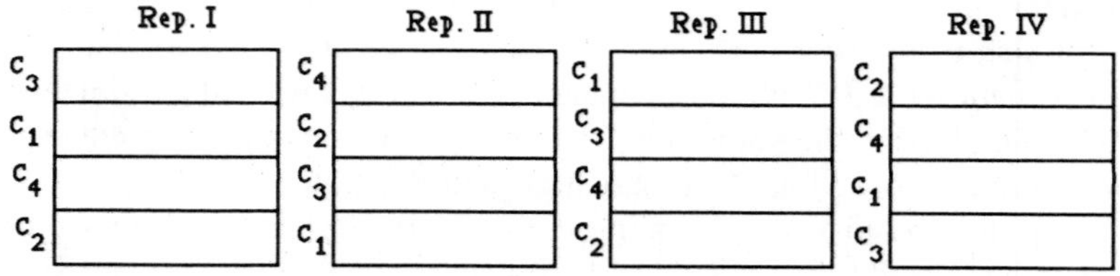

FIGURE 5.6.1.

Random allocation of the 4 cultivars (C_1, C_2, C_3, and C_4) to the horizontal strip with 4 replications.

STEP 2: Divide each block or replication into 4 vertical strips, and randomly assign the 4 nitrogen treatments ($b = 4$) following a randomization scheme discussed in Chapter 4. This is shown below in Figure 5.6.2.

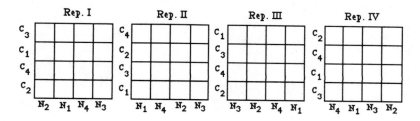

FIGURE 5.6.2.

A Strip-plot layout with random allocation of the 4 cultivars (C_1, C_2, C_3, and C_4) to the horizontal strip, and 4 nitrogen (N_1, N_2, N_3, and N_4) to the vertical strip with 4 replications.

Analysis of Variance: To perform the analysis of variance on a strip-plot experiment, we must analyze the data for the horizontal and vertical factors, and take account of the interaction. Thus, the analysis would involve the horizontal-factor analysis, vertical-factor analysis, and the interaction analysis. Each of these analyses will be performed using Example 5.6.1.

Example 5.6.1

In a study to determine the effect of N rate on soft winter wheat yields, agronomists have used a strip-plot design with 4 soft red winter wheat cultivars—Arthur 71, Auburn, Caldwell, and Compton—with 4 replications. The study was conducted on Parr silt loam soil, and the yield in bushels per acre are given in Table 5.6.1.

Solution

STEP 1: Calculate the replication totals (R), the grand total (G), and the cultivar total (A) by first constructing a table for the replication × horizontal factor (cultivar) totals (RA) as shown in Table 5.6.2, and then a second table for the replication × vertical factor (nitrogen) totals (RB) as shown in Table 5.6.3. Finally, construct a horizontal × vertical factor table of totals (AB) as shown in Table 5.6.4.

TABLE 5.6.1.

Grain Yields of 4 Soft Red Winter Wheat Cultivars Grown at 4 Nitrogen Fertilizer Rates

Cultivar	Nitrogen Rate (lb/acre)	Rep. I	Rep. II	Rep. III	Rep IV
				Grain Yield (bu/acre)	
Arthur 71	40	72	74	76	70
	80	76	75	74	78
	120	72	74	73	75
	160	74	76	82	86
Auburn	40	60	62	64	65
	80	61	63	69	68
	120	70	72	69	70
	160	72	70	82	86
Caldwell	40	75	73	72	80
	80	77	78	77	82
	120	80	82	86	88
	160	84	82	84	89
Compton	40	65	68	63	72
	80	68	72	74	76
	120	69	68	70	72
	160	73	75	74	76

TABLE 5.6.2.

The Replication × Cultivar (Factor A) Table of Yield Totals Computed from Data in Table 5.6.1

Cultivar	Rep. I	Rep. II	Rep. III	Rep. IV	Cultivar Total (A)
		Yield Total (RA)			
Arthur 71	294	299	305	309	1,207
Auburn	263	267	284	289	1,103
Caldwell	316	315	319	339	1,289
Compton	275	283	281	296	1,135
Rep. Total (R)	1,148	1,164	1,189	1,233	
Grand Total (G)					4,734

TABLE 5.6.3.

The Replication × Nitrogen (Factor B) Totals Computed from Data in Table 5.6.1

Nitrogen	Yield Total (RB)				Nitrogen Total (B)
	Rep. I	Rep. II	Rep. III	Rep. IV	
40	272	277	275	287	1,111
80	282	288	294	304	1,168
120	291	296	298	305	1,190
160	303	303	322	337	1,265

TABLE 5.6.4.

The Cultivar × Nitrogen Table of Totals Computed from Data in Table 5.6.1

Cultivar	Yield Total (AB)			
	N rate, lb/ac			
	40	80	120	160
Arthur 71	292	303	294	318
Auburn	251	261	281	310
Caldwell	300	314	336	339
Compton	268	290	279	298

STEP 2: Determine the degrees of freedom associated with each source of variation. The following sources of variation and the degrees of freedom are identified.

Replication $d.f.$	$= r - 1 = 3$
Horizontal factor $(A)d.f.$	$= a - 1 = 3$
Error (a) $d.f.$	$= (r - 1)(a - 1) = 9$
Vertical factor (B) $d.f.$	$= b - 1 = 3$
Error (b) $d.f.$	$= (r - 1)(b - 1) = 9$
$A \times B$ $d.f.$	$= (a - 1)(b - 1) = 9$
Error (c)	$= (r - 1)(a - 1)(b - 1) = 27$
Total $d.f.$	$= rab - 1 = 63$

STEP 3: Compute the correction factor and the total sum of squares as follows:

$$C = \frac{G^2}{rab} \tag{5-5}$$

$$= \frac{(4,734)^2}{4 \times 4 \times 4}$$

$$= 350,168.1$$

$$\text{Total } SS = \sum X^2 - C \tag{5-6}$$

$$= \left[(72)^2 + (74)^2 + \dots + (78)^2\right] - 350,168.1$$

$$= 353,034 - 350,168.1$$

$$= 2,865.90$$

STEP 4: Calculate the replication, cultivar, and error sum of squares for the horizontal factor analysis, using Equation 5-7 as given below:

$$\text{Replication } SS = \frac{\sum R^2}{ab} - C \tag{5-7}$$

$$= \frac{(1,148)^2 + (1,164)^2 + (1,189)^2 + (1,233)^2}{16} - 350,168.1$$

$$= \frac{5,606,810}{16} - 350,168.1$$

$$= 350,425.62 - 350,168.1$$

$$= 257.52$$

$$SS_A \text{ (Cultivar)} = \frac{\sum A^2}{rb} - C \tag{5-30}$$

$$= \frac{(1,207)^2 + (1,103)^2 + (1,289)^2 + (1,135)^2}{16} - 350,168.1$$

$$= 351,450.25 - 350,168.1$$

$$= 1,282.15$$

$$\text{Error } SS \ (a) = \frac{\sum (RA)^2}{b} - (C + \text{Rep. } SS + SS_A) \tag{5-31}$$

$$= \frac{(294)^2 + (263)^2 + .. + (296)^2}{4} - (350,168.1 + 257.52 + 1,282.15)$$

$$= 351,758 - 351,707.77$$

$$= 50.23$$

STEP 5: Compute the various sums of squares for the vertical factor analysis as shown below:

$$SS_B \ (\text{Nitrogen}) = \frac{\sum B^2}{ra} - C \tag{5-32}$$

$$= \frac{(1,111)^2 + (1,168)^2 + (1,190)^2 + (1,265)^2}{16} - 350,168.1$$

$$= 350,929.37 - 350,168.1$$

$$= 761.27$$

$$\text{Error } (b) \ SS = \frac{\sum (RB)^2}{b} - (C + \text{Rep. } SS + SS_B) \tag{5-33}$$

$$= \frac{(272)^2 + (282)^2 + ... + (337)^2}{4} - (350,168.1 + 257.52 + 761.27)$$

$$= 351,256 - 351,186.89$$

$$= 69.11$$

STEP 6: Compute the sums of squares for the interaction analysis as shown below:

$$SS_{AB} \text{ (Cultivar} \times \text{Nitrogen)} = \frac{\sum (AB)^2}{r} - (C + SS_A + SS_B) \qquad (5\text{-}34)$$

$$= \frac{(292)^2 + (251)^2 + \dots + (298)^2}{4}$$

$$- (350,168.1 + 1,282.15 + 761.27$$

$$= \frac{1,409,789}{4} - 352,211.52$$

$$= 352,449.50 - 352,211.52$$

$$= 237.98$$

$$\text{Error } SS\ (c) = TSS - \text{All other sum of squares} \qquad (5\text{-}35)$$

$$= 2,865.90 - (257.52 + 1,282.15 + 50.23 + 761.27$$

$$+ 69.11 + 237.98)$$

$$= 2,865.90 - 2,658.26$$

$$= 207.64$$

STEP 7: Calculate the mean square for each source of variation as:

$$\text{Replication} MS = \frac{\text{Rep. } SS}{r-1} = \frac{257.52}{3} = 85.84$$

$$MS_A = \frac{SS_A}{a-1} = \frac{1,282.15}{3} = 427.38$$

$$\text{Error } (a)\ MS = \frac{\text{Error } SS\ (a)}{(r-1)(a-1)} = \frac{50.23}{9} = 5.58$$

$$MS_B = \frac{SS_B}{b-1} = \frac{761.27}{3} = 253.76$$

$$\text{Error } (b) MS = \frac{\text{Error } SS\ (b)}{(r-1)(b-1)} = \frac{69.11}{9} = 7.68$$

$$MS_{AB} = \frac{SS_{AB}}{(a-1)(b-1)} = \frac{237.98}{9} = 26.44$$

$$\text{Error}(c)\,MS = \frac{\text{Error}(c)\,SS}{(r-1)(a-1)(b-1)} = \frac{207.64}{27} = 7.69$$

STEP 8: Compute the F value for each source of variation as shown in Table 5.6.5.

TABLE 5.6.5.

Analysis of Variance of Wheat Cultivar × N Response Experiment in a Strip-Plot Design

Source of Variation	Degree of Freedom	Sum of Squares	Mean Square	F
Replication	3	257.52	85.84	
Cultivar (A)	3	1,282.15	427.38	76.59**
Error (a)	9	50.23	5.58	
Nitrogen (B)	3	761.27	253.76	33.04**
Error (b)	9	69.11	7.68	
Cultivar × Nitrogen (A xB)	9	237.98	26.44	3.43**
Error (c)	27	207.64	7.69	
Total	63	2,865.90		

Note: cv (a) = 3.19%, cv (b) = 3.75%, cv (c) = 3.75%. ** = significant at 1% level.

STEP 9: Compare the computed F with the tabular F value given in Appendix E. For example, the tabular F for the effect of cultivar is 3.86 at 5% level of significance and 6.99 at 1% level of significance. This indicates that grain yield was significantly affected by the cultivar. The results also show that grain yield was significantly affected by nitrogen fertilizer. Similarly, The tabular F for the interaction between cultivar and nitrogen is 2.25 at the 5% level of significance and 3.14 at the 1% level of significance. Thus, the results show a significant interaction between N rate and cultivar, indicating that the N recommendation developed for one cultivar cannot be applied to the other cultivars tested in this experiment.

Finally, we will compute the coefficient of variation (cv) for the 3 error mean squares using Equation 5-17 as shown below:

$$cv\,(a) = \frac{\sqrt{\text{Error}\,(a)\,MS}}{\text{Grand Mean}} \times 100 \qquad\qquad (5\text{-}17)$$

$$= \frac{\sqrt{5.58}}{73.96} \times 100 = 3.19\%$$

$$cv\,(b) = \frac{\sqrt{\text{Error}\,(b)\,MS}}{\text{Grand Mean}} \times 100$$

$$= \frac{\sqrt{7.68}}{73.96} \times 100 = 3.75\%$$

$$cv\,(c) = \frac{\sqrt{\text{Error}\,(c)\,MS}}{\text{Grand Mean}} \times 100$$

$$= \frac{\sqrt{7.69}}{73.96} \times 100 = 3.75\%$$

As before, the value of the coefficient of variation (cv) indicates the degree of precision with which the factors are compared and is a good measure of the reliability of the experiment. The cv (a) is a measure of the precision associated with the horizontal factor, and the cv (b) with the vertical factor. The degree of precision in the measurement of the interaction between horizontal and vertical factor is shown by cv (c).

Missing Data: To illustrate how the value of a missing observation (a_ib_j) is estimated in a strip-plot experiment, assume that the value is missing on the Auburn cultivar in Example 5.6.1, receiving 120 lb of nitrogen per acre in replication II, which is 72. To find the missing value, the following equation is used.

$$\widehat{M} = \frac{a\,\{b\,[A_iB_j]\,\text{-}A_i\} + r\,(aH + bV\,\text{-}B) - bv + G}{(a-1)(b-1)(r-1)} \qquad\qquad (5\text{-}36)$$

where

$\widehat{M}$ = estimated value of the missing observation

a = level of horizontal factor

b = level of vertical factor

r = number of replications

A_iB_j = total of observed values of the treatment that contains the missing observation

A_i = total of all observations that receive the ith level of horizontal factor A

H = total of observed values of the horizontal strip that contain the missing observation

B = total of observed values of the replication that contain the missing observation

V = total of observed values of the vertical strip that contain the missing observation

v = total of observed value of the specific level of the vertical factor that contains the missing observation

G = total of all observed values.

To estimate this missing value, the parameters to work with are:

$$a = 4$$
$$b = 4$$
$$r = 4$$
$$A_iB_j = 281 - 72 = 209$$
$$A_i = 1{,}103 - 72 = 142$$
$$H = 267 - 72 = 142$$
$$B = 1{,}164 - 72 = 1{,}092$$
$$V = 296 - 72 = 224$$
$$v = 1{,}190 - 72 = 1{,}118$$
$$G = 4{,}734 - 72 = 4{,}662$$

Thus, we have:

$$\widehat{M} = \frac{4\left\{\left[4(209)-1{,}031\right]\right\} + 4\left[4(142)+4(224)-1{,}092\right] - (4)(1{,}118)+4{,}662}{(4-1)(4-1)(4-1)}$$
$$= \frac{13{,}862 - 12{,}712}{27} = 42.6$$

Once estimated, the missing value is placed in the table of observed values, and the analysis of variance is performed in the usual way. Once again, remember to subtract 1 degree of freedom from the total and error before computing the F value.

REFERENCES AND SUGGESTED READINGS

Anderson, R. L. 1946. "Missing-plot techniques." *Biomed. Bull.* 2: 41-47.

Bennett, C. A. and Franklin, N. L. 1954. *Statistical Analysis in Chemistry and the Chemical Industry.* New York: John Wiley & Sons.

Bundy, L. G. and Carter, P. R. 1988. "Corn hybrid response to nitrogen fertilization in the northern corn belt." *J. Prod. Agric.* 1(2):99 -104.

Cochran, W. G. and Cox, G.M. 1957. *Experimental Designs.* 2nd ed. New York: John Wiley & Sons, Chapter 5, and 168-170.

Goulden, C.H. 1952. *Methods of Statistical Analysis.* 2nd ed. New York: John Wiley & Sons, 94.

Tocher, K. D. 1952. "The design and analysis of block experiments." *J. R. Stat. Soc. Ser. B*, 14: 45-100.

EXERCISES

1. An animal scientist wishes to conduct a split-plot design experiment where 2 factors (2 breeds of swine and 4 different diets) are to be arranged in 4 blocks. The objective of the experiment is to see if the 4 diets are different from one another. Show the layout of this experiment, keeping the objective in mind.

2. The omnivorous looper, larva of the moth *Sabulodes aegrotata* (Guenee) is a sporadic pest of California's avocado. Researchers are testing 4 chemicals to control the omnivorous looper in avocado orchards. The treatments were assigned to the subplots, whereas the concentration rates of active ingredients were assigned to the main plots. The results of the experiment are shown below.

	No. of Larvae per Tree Sample at 14-day Posttreatment Interval		
Treatment	Rep. I	Rep. II	Rep. III
	Active Ingredient 1 lb/ac		
Dylox 80SP	8	10	11
Kryocide 8F	12	9	10
Lannate L	2	4	3
Orthene 75SP	1	2	4
Control	14	18	20
	Active Ingredient 2 lb/ac		
Dylox 80SP	5	9	8
Kryocide 8F	6	4	5
Lannate L	1	3	4
Orthene 75SP	1	2	4
Control	17	19	24
	Active Ingredient 4 lb/ac		
Dylox 80SP	3	3	5
Kryocide 8F	5	4	7
Lannate L	2	1	3
Orthene 75SP	1	2	1
Control	12	15	22

Insecticide Used for Omnivorous Looper Control in San Diego County

(a) Perform the analysis of variance on the data.
(b) What conclusions can be drawn from this analysis.
(c) Determine the degree of precision with which the treatments are compared.

3. In a study to determine the viability of triticale as a feedstuff for poultry, animal researchers conducted a split-plot experiment. They gathered the following data on the weight gain of the birds fed triticale and control.

Cumulative Growth of Broilers Fed Control or Triticale Diets

| Diet | Weight gain over days(lb/bird) | | | |
	Rep. I	Rep. II	Rep. III	Rep. IV
			7 Days	
Control	0.25	0.31	0.22	0.28
Low-triticale	0.26	0.28	0.29	0.27
Medium-triticale	0.22	0.21	0.24	0.22
			14 Days	
Control	0.65	0.70	0.59	0.68
Low-triticale	0.73	0.75	0.79	0.69
Medium-triticale	0.70	0.69	0.65	0.68
			21 Days	
Control	1.25	1.31	1.24	1.27
Low-triticale	1.29	1.28	1.29	1.30
Medium-triticale	1.22	1.21	1.23	1.22
			28 Days	
Control	1.35	1.33	1.29	1.28
Low-triticale	1.46	1.38	1.40	1.39
Medium-triticale	1.42	1.31	1.34	1.36

(a) Perform the analysis of variance on the data.
(b) Compute the 2 coefficients of variation, one corresponding to the main plot and another to the subplot analysis.

4. The western flower thrip, *Frankliniella occidentalis* (Pergande), is a major threat to the floricultural crops around the world. Researchers have conducted a split-plot experiment in which biological control has been used to control the thrips. In this experiment the objective is to determine if using predaceous mites such as *A. cucumeris* and *A.barkeri*, separately and in combination, would reduce the thrip population. Using chrysanthemum plants, the researchers have gathered the following data.

Impact of Predaceous Mites on the Number of Western Flower Thrips Infesting Chrysanthemum

Replication	Day	Amblyseius cucumeris	Amblyseius barkeri	A. cucumeris + A. barkeri	Control
I	Day 10	3	2	1	7
	Day 17	2	4	2	9
	Day 24	1	2	2	21
	Day 31	6	3	8	38
II	Day 10	2	1	2	9
	Day 17	3	4	1	10
	Day 24	5	2	3	25
	Day 31	3	5	7	34
III	Day 10	1	3	1	8
	Day 17	3	5	1	14
	Day 24	2	1	3	23
	Day 31	5	2	9	32
IV	Day 10	2	4	2	10
	Day 17	1	2	2	11
	Day 24	4	5	5	21
	Day 31	6	6	10	39

(a) Analyze the data.
(b) Compute the coefficient of variation for the 2 categories of the data.

5. Suppose in the above experiment, the researchers lost some of the observations as shown in the table below. The missing data is shown as (-) in
each column of the data set.

Impact of Predaceous Mites on the Number of Western Flower Thrips Infesting
Chrysanthemum

Replication	Day	Amblyseius cucumeris	Amblyseius barkeri	A. cucumeris + A. barkeri	Control
I	Day 10	3	2	1	7
	Day 17	2	4	2	9
	Day 24	1	2	2	-
	Day 31	-	3	8	38
II	Day 10	2	1	2	9
	Day 17	3	4	1	10
	Day 24	5	2	-	25
	Day 31	3	5	7	34
III	Day 10	1	3	1	8
	Day 17	3	-	1	14
	Day 24	2	1	3	23
	Day 31	5	2	9	32
IV	Day 10	2	4	2	10
	Day 17	1	2	-	11
	Day 24	4	5	5	21
	Day 31	6	6	10	39

(a) How would you estimate the missing values, and perform the analysis of variance?

(b) Is there a significant difference between the results obtained in problem 3 and the present case?

6. In a strip-plot design experiment, researchers were interested more in the interaction between the amount of fertilizer and rainfall than in either of these factors alone. They have hypothesized that fertilizers produce more range forage in drought than normal years. The data collected from the experiment is shown below.

Range Forage Yields per acre from Fertilizer Application in 2 Normal Precipitation Years and 2 Drought Years

Rainfall (in)	Fertilizer Application (lb/ac)	Forage Yield (lb/ac)				
		Rep. I	Rep. II	Rep. III	Rep. IV	Rep. V
	Ammonium sulfate					
17.5	(300 lb)	4500	4355	4100	4600	4250
12.8	" "	4100	4235	4005	4095	4050
6.0	" "	1400	1325	1200	1375	1390
6.5	" "	1000	1025	995	1020	1000
	Ammonium nitrate					
17.5	(180 lb)	4300	4235	4025	4165	4120
12.8	" "	4050	4110	4033	4195	4250
6.0	" "	1630	1624	1595	1675	1595
6.5	" "	1060	1028	1000	1029	1016
	16-20-0					
17.5	(375 lb)	4250	4151	4170	4300	4290
12.8	" "	3700	3935	3205	3495	3850
6.0	" "	1400	1302	1296	1315	1390
6.5	" "	1100	1025	991	1022	1019
	Control					
17.5	(None)	3950	3801	3905	3390	3890
12.8	" "	4101	4035	4205	4007	4100
6.0	" "	901	897	906	942	899
6.5	" "	698	674	688	700	645

(a) Perform the analysis of variance.
(b) Compute the coefficient of variation for the problem.
(c) Assume that the observation for the plot receiving ammonium nitrate when rainfall is 12.8 inches in replication II (which is 4110) is missing. How do you estimate this missing value before the analysis is performed?

Chapter 6

THREE (OR MORE)-FACTOR EXPERIMENTAL DESIGNS

6.1 INTRODUCTION

In the previous chapter, we discussed how 2 factors could be studied in the same environment simultaneously. The concepts developed in the previous 2 chapters could be extended to include 3 or more factors. Studying several factors together offers the advantage of simultaneously examining the interaction between factors. Such interactions, if statistically significant, take precedent over the main effects when reporting the result of an experiment. It should be kept in mind, however, that the number of treatments to study increases rapidly as more factors are added in the experiment. This point was made earlier in Chapter 5 when discussing factorial experiments. Additionally, as the number of factors increases in an experiment, so does the interaction between factors. This complicates the analysis and requires great care in interpreting the results.

In this chapter, we shall consider the simplest arrangement of a factorial experiment, or 2^3 experiment, where the experiment includes 3 factors (A, B, and C) each at 2 levels (a_1, a_2, b_1, b_2 and c_1, c_2). In such a 3-factor experiment, we have to estimate and test 3 main effects A, B, and C ; 3 first-order (2-factor) interactions, AB, BC, and AC ; and 1 second-order (3-factor) interaction ABC. Additionally, we shall consider blocking in the case in which the blocks are not large enough to contain all the treatments, as was the case with split-plot design where each whole plot contained only 1 level of each of the main treatments.

177

Inclusion of an additional factor to a 2^2 factorial experiment does not change
the principle of design or the randomization and layout of the experiment. The
data matrix for a 3-factor design is presented in Table 6.1.

TABLE 6.1.

Data Matrix for a 3-Factor Design

Factor A	Factor B	1	2	3	. . .	c
				Factor C		
1	1	X_{1111}	X_{1112}	X_{1113}	. .	X_{111c}
	2	X_{1121}	X_{1122}	X_{1123}	. .	X_{112c}
	3	X_{1131}	X_{1132}	X_{1133}	. .	X_{113c}
	.	.	.	.	. .	.
	.	.	.	.	. .	.
	b	X_{11b1}	X_{11b2}	X_{11b3}		X_{11bc}
2	1	X_{1211}	X_{1212}	X_{1213}	.	X_{121c}
	2	X_{1221}	X_{1222}	X_{1223}	.	X_{122c}
	3	X_{1231}	X_{1232}	X_{1233}	.	X_{123c}
	.	.	.	.	.	.
	.	.	.	.	.	.
	b	X_{12b1}	X_{12b2}	X_{12b3}		X_{12bc}
a	1	X_{1a11}	X_{1a12}	X_{1a13}	.	X_{1a1c}
	2	X_{1a21}	X_{1a22}	X_{1a23}	.	X_{1a2c}
	3	X_{1a31}	X_{1a32}	X_{1a33}	.	X_{1a3c}
	.	.	.	.	.	.
	.	.	.	.	.	.
	b	X_{1ab1}	X_{1ab2}	X_{1ab3}	.	X_{1abc}

The underlying model of the 3-factor experiment is of the form

$$X_{ijkl} = \mu + \alpha_i + \beta_j + \gamma_k + (\alpha\gamma)_{ik} + (\beta\gamma)_{jk} + (\alpha\beta\gamma)_{ijk} + \varepsilon_{ijk.}$$

For $i = 1, . . a; j = 1, .. b; k = 1, ...c, l = 1,...n$

where

X_{ijkl} = The lth yield observation of the ijkth treatment combination

μ = overall mean; that is, the average of the population mean responses for the abc treatments

α_i = effect of the i th level of the first factor (A); averaged over the b levels of the second factor (B). The ith level of the first factor adds α_i to the overall mean μ

β_j = average effect of the jth level of factor B

γ_k = average effect of the kth level of factor C

$(\alpha\beta)_{ij}$ = interaction between the ith level of factor A and the jth level of factor B; average of the population means for the c treatments which involve the ith level of A and jth level of B minus $\mu + \alpha_i + \beta_j$

$(\alpha\gamma)_{ik}$ = interaction of the ith level of factor A with the kth level of factor C; the average of the population means for the b treatments involving the ith level of A and the kth level of C minus $\mu + \alpha_i + \beta_j$

$(\beta\gamma)_{jk}$ = interaction of the jth level of factor B with the kth level of factor C; the average of the population means for the a treatments which involve jth level of B and the kth level of C minus $\mu + \alpha_i + \beta_j$

$(\alpha\beta\gamma)_{ijk}$ = interaction between the ith level of factor A, the jth level of factor B, and the kth level of factor C; population mean response for the ijkth treatment minus $\mu + \alpha_i + \beta_j + \gamma_k + (\alpha\beta)_{ij} + (\alpha\gamma)_{ik} + (\beta\gamma)_{jk}$

ε_{ijkl} = the random error component.

The above equation specifies that there are n observations on each of the abc treatment combinations, a levels of factor A, b levels of factor B, and c levels of factor C. The error component, ε_{ijkl}, is assumed to be independently and normally distributed, with 0 mean and variance σ_e^2 within each of the abc treatment combinations. The quantities defined above have all been encountered in Chapter 5 with the exception of $(\alpha\beta\gamma)_{ijk}$, which we designate as the second-order interaction (3 independent variables) to distinguish it from the first-order interactions that involve only 2 independent variables.

In the next sections of this chapter, we will illustrate the different designs with 3 or more factors, and how they are analyzed.

6.2   SPLIT-SPLIT-PLOT DESIGN

In Chapter 5 (Section 5.5) you were introduced to a 2-factor design called the split-plot design. An extension of this design is called the *split-split-plot design* where the subplot is further divided to include a third factor in the experiment. The design allows for 3 different levels of precision associated with the 3 factors. That is, the degree of precision associated with the main factor is lowest, while the degree of precision associated with the sub-subplot is the highest. Thus, the design is suited for those 3-factor experiments where different levels of precision are desired with the different factors.

Randomization: In the split-split-plot design, the main plot is divided into subplots, and the subplots are further divided into sub-subplots. The randomization process requires that the levels of the main plot factor be applied at random to the blocks. Similarly, the levels of the subplot factor should be applied randomly within the main plots, using separate randomization in each main plot. Finally, the levels of the last factor should be assigned to the sub-subplots at random, using a separate randomization in each subplot. To illustrate the randomization and layout, we will use a $2 \times 2 \times 3$ factorial experiment replicated 4 times in a randomized complete block design arranged in a split-split-plot layout. Factor A (2 hybrids, denoted as H_1 and H_2) is assigned to the main plot. Factor B (2 row spacings, denoted as S_1 and S_2) is assigned by the experimenter to the subplot. Factor C (3 plant densities, denoted as D_1, D_2, and D_3) is assigned to the sub-subplot. The steps followed are given below.

STEP 1:   Divide the experimental area into 4 replications and each replication into 2 main plots. Factor A (2 hybrids- H_1 and H_2) is randomly assigned to the main plot as shown in Figure 6.2.1.

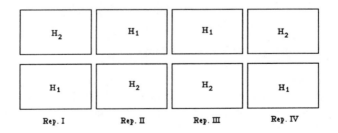

FIGURE 6.2.1.

Random assignment of factor A (hybrids, H_1 and H_2) to the 2 main plots in each of the 4 replications.

STEP 2:   Subdivide each main plot into 2 subplots, and assign randomly the 2 levels of factor B (row spacings - S_1 and S_2) to the subplots within the main plots using separate randomization in each main plot as is shown in Figure 6.2.2.

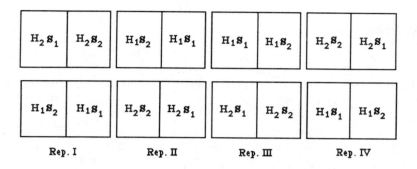

Rep. I   **Rep. II**   **Rep. III**   **Rep. IV**

FIGURE 6.2.2.

Random assignment of factor B (row spacing, S_1 and S_2) to the 2 subplots in each of the 2 main plots in the 4 replications.

STEP 3: Divide each subplot into 3 sub-subplots, and assign at random the 3 levels of factor C (plant densities - D_1, D_2 and D_3) to the sub-subplots using separate randomization in each subplot as shown in Figure 6.2.3.

$H_2S_1D_3$	$H_2S_2D_1$		$H_1S_2D_1$	$H_1S_1D_2$		$H_1S_1D_2$	$H_1S_2D_2$		$H_2S_2D_1$	$H_2S_1D_1$
$H_2S_1D_1$	$H_2S_2D_2$		$H_1S_2D_3$	$H_1S_1D_1$		$H_1S_1D_1$	$H_1S_2D_1$		$H_2S_2D_3$	$H_2S_1D_2$
$H_2S_1D_2$	$H_2S_2D_3$		$H_1S_2D_2$	$H_1S_1D_3$		$H_1S_1D_3$	$H_1S_2D_3$		$H_2S_2D_2$	$H_2S_1D_3$

$H_1S_2D_1$	$H_1S_1D_2$		$H_2S_2D_3$	$H_2S_1D_1$		$H_2S_1D_3$	$H_2S_2D_1$		$H_1S_1D_1$	$H_1S_2D_2$
$H_1S_2D_3$	$H_1S_1D_1$		$H_2S_2D_2$	$H_2S_1D_3$		$H_2S_1D_2$	$H_2S_2D_3$		$H_1S_1D_3$	$H_1S_2D_1$
$H_1S_2D_2$	$H_1S_1D_3$		$H_2S_2D_1$	$H_2S_1D_2$		$H_2S_1D_1$	$H_2S_2D_2$		$H_1S_1D_2$	$H_1S_2D_3$

Rep. I   **Rep. II**   **Rep. III**   **Rep. IV**

FIGURE 6.2.3.

A $2 \times 2 \times 3$ factorial experiment layout plan arranged in a split-split-plot design with 2 hybrids assigned to the main plots, 2 row spacing assigned to the subplots, and 3 plant densities assigned to the sub-subplots with 4 replications.

Analysis of Variance: The analysis of variance performed on the data from a split-split-plot design is similar to the one carried out with the split-plot design. However, the number of types of comparisons has increased because of the added factor and a second split to obtain the split-split-plot design. To illustrate the steps in computation, the following example is given.

Example 6.2.1

A study was conducted in Michigan to determine the influence of plant density and hybrids on corn (*Zea mays* L.) yield. The experiment was a $2 \times 2 \times 3$ factorial replicated 4 times in a randomized complete block design arranged in a split-split-plot layout. In this experiment, factor A is the 2 corn hybrids (P3730 and B70 × LH55) assigned to the mainplots, factor B is the 2 row spacings (12 and 25 in.) assigned to the subplots, and factor C is the 3 target plant densities (12,000, 16,000, and 20,000 plants per acre) assigned to the sub-subplots. The data gathered for the experiment are shown in Table 6.2.1.

TABLE 6.2.1.
Yield of 2 Corn Hybrids (Bu/ac) with 2 Row Spacings and 3 Plant Densities

Hybrid	Row Spacing (inches)	Plant Density (plants/ac)	I	II	III	IV
P3730	12	12,000	140	138	130	142
		16,000	145	146	150	147
		20,000	150	149	146	150
			435	433	426	439
	25	12,000	136	132	134	138
		16,000	140	134	136	140
		20,000	145	138	138	142
			421	404	408	420
B70 × LH55	12	12,000	142	132	128	140
		16,000	146	136	140	141
		20,000	148	140	142	140
			436	408	410	421
	25	12,000	132	130	136	134
		16,000	138	132	130	132
		20,000	140	134	130	136
			410	396	396	402

Column headers: Grain Yield (Bu/ac), Replications (I, II, III, IV)

Solution _____

The statistical analyses are performed on the mainplot, the subplot, and the sub-subplot. The steps are:

STEP 1: Perform the main-plot analysis by computing the replication totals (R), the grand total (G), and the hybrid total (A) by first constructing a table for the replication $\times$ hybrid totals (RA) as shown in Table 6.2.2.

TABLE 6.2.2.

The Replication × Hybrid (factor A) Table of Yield Totals Computed from Data in Table 6.2.1

Hybrid	Yield Total (RA)				Hybrid Total (A)
	Rep. I	Rep. II	Rep. III	Rep. IV	
P3730	856	837	834	859	3,386
B70 × LH55	846	804	806	823	3,279
Rep. Total (R)	1,702	1,641	1,640	1,682	
Grand Total (G)					6,665

STEP 2: Determine the degrees of freedom associated with each source of variation. The following sources of variation and the degrees of freedom are identified.

Replication $d.f.$ $= r - 1 = 3$

Main-plot factor (A) $d.f.$ $= a - 1 = 1$

Error (a) $d.f.$ $= (r - 1)(a - 1) = 3$

Subplot factor (B) $d.f.$ $= b - 1 = 1$

$A \times B$ $d.f.$ $= (a - 1)(b - 1) = 1$

Error (b) $d.f.$ $= a\,(r - 1)(b - 1) = 6$

Sub-subplot factor (C) $= c - 1 = 2$

$A \times C$ $d.f.$ $= (a - 1)(c - 1) = 2$

$B \times C$ $d.f.$ $= (b - 1)(c - 1) = 2$

$A \times B \times C$ $d.f.$ $= (a - 1)(b - 1)(c - 1) = 2$

Error (c) $= ab \ (r - 1)(c - 1) = 24$

Total $d.f.$ $= rabc - 1 = 47$

STEP 3: Compute the correction factor and the total sum of squares for the main plot as:

$$C = \frac{G^2}{rabc} \tag{6-1}$$

$$= \frac{(6,665)^2}{4 \times 2 \times 2 \times 3}$$

$$= 44,422,225/48$$

$$= 925,463.02$$

$$Total \ SS = \sum X^2 - C \tag{6-2}$$

$$= \left[(140)^2 + (138)^2 + \ldots + (136)^2\right] - 925,463.02$$

$$= 927,177 - 925,463.02$$

$$= 1,713.98$$

STEP 4: Calculate the replication, hybrid, and error sum of squares for factor A, for the main-plot analysis as:

$$Replication \, SS = \frac{\sum R^2}{abc} - C \tag{6-3}$$

$$= \frac{(1,702)^2 + (1,641)^2 + (1,640)^2 \, (1,682)^2}{12} - 925,463.02$$

$$= 11,108,409/12 - 925,463.02$$

$$= 925,700.75 - 925,463.02$$

$$= 237.73$$

$$SS_A \text{ (Hybrid)} = \frac{\sum A^2}{rbc} - C \tag{6-4}$$

$$= \frac{(3,386)^2 + (3,279)^2}{(4)(2)(3)} - 925,463.02$$

$$= 925,701.54 - 925,463.02$$

$$= 238.52$$

$$\text{Error } SS\,(a) = \frac{\sum (RA)^2}{bc} - \left(C + \text{Rep. } SS + SS_A\right) \tag{6-5}$$

$$= \frac{(856)^2 + (846)^2 + \dots + (823)^2}{(2)(3)} - (925,463.02 + 237.73 + 238.52)$$

$$= 5,555,839/6 - 925,939.27$$

$$= 33.90$$

STEP 5: Compute the various sums of squares for the subplot analysis by first constructing a summary table of yield totals for factor $A \times$ factor B or (AB), and a 3-way table of totals for the replication $\times$ factor $A \times$ factor B or (RAB) as shown in Tables 6.2.3 and 6.2.4.

TABLE 6.2.3.

Summary Table of Yield Totals for the Hybrid × Row Spacing (Factor A × Factor B) Computed from Data in Table 6.2.1

	Yield Total (AB)	
	Row Spacing (in.)	
Hybrid	12	25
P3730	1,733	1,653
B70 × LH55	1,675	1,604
Row Spacing Total (B)	3,408	3,257

TABLE 6.2.4.

Summary Table of Yield Totals for the Replication × Hybrid × Row Spacing or (*RAB*) Computed from Data in Table 6.2.1

Hybrid	Row Spacing (in.)	Rep. I	Rep. II	Rep. III	Rep. IV
				Yield Total (*RAB*)	
P3730	12	435	433	426	439
	25	421	404	408	420
B70 × LH55	12	436	408	410	421
	25	410	396	396	402

$$SS_B \text{ (Row Spacing)} = \frac{\sum B^2}{rac} - C \tag{6-6}$$

$$= \frac{(3{,}408)^2 + (3{,}257)^2}{(4)(2)(3)} - 925{,}463.02$$

$$= 22{,}222{,}513/24 - 925{,}463.02$$

$$= 925{,}938.04 - 925{,}463.02$$

$$= 475.02$$

$$SS_{AB} \text{ (Hybrid} \times \text{Row Spacing)} = \frac{\sum (AB)^2}{rc} - (C + SS_A + SS_B) \tag{6-7}$$

$$= \frac{(1{,}733)^2 + (1{,}675)^2 + (1{,}653)^2 + (1{,}604)^2}{(4)(3)}$$

$$- (925{,}463.02 + 238.52 + 475.02)$$

$$= 926{,}178.25 - 926{,}176.56$$

$$= 1.69$$

$$\text{Error }(b)\ SS = \frac{\sum (RAB)^2}{c} - (C + \text{Rep. } SS + SS_A + \text{Error }(a)\ SS + SS_B + SS_{AB})\ (6\text{-}8)$$

$$= \frac{(435)^2 + (412)^2 + \ldots + (402)^2}{3}$$

$$- (925{,}463.02 + 237.73 + 238.52 + 33.90 + 475.02 + 1.69)$$

$$= 2{,}779{,}469/3 - 926{,}449.88$$

$$= 926{,}489.67 - 926{,}449.88$$

$$= 39.79$$

STEP 6: Perform the sub-subplot analysis by constructing 3 summary tables of yield totals for (1) factor $A \times$ factor C or (AC), (2) factor $B \times$ factor C or (BC), and (3) for factor $A \times B \times C$ or (ABC) as shown in Tables 6.2.5, 6.2.6, and 6.2.7, respectively.

TABLE 6.2.5.

Summary Table of Yield Totals for Hybrid × Plant Density (Factor A × Factor C) Computed from Data in Table 6.2.1

Hybrid	Yield Total (AC) Plant Density/ac		
	12,000	16,000	20,000
P3730	1,090	1,138	1,158
B70 × LH55	1.074	1.095	1,110
Plant Density Total (C)	2,164	2,233	2,268

In computing the sum of squares for factor C it should be kept in mind that the squared C value is different from the correction factor notation (C) as shown below:

TABLE 6.2.6.

Summary Table of Yield Totals for Row Spacing × Plant Density (factor B × factor C) Computed from Data in Table 6.2.1

| | Yield Total (BC) | | |
| | Plant Density/ac | | |
Row Spacing	12,000	16,000	20,000
12	1,092	1,151	1,165
25	1,072	1,082	1,103

TABLE 6.2.7.

Summary Table of Yield Totals for Hybrid × Row Spacing × Plant Density (factor A × factor B × factor C) Computed from Data in Table 6.2.1

| | | Yield Total (ABC) | | |
| | | Plant Density/ac | | |
Hybrid	Row Spacing	12,000	16,000	20,000
P3730	12	550	588	595
	25	540	550	563
B70 × LH55	12	542	563	570
	25	532	532	540

$$SS_C(Plant\ Density) = \frac{\sum C^2}{rab} - C \qquad (6\text{-}9)$$

$$= \frac{(2,164)^2 + (2,233)^2 + (2,268)^2}{(4)(2)(2)} - 925,463.02$$

$$= 14,813,009/16 - 925,463.02$$

$$= 925,813.06 - 925,463.02$$

$$= 350.04$$

$$SS_{AC} = \frac{\sum (AC)^2}{rb} - (C + SS_A + SS_C)$$ (6-10)

$$= \frac{(1,090)^2 + (1,074)^2 + (1,110)^2}{(4)(2)} - (925,463.02 + 238.52 + 350.04)$$

$$= 7,408,709/8 - 926,051.58$$

$$= 926,088.626 - 926,051.58$$

$$= 37.04$$

$$SS_{BC} = \frac{\sum (BC)^2}{ra} - (C + SS_B + SS_C)$$ (6-11)

$$= \frac{(1,092)^2 + (1,072)^2 + \dots + (1,103)^2}{(4)(2)} - (925,463.02 + 275.02 + 350.04)$$

$$= 7,411,007/8 - 926,288.08$$

$$= 926,375.86 - 926,288.08$$

$$= 87.80$$

$$SS_{ABC} = \frac{\sum (ABC)^2}{r} - (C + SS_A + SS_B + SS_C + SS_{AB} + SS_{AC} + SS_{BC})$$ (6-12)

$$= \frac{(550)^2 + (540)^2 + \dots + (540)^2}{4}$$

$$- (925,463.02 + 238.52 + 475.02 + 350.04 + 1.69 + 37.04 + 87.80)$$

$$= 3,706,619/4 - 926,653.13$$

$$= 926,654.75 - 926,653.13$$

$$= 1.62$$

$$\text{Error }(c)\ SS = \text{Total }SS - \text{All other sum of squares} \qquad (6\text{-}13)$$
$$= 1,713.98 - (237.73 + 238.52 + 33.90 + 475.02 + 1.69$$
$$+ 39.79 + 350.04 + 37.04 + 87.80 + 1.62)$$
$$= 1,713.98 - 1,503.15$$
$$= 210.83$$

STEP 7: Calculate the mean square for each source of variation as:

$$\text{Replication} MS = \frac{\text{Rep. }SS}{r-1} = \frac{273.73}{3} = 91.24 \qquad (6\text{-}14)$$

$$MS_A = \frac{SS_A}{a-1} = \frac{238.53}{2-1} = 238.53 \qquad (6\text{-}15)$$

$$\text{Error }(a)\ MS = \frac{\text{Error }(a)\ SS}{(r-1)(a-1)} = \frac{33.90}{(3)(1)} = 11.30 \qquad (6\text{-}16)$$

$$MS_B = \frac{SS_B}{b-1} = \frac{475.02}{2-1} = 475.02 \qquad (6\text{-}17)$$

$$MS_{AB} = \frac{SS_{AB}}{(a-1)(b-1)} = \frac{1.69}{1} = 1.69 \qquad (6\text{-}18)$$

$$\text{Error }(b) MS = \frac{\text{Error }SS\ (b)}{a\ (r-1)(b-1)} = \frac{39.79}{2(3)(1)} = 6.63 \qquad (6\text{-}19)$$

$$MS_C = \frac{SS_C}{c-1} = \frac{350.40}{3-1} = 175.20 \qquad (6\text{-}20)$$

$$MS_{AC} = \frac{SS_{AC}}{(a-1)(c-1)} = \frac{37.04}{(1)(2)} = 18.52 \qquad (6\text{-}21)$$

$$MS_{BC} = \frac{SS_{BC}}{(b-1)(c-1)} = \frac{87.80}{(1)(2)} = 43.90 \tag{6-22}$$

$$MS_{ABC} = \frac{SS_{ABC}}{(a-1)(b-1)(c-1)} = \frac{1.62}{(1)(1)(2)} = 0.81 \tag{6-23}$$

$$\text{Error}(c)\,MS = \frac{\text{Error}(c)\,SS}{ab\,(r-1)(c-1)} = \frac{210.83}{(4)(3)(2)} = 8.78 \tag{6-24}$$

STEP 8: Compute the F value for each source of variation as shown in Table 6.2.8.

TABLE 6.2.8.
Analysis of Variance of a 3-Factor Corn Experiment in a Split-Split-Plot Design

Source of Variation	Degree of Freedom	Sum of Squares	Mean Square	F
Replication	3	273.73	91.24	
Hybrid (A)	1	238.53	238.53	21.11**
Error (a)	3	33.90	11.30	
Row Spacing (B)	1	475.02	475.02	71.64**
AB	1	1.69	1.69	<1
Error (b)	6	39.79	6.63	
Plant Density (C)	2	350.40	175.20	19.95**
AC	2	37.04	18.52	2.11^{ns}
BC	2	87.80	43.90	5.00*
ABC	2	1.62	0.81	<1
Error (c)	24	210.83	8.78	
Total	47	1,713.98		

Note: cv (a) = 2.42%, cv (b) = 1.85%, cv (c) = 2.13%. ** = significant at 1% level, * = significant at 5%, ^{ns} = nonsignificant.

STEP 9: Compare the computed F value of Table 6.2.8 with the tabular F value given in Appendix E. The results indicate that the interaction between row spacing and plant density is significant while the other interactions are all nonsignificant.

Finally, we will compute the coefficient of variation (*cv*) for the 3 error mean squares as shown below and reported at the bottom of Table 6.2.8.

$$cv\,(a) = \frac{\sqrt{\text{Error}\,(a)\,MS}}{\text{Grand Mean}} \times 100 \qquad\qquad (6\text{-}25)$$

$$= \frac{\sqrt{11.30}}{138.85} \times 100 = 2.42\%$$

$$cv\,(b) = \frac{\sqrt{\text{Error}\,(b)\,MS}}{\text{Grand Mean}} \times 100$$

$$= \frac{\sqrt{6.63}}{138.85} \times 100 = 1.85\%$$

$$cv\,(c) = \frac{\sqrt{\text{Error}\,(c)\,MS}}{\text{Grand Mean}} \times 100$$

$$= \frac{\sqrt{8.78}}{138.85} \times 100 = 2.13\%$$

As before, the value of the coefficient of variation (*cv*) indicates the degree of precision with which the factors are compared and is a good measure of the reliability of the experiment. The *cv* (*a*) is a measure of the degree of precision associated with the main effect of the main-plot factor, and the *cv* (*b*) shows the degree of precision of the subplot factor in relation to the mainplot factor and their interactions. The degree of precision in the measurement of the main effect of the sub-subplot factor and its interaction with all other factors is given by the *cv* (*c*). As was indicated in the earlier sections of this chapter, the degree of precision of the main and subplot factors is sacrificed in order to have greater precision in the measurement of the sub-subplot factor. Thus, we would expect *cv* (*a*) to have the largest value and *cv*(*c*), the smallest value. However, in this particular example, while *cv*(*a*) value is the largest, *cv*(*c*) is not the smallest. Such unexpected results are not uncommon.

6.3   STRIP-SPLIT-PLOT DESIGN

This design, which is an extension of the strip-plot design discussed in Section 5.6, is a 3-factor design that accommodates the inclusion of the third factor by dividing the intersection plot into subplots. Essentially, the design has 4 plot sizes—the horizontal strip, the vertical strip, the intersection plot, and the subplot each measured with different levels of precision. As in the case of the strip-plot design, the precision associated with the subplot factor and its interactions with the other factor is the highest.

Randomization and Layout: To illustrate the randomization and layout, let us take an example of a $4 \times 2 \times 4$ factorial experiment replicated 4 times in a randomized complete block design arranged in a strip-split-plot layout. Factor A (first harvest time, denoted as T_1, T_2, T_3, and T_4) is assigned to the horizontal plot. Factor B (herbicide application, denoted as H_0, and H_1) is assigned to the vertical plot. Factor C (4 alfalfa cultivars denoted as C_1, C_2, C_3 and C_4) is assigned to the subplot. As before, the treatment levels and the number of replications are denoted as a, b, c, and r. The steps followed are given below.

STEP 1: Divide the experimental area into 4 blocks or replications ($r = 4$) where each block is further divided into 4 horizontal strips. Randomly assign the 4 treatments ($a = 4$) to each of the strips, following any of the randomization schemes outlined in Chapter 4. Figure 6.3.1 shows the layout for this step.

STEP 2: Divide each block or replication into 2 vertical strips, and randomly assign the 2 herbicide treatments ($b = 2$), following a randomization scheme discussed in Chapter 4. This is shown below in Figure 6.3.2.

STEP 3: Subdivide each of the intersection plots into 4 subplots and randomly assign the 4 alfalfa cultivars ($c = 4$ treatments) to them separately and independently. The final layout of the experiment is shown in Figure 6.3.3.

Analysis of Variance: As in the split-plot design, we need to perform separate analyses for the vertical, horizontal, and intersection factors. These analyses are performed for each factor as illustrated in the following example.

Example 6.3.1

Four cultivars of alfalfa (*Medicago sativa* L.) were studied in newly planted alfalfa plots, with or without the use of pre-emergence herbicide, to determine whether or not variable dates of first cutting of alfalfa, followed by subsequent harvests after the alfalfa had reached the bloom stage, would influence yield. The forage dry matter production results are shown in Table 6.3.1. Perform the analysis of variance.

Solution

STEP 1: Calculate the treatment totals (ABC) as shown in Table 6.3.1.

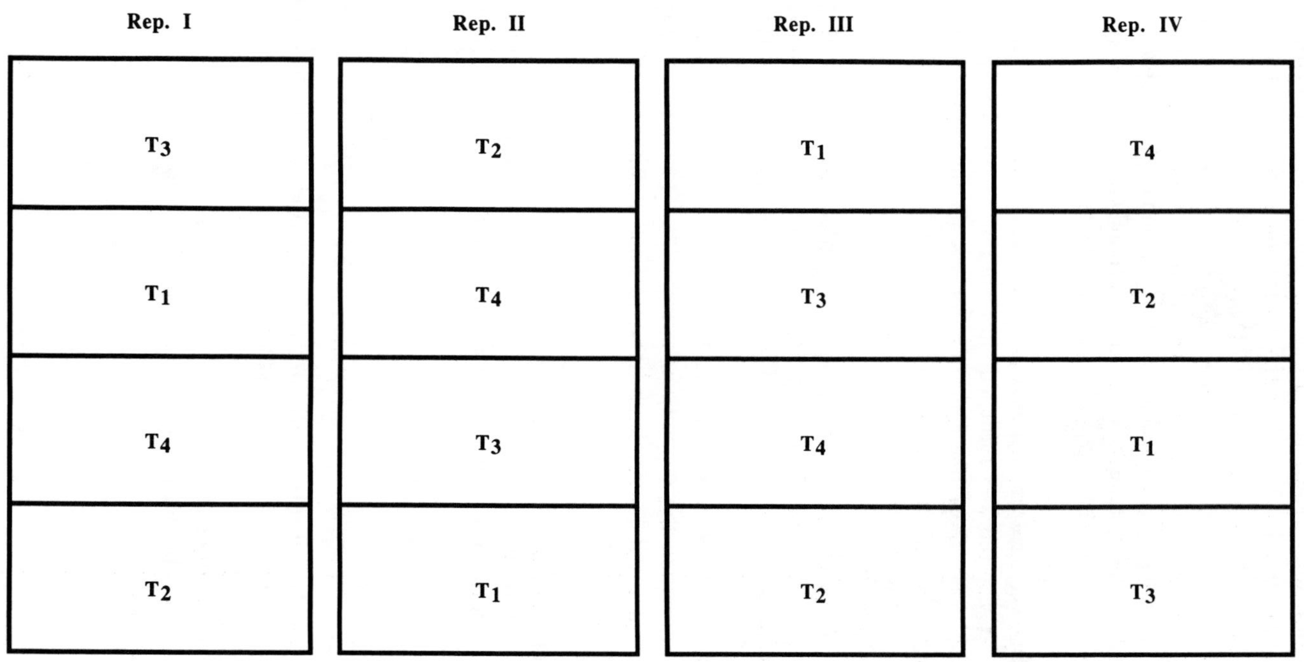

FIGURE 6.3.1.

Random assignment of the horizontal factor (first harvest time T_1, T_2, T_3, and T_4) to the horizontal plots with 4 replications.

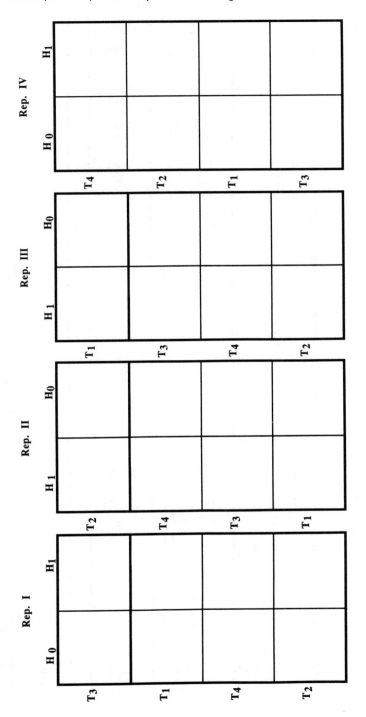

FIGURE 6.3.2.

Random assignment of the horizontal factor (first harvest time T_1, T_2, T_3, and T_4) to the horizontal plots and the vertical factor (herbicide, H_0 and H_1) to the vertical plots with 4 replications.

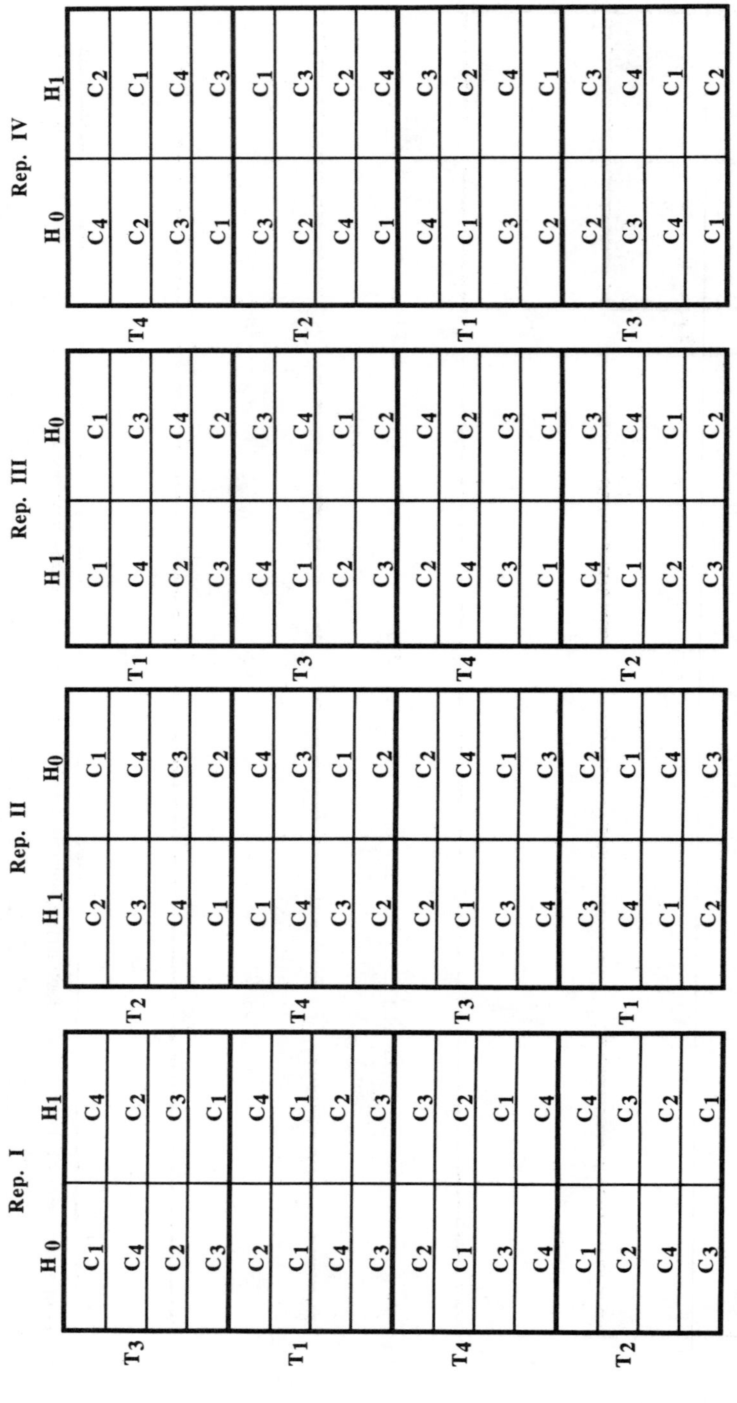

FIGURE 6.3.3.

Layout of a $4 \times 2 \times 4$ factorial experiment arranged in a strip-split-plot design with 4 replications.

TABLE 6.3.1.

Yield of Dry Matter Forage from 4 Alfalfa Cultivars with Varied First-Time Harvest in the Absence and Presence of Herbicide

First Harvest Time	Herbicide	Cultivar	I	II	III	IV	Total (*ABC*)

First Harvest Time	Herbicide	Cultivar	I	II	III	IV	Total (*ABC*)
Early bud	Absence	Kanza	1.9	1.8	1.6	1.8	7.1
Late bud			2.0	2.1	1.9	2.2	8.2
Early bloom			2.4	2.3	2.1	2.2	9.0
Late Bloom			2.6	2.4	2.5	2.4	9.9
Early bud	Presence		5.5	5.2	5.7	5.3	21.7
Late bud			5.8	5.2	5.6	5.1	21.7
Early bloom			6.4	6.4	6.2	6.1	25.1
Late Bloom			6.2	6.0	6.4	6.1	24.7
Early bud	Absence	Liberty	1.8	1.8	1.0	1.5	6.1
Late bud			2.4	2.5	2.8	2.4	10.1
Early bloom			2.6	2.3	2.5	2.4	9.8
Late Bloom			2.9	2.7	2.6	2.4	10.6
Early bud	Presence		5.1	5.4	5.2	5.5	21.2
Late bud			5.0	5.1	5.1	4.9	20.1
Early bloom			5.3	5.6	5.2	5.3	21.4
Late Bloom			5.4	5.7	5.5	5.3	21.9
Early bud	Absence	Arc	2.0	2.2	2.1	2.0	8.3
Late bud			2.4	2.1	2.3	2.2	9.0
Early bloom			2.4	2.2	2.6	2.4	9.6
Late Bloom			2.6	2.8	2.4	2.2	10.0
Early bud	Presence		5.1	5.5	5.3	5.2	21.1
Late bud			5.7	5.9	5.1	5.4	22.1
Early bloom			5.3	5.5	5.1	5.3	21.2
Late Bloom			5.7	5.6	5.7	5.5	22.5
Early bud	Absence	CA90	1.6	1.2	1.5	1.3	5.6
Late bud			2.5	2.2	2.1	2.6	9.4
Early bloom			2.3	2.2	2.6	2.1	9.2

Dry Matter Yield (tons/ac) — Replications

TABLE 6.3.1. (Continued)

First Harvest Time	Herbicide	Cultivar	Dry Matter Yield (tons/ac)				
			Replications				Total (ABC)
			I	II	III	IV	
Late Bloom			2.8	2.9	2.4	2.3	10.4
Early bud	Presence		7.4	7.1	7.7	7.3	29.5
Late bud			7.8	7.3	7.9	7.2	30.2
Early bloom			7.5	7.1	7.6	7.3	29.5
Late Bloom			7.7	7.1	7.4	7.9	30.1

STEP 2: Determine the degrees of freedom associated with each source of variation. The following sources of variation and the degrees of freedom are identified.

Replication $d.f.$ $= r - 1 = 3$

Horizontal factor (A) $d.f.$ $= a - 1 = 3$

Error (a) $d.f.$ $= (r - 1)(a - 1) = 9$

Vertical factor (B) $d.f.$ $= b - 1 = 1$

Error (b) $d.f.$ $= (r - 1)(b - 1) = 3$

$A \times B$ $d.f.$ $= (a - 1)(b - 1) = 3$

Error (c) $= (r - 1)(a - 1)(b - 1) = 9$

Subplot factor (C) $d.f.$ $= c - 1 = 3$

$A \times C$ $d.f.$ $= (a - 1)(c - 1) = 9$

$B \times C$ $d.f.$ $= (a - 1)(b - 1) = 3$

$A \times B \times C$ $d.f.$ $= (a - 1)(b - 1)(c - 1) = 9$

Error (d) $= ab (r - 1)(c - 1) = 72$

Total $d.f.$ $= rabc - 1 = 127$

STEP 3: Perform the vertical analysis by calculating the replication totals (R), the grand total (G), and the first harvest time total (A). To do this, we need to construct a 2-way table for the replication × first harvest time totals (RA) as shown in Table 6.3.2. To show how we have computed the yield total for all the replications in Table 6.3.2, we have used the early bud in replication I as an example, and the total is computed as:

$$1.9 + 5.5 + 1.8 + 5.1 + 2.0 + 5.1 + 1.6 + 7.4 = 30.4$$

TABLE 6.3.2.

Summary Table of Replication × First Harvest Time (Factor A) Yield Totals Computed from Data in Table 6.3.1

First Harvest	Yield Total (RA)				First-Time Harvest Total
Time	Rep. I	Rep. II	Rep. III	Rep. IV	(A)
Early bud	30.4	30.2	30.1	29.9	120.6
Late bud	33.6	32.4	32.8	32.0	130.8
Early bloom	34.2	33.6	33.9	33.1	134.8
Late bloom	35.9	35.2	34.9	34.1	140.1
Rep. Total (R)	134.1	131.4	131.7	129.1	
Grand Total (G)					526.3

STEP 4: Compute the correction factor and the various sums of squares as:

$$C = \frac{G^2}{rabc} \tag{6-1}$$

$$= \frac{(526.3)^2}{4 \times 4 \times 2 \times 4}$$

$$= 277{,}991.68/128$$

$$= 2{,}164.00$$

$$\text{Total } SS = \sum X^2 - C \tag{6-2}$$

$$= \left[(1.9)^2 + (1.8)^2 + \ldots + (7.9)^2 \right] - 2{,}164.00$$

$$= 2{,}684.39 - 2{,}164.00$$

$$= 520.39$$

STEP 5: Calculate the replication, factor A (first harvest time), and the error sum of squares for factor A as:

$$\text{Replication} SS = \frac{\sum R^2}{abc} - C \tag{6-3}$$

$$= \frac{(134.1)^2 + (131.4)^2 + (131.7)^2 (129.1)^2}{32} - 2,164$$

$$= 69,260.47/32 - 2,164$$

$$= 2,164.39 - 2,164$$

$$= 0.39$$

$$SS_A \text{ (First Time Harvest)} = \frac{\sum A^2}{rbc} - C \tag{6-4}$$

$$= \frac{(120.6)^2 + (130.8)^2 + (134.8)^2 + (140.1)^2}{(4)(2)(4)} - 2,164$$

$$= 69,452.05/32 - 2,164$$

$$= 2,170.38 - 2,164$$

$$= 6.38$$

$$\text{Error } SS\,(a) = \frac{\sum (RA)^2}{bc} - (C + \text{Rep. } SS + SS_A) \tag{6-5}$$

$$= \frac{(30.4)^2 + (33.6)^2 + ... + (34.1)^2}{(2)(4)} - (2,164 + 0.39 + 6.38)$$

$$= 17,366.87/8 - 2,170.77$$

$$= 2,170.86 - 2,170.77$$

$$= 0.09$$

STEP 6: Perform the horizontal analysis by constructing a two-way table for the Replication $\times$ Herbicide totals (RB) as shown in Table 6.3.3, and then compute the various sums of squares as:

$$SS_B \text{ (Herbicide)} = \frac{\sum B^2}{rac} - C \qquad (6\text{-}6)$$

$$= \frac{(142.3)^2 + (384.0)^2}{(4)(4)(4)} - 2{,}164$$

$$= 167{,}705.29/64 - 2{,}164$$

$$= 2{,}620.4 - 2{,}164$$

$$= 456.4$$

TABLE 6.3.3.

Summary Table of Yield Totals for the Replication × Herbicide (RB) Computed from Data in Table 6.3.1

Herbicide	Yield Total (RB)				Herbicide
	Rep. I	Rep. II	Rep. III	Rep. IV	Total (B)
Absent	37.2	35.7	35.0	34.4	142.3
Present	96.9	95.7	96.7	94.7	384.0

$$\text{Error } (b) \, SS = \frac{\sum (RB)^2}{ac} - (C + \text{Rep. } SS + SS_B) \qquad (6\text{-}26)$$

$$= \frac{(37.2)^2 + (96.9)^2 + \ldots + (94.7)^2}{16} - (2{,}164 + 0.39 + 456.4)$$

$$= 41{,}933.77/16 - 2{,}620.79$$

$$= 2{,}620.86 - 2{,}620.79$$

$$= 0.07$$

STEP 7: Perform the interaction analysis by constructing a 2-way table for the vertical × horizontal factor totals (AB), and a 3-way table of totals for the replication × vertical × horizontal factors (RAB) as shown in Tables 6.3.4 and 6.3.5. Using these summary tables, compute the various sums of squares as:

TABLE 6.3.4.

Summary Table of Yield Totals for First Harvest Time ×
Herbicide(factor A × factor B) Computed from Data in Table 6.3.1

Herbicide	Yield Total (AB)			
	Early bud	Late bud	Early bloom	Late bloom
Absent	27.1	36.7	37.6	40.9
Present	93.5	94.1	97.2	99.2

TABLE 6.3.5.

Summary Table of Yield Totals for the Replication × First Harvest Time
× Herbicide or (RAB) Computed from Data in Table 6.3.1

First Harvest Time	Herbicide	Yield Total (RAB)			
		Rep. I	Rep. II	Rep. III	Rep. IV
Early bud	Absent	7.3	7.0	6.2	6.6
	Present	23.1	23.2	23.9	23.3
Late bud	Absent	9.3	8.9	9.1	9.4
	Present	24.3	23.5	23.7	22.6
Early bloom	Absent	9.7	9.0	9.8	9.1
	Present	24.5	24.6	24.1	24.0
Late bloom	Absent	10.9	10.8	9.9	9.3
	Present	25.0	24.4	25.0	24.8

$$SS_{AB} = \frac{\sum (AB)^2}{rc} - (C + SS_A + SS_B) \tag{6-7}$$

$$= \frac{(27.1)^2 + (93.5)^2 + ... + (99.2)^2}{(4)(4)} - (2,164 + 6.38 + 456.4)$$

$$= 42,053.41/16 - 2,626.78$$

$$= 2,628.34 - 2,626.78$$

$$= 1.56$$

$$\text{Error}(c)\,SS = \frac{\sum (RAB)^2}{c} - (C + \text{Rep. } SS + SS_A + \text{Error}(a)\,SS + SS_B \tag{6-27}$$

$$+ \text{Error}(b)\,SS + SS_{AB})$$

$$= 10{,}518.81/4 - (2{,}164 + 0.39 + 6.38 + 0.09 + 456.4 + 0.07 + 1.56)$$

$$= 2{,}629.70 - 2{,}628.89$$

$$= 0.81$$

STEP 8: Perform the subplot analysis by constructing 2 summary tables of yield totals. The first table is for vertical factor $A \times$ subplot factor C and is presented as Table 6.3.6.

TABLE 6.3.6.

Summary Table of Yield Totals for Factor $A \times$ Factor C Computed from Data in Table 6.3.1

	Yield Total (AC)			
First Harvest Time	Kanza	Liberty	Arc	CA90
Early bud	28.8	27.3	29.4	35.1
Late bud	29.9	30.2	31.1	39.6
Early bloom	34.1	31.2	30.8	38.7
Late bloom	34.6	32.5	32.5	40.5
First-Time Harvest Total (C)	127.4	121.2	123.8	153.9

The second table is for the horizontal factor $B \times$ subplot factor C, which is shown in Table 6.3.7. From these tables compute the various sums of squares as given below:

TABLE 6.3.7.

Summary Table of Yield Totals for Factor $B \times$ Factor C Computed from Data in Table 6.3.1

	Yield Total (BC)			
Herbicide	Kanza	Liberty	Arc	CA90
Absent	34.2	36.6	36.9	34.6
Present	93.2	84.6	86.9	119.3

In computing the sum of squares for factor C it should be kept in mind that the squared C value is representing factor C and is different from the correction factor notation (C) as shown below:

$$SS_C \text{ (Cultivar)} = \frac{\sum C^2}{rab} - C \qquad\qquad (6\text{-}9)$$

$$= \frac{(127.4)^2 + (121.2)^2 + (123.8)^2 + (153.9)^2}{(4)(4)(2)} - 2,164$$

$$= 69,931.85/32 - 2,164$$

$$= 2,185.37 - 2,164$$

$$= 21.37$$

$$SS_{AC} = \frac{\sum (AC)^2}{rb} - (C + SS_A + SS_C) \qquad\qquad (6\text{-}10)$$

$$= \frac{(28.8)^2 + (27.3)^2 + \ldots + (40.5)^2}{(4)(2)} - (2,164 + 6.38 + 21.37)$$

$$= 17,545.01/8 - 2,191.75$$

$$= 2,193.13 - 2,191.75$$

$$= 1.38$$

$$SS_{BC} = \frac{\sum (BC)^2}{ra} - (C + SS_B + SS_C) \qquad\qquad (6\text{-}11)$$

$$= \frac{(34.2)^2 + (36.6)^2 + \ldots + (119.3)^2}{(4)(4)} - (2,164 + 456.4 + 21.37)$$

$$= 42,695.47/16 - 2,641.77$$

$$= 2,668.47 - 2,641.77$$

$$= 26.70$$

$$SS_{ABC} = \frac{\sum (ABC)^2}{r} - (C + SS_A + SS_B + SS_C + SS_{AB} + SS_{AC} + SS_{BC}) \quad (6\text{-}12)$$

$$= \frac{(7.1)^2 + (8.2)^2 + ... + (30.1)^2}{4}$$

$$- (2,164 + 6.38 + 456.4 + 21.37 + 1.56 + 1.38 + 26.70)$$

$$= 10,719.61/4 - 2,677.79$$

$$= 2,679.90 - 2,677.79$$

$$= 2.11$$

Error (d) SS = Total SS – All other sum of squares   (6-28)

$$= 520.39 - (0.39 + 6.38 + 0.09 + 456.4$$

$$+ 0.07 + 1.56 + 0.81 + 21.37 + 1.38 + 26.70 + 2.11)$$

$$= 520.39 - 517.26$$

$$= 3.13$$

STEP 9: Calculate the mean square for each source of variation as:

$$\text{Replication} MS = \frac{\text{Rep. } SS}{r - 1} = \frac{0.39}{3} = 0.13 \qquad (6\text{-}14)$$

$$MS_A = \frac{SS_A}{a - 1} = \frac{6.38}{4 - 1} = 2.13 \qquad (6\text{-}15)$$

$$\text{Error } (a) MS = \frac{\text{Error } (a) SS}{(r - 1)(a - 1)} = \frac{0.09}{(3)(3)} = 0.01 \qquad (6\text{-}16)$$

$$MS_B = \frac{SS_B}{b - 1} = \frac{456.4}{2 - 1} = 456.4 \qquad (6\text{-}17)$$

$$\text{Error} (b)MS = \frac{\text{Error } SS\ (b)}{(r-1)(b-1)} = \frac{0.07}{(3)(1)} = 0.02 \tag{6-29}$$

$$MS_{AB} = \frac{SS_{AB}}{(a-1)(b-1)} = \frac{1.56}{(3)(1)} = 0.52 \tag{6-18}$$

$$\text{Error} (c)MS = \frac{\text{Error } SS\ (c)}{(r-1)(a-1)(b-1)} = \frac{0.81}{(3)(3)(1)} = 0.09 \tag{6-30}$$

$$MS_C = \frac{SS_C}{c-1} = \frac{21.37}{4-1} = 7.12 \tag{6-20}$$

$$MS_{AC} = \frac{SS_{AC}}{(a-1)(c-1)} = \frac{1.38}{(3)(3)} = 0.15 \tag{6-21}$$

$$MS_{BC} = \frac{SS_{BC}}{(b-1)(c-1)} = \frac{26.70}{(1)(3)} = 8.9 \tag{6-22}$$

$$MS_{ABC} = \frac{SS_{ABC}}{(a-1)(b-1)(c-1)} = \frac{2.11}{(3)(1)(3)} = 0.23 \tag{6-23}$$

$$\text{Error} (d)\ MS = \frac{\text{Error} (d)\ SS}{ab\ (r-1)(c-1)} = \frac{3.13}{(8)(3)(3)} = 0.04 \tag{6-31}$$

STEP 10:   Compute the F value for each source of variation as shown in Table 6.3.8.

TABLE 6.3.8.

Analysis of Variance of Alfalfa Dry Matter Yield in Strip-Split-Plot Design

Source of Variation	Degree of Freedom	Sum of Squares	Mean Square	F
Replication	3	0.39	0.13	
First Harvest Time (A)	3	6.38	2.13	213.00**
Error (a)	9	0.09	0.01	
Herbicide (B)	1	456.40	456.40	a
Error (b)	3	0.07	0.02	
AB	3	1.56	0.52	5.78*
Error (c)	9	0.81	0.09	
Hybrid(C)	3	21.37	7.12	178.00**
AC	9	1.38	0.15	3.75**
BC	3	26.70	8.90	222.50**
ABC	9	2.11	0.23	5.75**
Error (d)	72	3.13	0.04	
Total	127	520.39		

Note: cv (a) = 2.43%, cv (c) = 7.30%, cv (d) = 4.87%. ** = significant at 1% level, * = significant at 5%. ^{a}Error (b) degrees of freedom is not adequate for valid test of significance.

STEP 11: Compare the computed F of Table 6.3.8 with the tabular F value given in Appendix E. The results indicate that all interactions were significant, implying that there are differences in the treatment combinations. To appropriately interpret large interactions, we need to supplement the analysis with detailed examination of the nature of the interactions. In the next chapter we will discuss mean comparisons that provide added information in interpreting the results.

Finally, we will compute the coefficient of variation (cv) for the 4 error mean squares. Because of inadequate degrees of freedom associated with factor B, we have not calculated the value of the $cv(b)$. The cv value for a, c, and d error mean squares are computed as shown below:

$$cv\,(a) = \frac{\sqrt{\text{Error}\,(a)\,MS}}{\text{Grand Mean}} \times 100 \tag{6-25}$$

$$= \frac{\sqrt{0.01}}{4.11} \times 100 = 2.43\%$$

$$cv\,(c) = \frac{\sqrt{\text{Error}\,(b)\,MS}}{\text{Grand Mean}} \times 100$$

$$= \frac{\sqrt{0.09}}{4.11} \times 100 = 7.30\%$$

$$cv\,(d) = \frac{\sqrt{\text{Error}\,(c)\,MS}}{\text{Grand Mean}} \times 100$$

$$= \frac{\sqrt{0.04}}{4.11} \times 100 = 4.87\%$$

As before, the value of the coefficient of variation (cv) indicates the degree of precision associated with the measurement of the effects of the different factors. The $cv\,(a)$ is a measure of the degree of precision associated with the vertical factor. The degree of precision associated with the interaction effect between factor A and B is given by the $cv\,(c)$. The $cv\,(d)$ captures the precision of all effects associated with the subplot factor.

6.4 FACTORIAL EXPERIMENTS IN FRACTIONAL REPLICATION

An advantage of using factorial experiments is that the experimenter is able to study several factors simultaneously. However, as the number of factors included in the experiment increases, so does the requirement for resources. This can be very costly and time-consuming. Additionally, technical considerations may dictate against the use of a fully replicated design. Consider, for example, an experiment in which 6 or 7 factors each at 2 levels are to be studied simultaneously. A full replicate for each case would require the use of 64 and 128 experimental units, respectively, whether confounding is to be used or not. In such a situation, it is worth considering whether estimates of the main effects and/ or interactions, with a given precision, can be obtained by using a fraction of the full replicate. Finney (1945), who described a half-replicate of a 4×2^4 agricultural experiment, proposed the use of *fractional replication* and outlined the methods of construction for 2^n and 3^n factorials.

The major advantage of using the design is that 5 or more factors, for example, can be included simultaneously in an exploratory experiment of a practicable size in which those factors that affect response are identified and can be separated from those factors whose level is unimportant. Another feature of the design is reduced block size. Since the design does not require that all treatment combinations be tested, the experimenter is able to use reduced block size and thus improve the homogeneity of the experimental unit.

The advantages of using fractional replication should be weighed against the loss of information from a reduced number of replications, especially when fewer factors are involved. Loss of information is not a major concern in situations where there are a large number of factors in the experiment, resulting in a large number of interactions. The larger the number of factors in the experiment, the greater the flexibility that the investigator may have in the choice of factor interactions to be sacrificed. Prior knowledge of factor effects and interactions is helpful in deciding which factor combinations are to be included and which are to be sacrificed. In practice, high-order interactions such as the 5 or 4-factor interactions are sacrificed. As suggested by Gomez and Gomez (1984, p. 168) in almost all cases, it is essential that the experimenter test a set of treatments that allows for estimation of all main effects and 2-factor interactions. Another concern with use of the fractional replication is misinterpretation of the results, a condition which is less of a problem in fully replicated designs. In Section 6.4.1 we will discuss reasons for the problems of misinterpretation.

6.4.1. Aliases and Defining Contrasts

In using fractional factorial designs, care must be taken in the interpretation of results that are affected by *aliases*. Aliases refer to 2 factorial effects that are represented by the same comparison. Another term that needs clarification is *defining contrasts*. A defining contrast refers to a contrast that is used to split the factorial into 2 half-replicates. To have a better understanding of these 2 terms, consider a 2^3 factorial experiment in which the following 4 combinations- a, b, c, and abc are tested. These 4 treatment combinations are only half of the complete replicate. We can estimate the main effects and interactions as contrasts, using the coefficients given in Table 6.4.1.1.

Notice that in Table 6.4.1.1 all of the treatment combinations for contrast ABC carry a + sign, indicating that this contrast cannot be estimated at all from the selected treatments. This is called a *defining contrast*. You should note that when using a half-replicate we lose 1 factorial effect *ABC* entirely, and each of the main effect is mixed with 1 of the 2 factor interactions.

When 2 factorial effects are represented by the same comparison, we refer to them as contrasts which are equivalent to each other, as is the case with $A = BC$, $B = AC$, and $C = AB$. These are called *aliases*. This means that the contrast which estimates A, for example, is the same as the contrast that estimates BC. Such equivalencies in contrasts make it difficult to distinguish whether the contrast is affected by A, or BC, or a mixture of both. In

TABLE 6.4.1.1.

Table 6.4.1.1. Coefficients for Factorial Effects Using 4 Treatments from a 2^3 Factorial Experiment

Contrast	Treatment Combination			
	a	b	c	abc
A	+	−	−	+
B	−	+	−	+
AB	−	−	+	+
C	−	−	+	+
AC	−	+	−	+
BC	+	−	−	+
ABC	+	+	+	+

fractional replication, every contrast has at least 1 other contrast as an alias. Difficulty in appropriately identifying the alias to which an effect is to be attributed often causes the problems, alluded to earlier, of misinterpretation of the results.

To minimize the risk of misinterpreting the results, we must know the aliases of the factorial effects we are interested in. To find the aliases, use is made of the following rule:

In the 2^n system, the alias of any factorial contrast is its generalized interaction with the defining contrast.

Using this rule, the alias for A in the defining contrast ABC is the interaction between A and ABC, which is $AABC$ or A^2BC. When interpreting a generalized interaction, squared terms are canceled (Cochran and Cox, 1957). Thus, A^2BC is read as BC, and the alias of B is AB^2C or AC.

To determine which factorial effect one should use as the defining contrast in a given situation, the following rule should be followed:

In order to get a fractional replicate, do not use as the defining contrast any factorial effect that must be estimated from the experiment run on that fractional replicate.

Following this rule, the experimenter should select as the defining contrast that factorial effect which has the largest number of interactions. Higher-order interactions are negligible and thus can be sacrificed.

Fractional factorial designs are widely used and can be constructed to fit most factorial experiments. However, the procedure for construction is complex and beyond the scope of this book. In the following example, we have presented a simple analysis of the fractional replication. The reader is referred to Cochran and Cox (1957), Finney (1945), Kempthorne (1952), and Brownlee, Kelly, and Loraine (1948) for further elaborations on—and more complex analyses of—fractional factorial designs. Appendix G presents several basic plans, each of which identifies the defining contrast(s) and the estimable factors that can be used for exploratory research. The next sections present the randomization and layout of a fractional factorial design and analysis.

Randomization and Layout: The randomization procedure requires several steps which are given below.

STEP 1: Using Appendix G, select a basic plan that meets the needs of the experiment in terms of factors and levels of the factors to be tested. Assume that we have an experiment in which 6 factors each at 2 levels are used. Thus from Appendix G, we select Plan 6 which meets the requirement for a 2^6 factorial experiment to be conducted in a half-replicate. The selected plan has the following treatment numbers assigned to each block as shown in Table 6.4.1.2.

TABLE 6.4.1.2.
A Half-Replicate of a 2^6 Factorial

Treatment No.	1	2	3	4	5	6	7	8
	9	10	11	12	13	14	15	16
Block I	(1)	ab	ac	bc	ae	af	ad	bd
	abef	ef	de	df	bf	be	ce	cf
Block II	acde	acdf	abdf	acef	cd	abcd	abcf	abce
	bcdf	bcde	bcef	abde	abcdef	cdef	bdef	adef

STEP 2: Randomly assign the block arrangement in the basic plan to the blocks in the field. This should be done in cases where there are more than 1 replication per block, to allow for appropriate randomization. Suppose we decide to have 2 replications, each with 2 blocks. This means we will have 32 experimental plots with each block having 16 plots in them. Follow any of the randomization schemes of Chapter 4 to assign the block numbers of the basic plan to the blocks in the field.

STEP 3: Randomly reassign the treatment in each block arrangement of the basic plan to the experimental plot separately and independently. This means that the 16 treatment numbers in each of the 4 blocks (2 in each replication) are randomly assigned to each block separately and independently. The results of this reassignment are shown in Table 6.4.1.3.

TABLE 6.4.1.3.

Randomly Reassigned Basic Plan Treatments to the Experimental Plot

Basic Plan	Assigned Field Plot Number			
Treatment	Rep. I		Rep. II	
Number	Block 1	Block 2	Block 1	Block 2
1	2	8	10	6
2	5	16	12	9
3	11	6	9	10
4	10	2	16	14
5	7	9	7	1
6	3	15	13	11
7	16	1	8	5
8	4	13	6	2
9	1	5	14	16
10	6	11	4	3
11	14	7	1	12
12	8	4	15	4
13	12	10	2	13
14	9	12	3	8
15	13	3	11	15
16	15	14	5	7

Based on this reassignment of the treatment numbers of the basic plan, we will assign treatment number 2 of the basic plan to plot 1 of block 1 of replication I, and treatment number 5 of the basic plan to plot 2 of block 1 of replication I. This is done for each plot in each block in each replication as shown in Figure 6.4.1.

Analysis of Variance: In performing the analysis, there are several methods to compute the sum of squares for main effects and 2-factor interactions. For simplicity, let us again assume we have a 2^6 factorial experiment with a half-replicate. To compute the sum of squares, we may construct fifteen 2-way tables for every pair of variables. Another approach would be to write down the combination of signs for each of the effects that are to be estimated. For this procedure, consult Cochran and Cox (1957). The quickest method of analyzing fractional replication of 2^n factorial is through an adaptation of Yates' algorithm. Such an adaptation proceeds as follows:

1. The fractional replication design which contains 2^{n-k} combination has a complete replication of $n-k$ factors. Arrange the $n-k$ factors in Yates' format, which ignores the k of the factors. A factor that is ignored is placed inside brackets. This is a reminder that the ignored factor has no

Block 1		Block 2	
1	2	1	2
ab	ae	abce	adef
3	4	3	4
de	ef	abcd	acdf
5	6	5	6
ad	ac	bcdf	bdef
7	8	7	8
cf	bc	acde	abcdef
9	10	9	10
(1)	af	cd	bcef
11	12	11	12
be	bd	abcf	acef
13	14	13	14
df	abef	bcde	abde
15	16	15	16
bf	ce	abdf	cdef

Block 1		Block 2	
1	2	1	2
ef	df	abcd	bcdf
3	4	3	4
abef	cf	bcde	cdef
5	6	5	6
ad	bf	acde	bcef
7	8	7	8
bd	af	cd	acdf
9	10	9	10
be	bc	adef	abdf
11	12	11	12
(1)	ce	abde	acef
13	14	13	14
ab	ac	abcdef	abce
15	16	15	16
de	ae	bdef	abcf

Replication I   Replication II

FIGURE 6.4.1.

Sample layout of a 2^6 factorial experiment with half of the treatments arranged in 2 blocks of 16 plots each.

part in the systematic order. For the 2^6 factorial experiment in which we have 6 factors (*ABCDEF*) each at 2 levels, there are 32 treatment combinations with a complete replication of the first 5 factors. This means that the last factor, or factor F, will be ignored in the arrangement process of treatment combinations.

2. Apply Yates' method as if the experiment were a complete replication of a 2^{n-k} factorial set. In our example, this would be a complete replicate of a 2^5 experiment.

3. Reintroduce the ignored factor (F) in the "Factorial Effect Identification" column as subsequently shown in Table 6.4.1.5. By doing so, each effect will have 1 or more aliases involving the ignored factor (F).
4. Finally, select the main effect and low-order interactions from the alias set to be identified with the contrast. For our example, we have used $ABCDEF$ as the defining contrast, hence the alias for A is $BCDEF$, and the alias for B is $ACDEF$, and so on. When we reintroduce the ignored factor, we have to identify its main and interaction effects for estimation purposes. The main effect of the ignored factor is F, and its interactions are AF, BF, CF, DF, and EF. The aliases for these effects are $F = ABCDE$, $AF = BCDE$, $BF = ACDE$, $CF = ABDE$, $DF = ABCE$, and $EF = ABCD$. The identification assigned to each of the 32 factorial effects appears under the column labeled "Contrast" in Table 6.4.1.5.

The following example is used to illustrate the analysis of variance for a fractional factorial design experiment.

Example 6.4.1

An animal scientist using a 2^6 factorial experiment in a half replicate is interested in knowing the impact of several factors on the daily weight gain of grazing beef steers. Specifically, the scientist is interested in knowing whether the 3-factor interactions are significant. Factors A, B, C, D, E, and F are known to have an effect on the performance of the animal. The experiment was conducted with 2 replications (to allow for the estimation of the 3-factor interactions) each with 32 treatment combinations. Each replication was further divided into 2 blocks each containing 16 plots. The daily weight gain data gathered from the experiment are given in Table 6.4.1.4.

Solution

STEP 1: Determine the number of factors each at 2 levels that is equal to the number of treatment combinations to be tested. As was stated in the previous section, in a fractional replication design of 2^{n-k} combinations, there is a complete replication of $n - k$ factors. The number of treatment combinations (t) equals 2^n, ignoring the k factors. Since the number of treatment combinations to be tested in the present example is 32, then $n = 5$ factors. This means that the complete replication contains the first 5 factors ($ABCDE$) and ignores the last factor (F).

STEP 2: Apply Yates' method as if the experiment were a complete replicate of a 2^5 factorial experiment, and arrange the treatments such that

TABLE 6.4.1.4.

Daily Weight Gain of Grazing Beef Steers from a 2^6 Factorial
Experiment Conducted as a Half-Replicate in Blocks of 16 Units with
2 Replications

Treatment Combination	Block 1			Treatment Combination	Block 2		
	Rep. I	Rep. II	Total		Rep. I	Rep. II	Total
(1)	2.13	2.18	4.31	ad	2.20	2.50	4.70
de	2.17	2.12	4.29	ae	2.35	2.24	4.59
df	2.40	2.11	4.51	af	2.34	2.34	4.68
ef	2.11	2.19	4.30	bd	1.50	1.89	3.39
ab	2.50	2.56	5.06	be	2.38	2.54	4.92
ac	2.33	2.43	4.76	bf	2.19	2.21	4.40
bc	2.15	2.30	4.45	cd	1.99	2.10	4.09
abde	1.99	2.00	3.99	ce	2.43	2.65	5.08
abdf	2.50	2.34	4.84	cf	2.00	2.10	4.10
abef	2.67	2.51	5.18	adef	1.50	2.35	3.85
acde	2.12	2.25	4.37	bdef	2.44	2.18	4.62
acdf	2.57	2.41	4.98	cdef	1.17	1.82	2.99
acef	2.00	2.28	4.28	abcd	2.67	2.41	5.08
bcde	2.82	2.69	5.51	abce	2.45	2.39	4.84
bcdf	2.16	2.19	4.35	abcf	1.90	1.85	3.75
bcef	1.56	1.89	3.45	abcdef	2.16	2.05	4.21
Total (RB)	36.18	36.45			33.67	35.62	

when treatment (1) is included in the set of treatments, it always is listed as the first treatment; then comes those treatments with fewer number of letters. For example, a comes before ab, and ac comes before ad and so on. When arranging the treatments in this fashion, omit the treatment identification letters corresponding to the ignored factor. In our example, the treatment identification letter af comes before ab because factor F is ignored and plays no role in this arrangement process. As a reminder, those treatments that involve factor f are placed in brackets as shown in Table 6.4.1.3.

STEP 3: Determine the degrees of freedom associated with each source of variation as given by the basic plan. The basic plan identifies the block, the main effect, and the 2 and 3-factor interactions. Since this experiment is performed with replications to allow for the measurement of the 3-factor interactions, we must account for the variability from the replications, and the block × replication interaction. Thus, the following sources of variation and degrees of freedom are associated with this experiment:

Effects	d.f.
Replication	= 1
Block	= 1
Block × Replication	= 1
Main	= 6
2-factor	= 15
3-factor	= 9
Error	= 30
Total	= 63

STEP 4: Calculate the replication total (R), the block total (B), replication × block totals (RB) as shown in Table 6.4.1.2, and the grand total (G) as follows:

Rep. I (R_1)   = 36.18 + 33.67 = 69.85

Rep. II (R_2)   = 36.45 + 35.62 = 72.07

Block 1 (B_1)   = 36.18 + 36.45 = 72.63

Block 2 (B_2)   = 33.67 + 35.62 = 69.29

G = 69.85 + 72.07 = 141.92

STEP 5: Calculate the correction factor and the total sum of squares as:

$$C = \frac{G^2}{rt} \tag{6-32}$$

where

$\quad\quad C \ = \$ Correction factor

$\quad\quad G \ = \$ Grand total

$$r = \text{Number of replications}$$

$$t = \text{Total number of treatments tested.}$$

Thus,

$$C = \frac{(141.92)^2}{2 \times 32} \tag{6-2}$$

$$= 314.71$$

$$\text{Total } SS = \sum X^2 - C$$

$$= \left[(2.13)^2 + (2.17)^2 + \ldots + (2.05)^2 \right] - 314.71$$

$$= 320.56 - 314.71$$

$$= 5.85$$

STEP 6: Calculate the replication, block, and block × replication sum of squares as:

$$\text{Replication } SS = \frac{\sum R^2}{t} - C \tag{6-33}$$

$$= \frac{(69.85)^2 + (72.07)^2}{32} - 314.71$$

$$= 314.78 - 314.71$$

$$= 0.07$$

$$\text{Block } SS = \frac{\sum B^2}{t} - C \tag{6-34}$$

$$= \frac{(72.63)^2 + (69.29)^2}{32} - 314.71$$

$$= 314.88 - 314.71$$

$$= 0.17$$

$$\text{Block} \times \text{Rep. } SS = \frac{\sum RB^2}{t/b} - (C + \text{Rep. } SS + \text{Block } SS) \tag{6-35}$$

where

b = Number of blocks in each replication.

Therefore, the block × replication sum of square is:

$$= \frac{(36.18)^2 + (36.45)^2 + (33.67)^2 + (35.62)^2}{32/2} - (314.71 + 0.07 + 0.17)$$

$$= 315.00 - 314.95$$

$$= 0.05$$

STEP 7: Calculate the t factorial effect totals by following the procedures outlined below.

1. Place the original 32 treatment combinations as the first set of t treatments or t_0 in column 2 of Table 6.4.1.5, following Yates' arrangement.
2. Divide the 32 t_0 treatment combinations into half ($t/2$), forming 16 successive pairs. Then place the sum of the values of successive pairs (t_1) into column 3 of Table 6.4.1.5. This task provides the first half of the t_1 values. To compute the values for the second half of t_1, the value of the first treatment in each of the 16 successive pairs is subtracted from the value of the second treatment combination respectively.

Computation of the values for the first and second half of t_1 is illustrated below:

1st pair = 4.31 + 4.68 = 8.99
2nd pair = 4.40 + 5.06 = 9.46
3rd pair = 4.10 + 4.76 = 8.86

.
.
.

16th pair = 5.51 + 4.21 = 9.72

The values for the second half of Column 3 are computed as:

9th pair = 4.68 - 4.31 = 0.37
10th pair = 5.06 - 4.40 = 0.66

TABLE 6.4.1.5.

Yates' Method of Computing the Sum of Squares for a 2^6 Factorial Experiment Conducted as a Half-Replicate in Blocks of 16 Units with 2 Replications Using the Daily Weight Gain Data

Treatment Combination	t_0	t_1	t_2	t_3	t_4	t_5	Factorial Effect Identification Contrast	Alias
(1)	4.31	8.99	18.45	35.51	71.45	141.92	(G)	(G)
$a\ (f)$	4.68	9.46	17.06	35.94	70.47	4.40	A	A
$b\ (f)$	4.40	8.86	17.44	36.64	4.25	2.16	B	B
ab	5.06	8.20	18.50	33.83	0.15	- 0.30	AB	AB
$c\ (f)$	4.10	9.21	18.99	0.99	- 0.81	- 1.34	C	C
ac	4.76	8.23	17.65	3.26	2.97	2.24	AC	AC
bc	4.45	9.07	16.75	1.14	0.03	- 0.18	BC	BC
$abc\ (f)$	3.75	9.43	17.08	- 0.99	- 0.33	- 3.72	ABC	ABC^a
$d\ (f)$	4.51	8.89	1.03	- 0.19	- 0.33	- 2.57	D	D
ad	4.70	10.10	- 0.04	- 0.62	- 1.01	0.14	AD	AD
bd	3.39	9.36	1.64	0.14	1.05	2.26	BD	BD
$abd\ (f)$	4.84	8.29	1.62	2.83	1.05	- 2.48	ABD	ABD
cd	4.09	8.14	0.55	- 1.07	0.21	4.12	CD	CD
$acd\ (f)$	4.98	8.61	0.59	1.10	- 0.32	0.02	ACD	ACD
$bcd\ (f)$	4.35	7.36	- 1.07	2.16	- 3.07	6.64	BCD	BCD
$abcd$	5.08	9.72	0.08	- 2.49	- 0.65	- 4.86	$ABCD$	EF
$e\ (f)$	4.30	0.37	0.47	- 1.39	0.43	- 0.98	E	E
ae	4.59	0.66	- 0.66	1.06	- 3.00	- 4.10	AE	AE
be	4.92	0.66	- 0.98	- 1.34	2.27	3.78	BE	BE
$abe\ (f)$	5.18	- 0.70	0.36	0.33	- 2.13	- 0.36	ABE	ABE
ce	5.08	0.19	1.21	1.07	- 0.43	- 0.68	CE	CE
$ace\ (f)$	4.28	1.45	- 1.07	- 0.02	2.69	0.14	ACE	ACE
$bce\ (f)$	3.45	0.89	0.47	0.04	2.17	- 0.53	BCE	BCE
$abce$	4.84	0.73	2.36	1.15	- 4.65	2.42	$ABCE$	DF
de	4.29	0.29	0.29	- 1.13	2.45	- 2.57	DE	DE
$ade\ (f)$	3.85	0.26	- 1.36	1.34	1.67	- 4.40	ADE	ADE
$bde\ (f)$	4.62	- 0.80	1.26	- 2.28	- 1.09	3.12	BDE	BDE
$abde$	3.99	1.39	- 0.16	1.89	1.11	- 6.82	$ABDE$	CF
$cde\ (f)$	2.99	- 0.44	- 0.03	-1.65	2.47	- 0.78	CDE	CDE
$acde$	4.37	- 0.63	2.19	- 1.42	4.17	2.20	$ACDE$	BF
$bcde$	5.51	1.38	0.19	2.22	0.23	1.70	$BCDE$	AF
$abcde\ (f)$	4.21	- 1.30	- 2.68	- 2.87	- 5.09	- 5.32	$ABCDE$	F

^{a} (block)

11th pair = 4.76 - 4.10 = 0.66

.

.

.

16th pair = 4.21 - 5.51 = -1.30

3. Compute the values for t_2 following the same procedures outlined in the previous step, except now we use t_1 instead of t_0 in our computation. Apply the same process to compute t_3, t_4, and t_5, each time using the t values of the previous step. With n factors, the process is continued until we reach column (n).

STEP 8: Label each of the factorial effect totals computed in the last step with the letters of the corresponding treatments, and place them under the heading "Contrast" as shown in column 8 of Table 6.4.1.5. Note that the first value of column 7 or t_5 is the grand total (G). The second value refers to the treatment combination a (f). Given that factor F was ignored in arranging the treatment combinations, this value is therefore assigned to the A main effect. The assignment of the ignored factor was discussed earlier. Proceed with the identification of each of 32 specific factorial effects.

STEP 9: Identify the aliases for the factorial effects and place them under the heading "Alias" in column 9 of Table 6.4.1.5. As was explained earlier, when we reintroduce the ignored factor we have to identify its main and interaction effects for estimation purposes. The main effect of the ignored factor is F, and its interactions are AF, BF, CF, DF, and EF. The aliases for these effects are $F = ABCDE$, $AF = BCDE$, $BF = ACDE$, $CF = ABDE$, $DF = ABCE$, and $EF = ABCD$. Since we used blocking in this experiment, $ABC = DEF$ is lost by confounding.

STEP 10: Calculate the sum of square for the main effect, the 2-factor interactions, the 3-factor interactions, and the error, using the values given in column 7 of Table 6.4.1.5, in the following manner:

$$\text{Main effect } SS = \frac{(A)^2 + (B)^2 + (C)^2 + (D)^2 + (E)^2 + (F)^2}{(r)(2)^k} \tag{6-36}$$

$$= [(4.40)^2 + (2.16)^2 + (-1.34)^2 + (-2.57)^2 + (-0.98)^2$$
$$+ (-5.32)^2]/(2)(32)$$

$$= 61.69/64$$

$$= 0.96$$

$$2\text{-factor interaction} SS = \left[(AB)^2 + (AC)^2 + ... + (BF)^2 + (AF)^2\right]/(r)(2)^k \quad (6\text{-}37)$$

$$= \left[(-0.30)^2 + (2.24)^2 + ... + (2.20)^2 + (1.70)^2\right]/(2)(32)$$

$$= 149.13/64$$

$$= 2.33$$

$$3\text{-factor interaction} SS = \left[(ABD)^2 + (ACD)^2 + ... + (BFE)^2 + (AFE)^2\right]/(r)(2)^k$$

$$= 80.37/(2)(32) \quad (6\text{-}38)$$

$$= 1.26$$

To compute the error sum of squares, we subtract all the sum of squares from the total sum of squares computed in Step 5. This is shown below:

$$\text{Error } SS = \text{Total } SS - (\text{the sum of all other } SS) \quad (6\text{-}39)$$

$$= 5.85 - (0.07 + 0.17 + 0.05 + 0.96 + 2.33 + 1.26)$$

$$= 5.85 - 4.84$$

$$= 1.01$$

STEP 11:   Calculate the mean square for each source of variation as:

Replication MS $= \text{Replication } SS /1$

$= 0.07/1$

$= 0.07$

Block MS $= \text{Block } SS / 1$

$= 0.17/1$

$= 0.17$

Block $\times$ Replication MS $= \text{Block} \times \text{Rep. } SS / 1$

$= 0.05/1$

$$= 0.05$$

Main effect MS $= $ Main effect SS /6

$$= 0.96/6$$

$$= 0.16$$

2-factor interaction MS $= $ 2-factor interaction SS /15

$$= 2.33/15$$

$$= 0.16$$

3-factor interaction MS $= $ 3-factor interaction SS / 9

$$= 1.26/9$$

$$= 0.14$$

Error MS $= $ Error SS / 30

$$= 1.01/30$$

$$= 0.03$$

STEP 12: Compute the F value for each source of variation as shown in Table 6.4.1.6.

STEP 13: Compare the computed F ozf Table 6.4.1.6 with the tabular F value given in Appendix E. The results indicate that the main effects and all interactions were highly significant. Similarly, the results show that the block effect was significant at the 5% level, implying that there are differences in the treatment combinations among blocks. It should be noted once again that without the use of replications in the experiment, it would not have been possible to estimate the 3-factor interactions. Therefore, in experiments where 3-factor interactions may not be the primary concern of the experimenter, the experiment could be conducted without replications, following the same procedures outlined above.

In summary, there are several points that should be considered when using fractional replication. First, the experimenter would have to make a choice of

TABLE 6.4.1.6.

Analysis of Variance of Daily Weight Gain Data From a Fractional Replication Design

Source of Variation	Degree of Freedom	Sum of Squares	Mean Square	F
Replication	1	0.07	0.07	2.33^{ns}
Block	1	0.17	0.17	$5.67*$
Block × replication	1	0.05	0.05	1.67^{ns}
Main effect	6	0.96	0.16	$5.33**$
2-factor interaction	15	2.33	0.16	$5.33**$
3-factor interaction	9	1.26	0.14	$4.67**$
Error	30	1.01	0.03	
Total	63	5.85		

Note: ^{ns} = nonsignificant, * = significant at 5% level, ** = significant at 1% level.

the number of levels for factors to be included in the experiment. Generally, use is made of 2 levels, with the exception of the cases where fitting a second-degree response surface for a quantitative factor is desired. Second, experimenters often desire to know the direction and amount of the main effect of a factor and whether there is interaction between 2 factors. Under such circumstances, use of a 2-level design that estimates the main effect free of a 2-factor or a 3-factor interaction is warranted. Third, when use is made of smaller designs, estimation of the residual standard deviation is quite difficult—and at times impossible—from the observations. In such conditions, knowledge of a prior estimate is desirable. This is not a problem with larger designs where an estimate can be found if 3-factor or higher-order interactions are negligible. Fourth, doubtful results may be obtained because of the presence of aliases. When there is difficulty in discerning some important effects, further investigation is a desirable alternative. In some circumstances it may prove helpful to conduct an experiment of similar size to the initial one, perhaps converting a 1/16th-replicate into a 1/8th-replicate to provide for a more favorable alias structure and, in particular, separating out the pair or pairs of contrasts that have given rise to interpretation difficulties. Finally, fractional replication designs allow for the inclusion of many factors that an experimenter may deem appropriate, based on prior knowledge and theoretical considerations. Such flexibility in the design is the hallmark of fractional replication designs.

REFERENCES AND SUGGESTED READINGS

Brownlee, K. A., Kelly, B. K., and Loraine, P. K. 1948. "Fractional replication arrangement for factorial experiments with factors at two levels." B*iometrika* 35: 268-276.

Finney, D. J. 1945. "The fractional replication of factorial arrangements." *Ann. Eugen* 12:291-301.

Kempthorne, O. 1952. *The Design and Analysis of Experiments*. New York: John Wiley & Sons.

Kempthorne, O. and Tischer, R. G. 1953. "An example of the use of fractional replication." *Biometrics* 9:295-303.

Cochran, W. G. and Cox, G. M. 1957. *Experimental Designs*. 2nd ed. New York: John Wiley & Sons, chap. 6A.

Gomez, K. A., and Gomez A. A. 1984. *Statistical Procedures for Agricultural Research* 2nd ed., New York: John Wiley & Sons, chap. 4.

Petersen, R. G. 1985. *Design and Analysis of Experiments,* New York: Marcel Dekker, chap. 8.

EXERCISES

1. A crop scientist is interested in using a split-split-plot design experiment where there are 3 factors of interest. He wishes to assign the 4 planting dates (P_1, P_2, P_3, P_4) to the main plot arranged in randomized complete blocks (I, II, III, and IV). Subplots are to be Fumigated (F_1) and nonfumigated (F_2) for mite control. The 4 harvest dates (H_1, H_2, H_3, H_4) are to be assigned to the sub-subplots. Show how you would lay out this experiment in the field.

2. Certain varieties of cotton such as Acala SJ-2 show potassium deficiency even when the soils, by test, do not show any deficiency. This problem is severe on soils with high levels of Verticillium wilt. In a 2 × 2 × 3 split-split-plot design experiment to determine whether potassium deficiency in cotton is disease-induced, the researchers have assigned the fertilizers (NPK) to the main plot, the treatments to the subplots, and the depth soil as a sub-subplot factor. They have collected soil samples from 2 depths in fumigated and nonfumigated plots and recorded the following data.

Effect of soil fumigation on levels of N, P, and K at 2 sampling depths.

| | Parts Per Million | | | | | | | | |
| | N | | | P | | | K | | |
Treatment	Rep.1	Rep.2	Rep.3	Rep.1	Rep.2	Rep.3	Rep.1	Rep.2	Rep.3
				0-12 in. depth					
Fumigated	30.2	30.3	31.0	24.2	24.5	23.8	92.0	94.0	93.3
Nonfumigated	68.0	67.8	68.2	19.0	19.3	19.5	104.8	105.1	104.6
				12-24 in. depth					
Fumigated	28.2	30.3	28.5	20.2	21.5	20.8	89.0	88.4	88.7
Nonfumigated	30.2	29.8	30.1	17.9	18.3	18.5	93.8	93.0	92.9

(a) Do a main-plot analysis.
(b) Do a subplot analysis.
(c) Do a sub-subplot analysis.
(d) Compute the coefficient of variation.

3. An environmental horticulturist is interested in finding out whether (1)
stress-adapted landscapes save water, (2)whether irrigation equal to 15%
or less reference evapotranspiration (ET_0) can be applied to established
shrubs and ground cover without any drought related injury. This is a 3-
factor experiment designed to test the effect of 3 irrigation regimes (no
irrigation, 12.0 inches, and 24.0 inches of water) and 2 different irrigation
methods (drip and furrow) on the growth of shrubs and ground covers such
as Xylosma, Oleander, Coton-easter, Juniper, Ice plant, and Hedera. The
experiment is strip-split-plot design replicated 3 times. The data collected
at the end of a 2-year period is given below.

Growth of shrubs and ground cover as a function of irrigation water received from
April to August.

Plantings	Irrigation Method	Inches of Water Applied	Growth in inches Rep. I	Rep. II	Rep. III
Xylosma	Drip	0.0	8.0	8.4	9.5
"	"	12.0	19.5	20.1	20.2
"	"	24.0	30.6	31.0	31.4
"	Furrow	0.0	6.0	5.4	5.8
"	"	12.0	12.8	16.9	17.4
"	"	24.0	28.2	27.6	29.4
Oleander	Drip	0.0	18.0	19.4	19.5
"	"	12.0	39.5	40.1	40.3
"	"	24.0	60.6	59.0	61.4
"	Furrow	0.0	16.0	15.4	15.7
"	"	12.0	22.8	36.9	37.4
"	"	24.0	48.2	47.6	49.4
Coton easter	Drip	0.0	6.0	6.4	6.5
"	"	12.0	35.5	31.1	30.6
"	"	24.0	40.6	41.0	41.3
"	Furrow	0.0	4.0	4.4	4.8
"	"	12.0	19.8	16.9	18.4
"	"	24.0	25.2	27.6	29.5
Juniper	Drip	0.0	12.0	12.4	12.7
"	"	12.0	10.5	10.1	10.2
"	"	24.0	20.6	18.0	19.3
"	Furrow	0.0	10.0	11.1	10.8
"	"	12.0	9.8	6.9	7.4
"	"	24.0	13.2	14.8	15.4
Ice Plant	Drip	0.0	22.0	18.8	19.5

Table Continued

Plantings	Irrigation Method	Inches of Water Applied	Growth in inches		
			Rep. I	Rep. II	Rep. III
"	"	12.0	26.5	28.1	27.2
"	"	24.0	33.6	33.0	32.4
"	Furrow	0.0	16.0	15.4	15.8
"	"	12.0	22.5	26.9	25.4
"	"	24.0	28.8	29.6	29.9
Hedera	Drip	0.0	16.0	17.4	18.5
"	"	12.0	20.5	22.1	20.7
"	"	24.0	40.6	41.0	41.4
"	Furrow	0.0	12.0	13.2	14.6
"	"	12.0	15.8	14.9	14.6
"	"	24.0	28.2	27.6	29.4

(a) Perform the analysis of variance.
(b) What conclusions can you draw from the analysis?
(c) Compute the coefficient of variation for the factors.

4. Irrigation specialists have found that subsurface drip irrigation has a number of potential advantages over conventional surface irrigation. However, the depth at which a subsurface irrigation system is placed has implications for seed germination. The researcher is interested in the depth of burial of the subsurface drip tape, the depth of seed placement, and irrigation frequency reference evapotranspiration (ET_0). Using a strip-split-plot design experiment with melon seeds, the following data were collected when the experiment was replicated 4 times:

Number of emerged seedlings out of 100 seeds planted 14 days after initial irrigation

Planting Depth (in.)	Burial Depth of Irrigation Tape	Amount of Water Applied	Number of Emerged Seedlings			
			Rep. I	Rep. II	Rep. III	Rep. IV
0.5	6	0.25 ET_0	35	32	34	36
1.5	"	"	53	54	52	56
2.5	"	"	1	0	0	2
0.5	9	"	40	38	41	36
1.5	"	"	38	37	35	38

Table Continued

Planting Depth (in.)	Burial Depth of Irrigation Tape	Amount of Water Applied	Number of Emerged Seedlings			
			Rep. I	Rep. II	Rep. III	Rep. IV
2.5	"	"	0	0	1	1
0.5	12	"	9	10	12	8
1.5	"	"	30	32	28	30
2.5	"	"	0	0	0	0
0.5	6	0.50 ET$_0$	58	60	62	58
1.5	"	"	45	47	48	52
2.5	"	"	6	4	5	3
0.5	9	"	40	38	42	36
1.5	"	"	58	57	55	58
2.5	"	"	3	1	5	4
0.5	12	"	30	30	32	28
1.5	"	"	60	52	58	60
2.5	"	"	5	6	3	4
0.5	6	0.75 ET$_0$	50	50	52	46
1.5	"	"	55	57	58	52
2.5	"	"	1	1	2	2
0.5	9	"	48	48	44	46
1.5	"	"	45	47	42	40
2.5	"	"	4	3	1	2
0.5	12	"	28	30	31	26
1.5	"	"	36	34	38	32
2.5	"	"	2	2	4	2

(a) What do you conclude from the analysis?
(b) What do the coefficients of variation in this problem show?

5. In a fractional factorial design experiment (half-replicate), a crop scientist is interested in knowing the impact of 5 different fertilizers each at 2 levels on yield of corn. The scientist wishes to have 2 blocks in each of the 2 replications. Show how you would lay out the experiment for the scientist.

6. Suppose that in the previous example, the data collected from the experiment are as shown below.

Corn yield from a 2^5 factorial experiment conducted as a half-replicate with 2 blocks each containing 2 replicates.

| | Grain Yield (bu/ac) | | | Grain Yield (bu/ac) | |
| | Block 1 | | | Block 2 | |
Treatment	Rep. I	Rep. II	Treatment	Rep. I	Rep. II
(1)	132	125	ac	112	114
ae	151	145	ad	122	132
ab	136	138	bc	145	148
be	144	142	bd	161	158
acde	171	162	de	145	138
cd	154	159	ce	162	168
bcde	143	138	abde	144	140
abcd	132	136	abce	155	159

(a) Perform the analysis of variance.

(b) Which one of the main effects and interactions were highly significant?

Chapter 7

TREATMENT MEANS COMPARISONS

7.1 INTRODUCTION

Researchers have at their disposal various methods to compare means of treatments from experiments. Deciding which treatment means are significantly different is called *mean separation*. In conducting experiments to determine which treatment mean is significant, it is essential to keep in mind that the best method is one that meets the objectives of the experiment, is simple, and does not require assumptions that are difficult to meet. Given these criteria, a researcher can design an appropriate experiment that compares treatment means and provides answers to his/her queries.

Researchers often use 2 or more factors in a factorial experiment, to determine the average effect of the factors over a wide range of conditions. For example, a horticulturist may be interested in knowing how a variety of beans responds to the different rates of fertilizer application. Depending on what question the investigator is trying to answer, the experiment can take several forms. For instance, if the horticulturist is interested in knowing whether there is any response to the fertilizer, such a question can easily be answered by comparing the mean yield from the fertilized fields with the mean yield from the control plots. If, on the other hand, the horticulturist wants to know if there are differences between the different rates of fertilizer applications, the mean yield of each treatment (rate of application) can be compared with the others to determine the best among them.

Additional factors can be introduced in this simple mean comparison example to illustrate the diversity of the mean comparisons. For instance, if the horticulturist decides to introduce an added criterion, such as band or broadcast application of fertilizer, in conducting the experiment, the question that the experimenter is asking is whether there is a difference in yield if fertilizer is applied through band application or broadcast. A comparison of means between those fields with band and broadcast applications should provide the answer.

The above comparisons are classified as either *paired* or *group comparisons*. Paired comparisons are the simplest and are used extensively in agricultural research. Group comparisons are made when treatments are classified according to some common characteristics. Because planned group comparisons allow for a more precise mean separation than multiple comparison tests, the *F* test is a much more meaningful test among the groups. Once it is determined that a significant *F* exists, we then ask which of the mean values are significantly different. The ordinary *t* test for the difference between means is applied to every pair of means. To save time in such a comparison, the least significant difference is computed. This will be explained in the next section.

7.2 COMPARISONS OF PAIRED MEANS

As was mentioned earlier, agricultural researchers often use this simple method of comparing treatment means. Paired comparisons can be carried out either as *planned* or *unplanned*. When planned, the researcher identifies, prior to conducting the experiment, a specific pair of treatments to be compared . The simplest comparison of this kind usually involves comparing each of the treatments with the control treatment. In the unplanned comparison, on the other hand, the researcher attempts to compare every possible pair of treatment means instead of choosing a particular set of treatments to be compared.

The procedures used in making a paired comparison are the *least significant difference test* (*LSD*) and the *Duncan's multiple range test* (*DMRT*). When using a planned paired comparison, *LSD* is more appropriate, whereas in an unplanned paired comparison, use should be made of the *DMRT*.

7.2.1 Least Significant Difference Test

The least significance test is a form of *t* test that is used to determine whether the difference between 2 treatment means is significant at a prescribed level of α. The *LSD*, while a suitable test for such a comparison, should not be used for comparing all possible means—especially when the *F* test is not significant. Its indiscriminate use can lead to misleading interpretation of results. Appropriately, *LSD* should be used for planned comparisons when the number of comparisons is not large. As the number of comparisons increases, the chance of spuriously significant results increases. This means that the researcher will observe some differences that are significant, but not at the level of significance chosen. It has been pointed out by Cochran and Cox (1957) that . . .if several *t*-tests have been performed, the probability that at least one of these is apparently significant is greater than 0.05. If the *t*-test are independent, this probability is 0.23 for 5 tests, 0.40 for 10 tests, and 0.64 for 20 tests.

What this means is that comparison between means further apart than any 2 in an array will be made at lower levels of significance. Therefore, it is wise not to use *LSD* for comparing all possible pairs of means. Gomez and Gomez (1984) suggested that such comparisons should be limited to less than 6 for the results to be meaningful.

As an *LSD* test is similar to a *t* test, the test for the statistical significance of the difference between 2 means follows the same reasoning. Suppose we have 2 treatments *i* and *j*, and are interested in using a *t* test to compare the difference between their means; we would have:

$$t = \frac{\bar{d} - \mu_{\bar{d}}}{s_{\bar{d}}} \tag{7-1}$$

where

$\bar{d} = \bar{X}_1 - \bar{X}_2$ or the difference between two means

$\mu_{\bar{d}}$ = population of mean differences where the mean is 0

$s_{\bar{d}}$ = the standard error of the mean difference.

To perform the *t* test using *LSD*, we would simply replace $\bar{d}$ with *LSD* and $\mu_{\bar{d}}$ with 0 (as it is assumed that the mean from a population of mean differences is 0), and our equation becomes:

$$t_\alpha = \frac{LSD - 0}{s_{\bar{d}}} \tag{7-2}$$

or

$$t_\alpha = \frac{LSD}{s_{\bar{d}}} \tag{7-3}$$

Using Equation 7-3 to solve for *LSD*, we will have:

$$LSD = (t_\alpha)(s_{\bar{d}}) \tag{7-4}$$

We can use Equation 7-4 to test the *LSD* at a prescribed level of significance. The level of significance (α) serves as the boundary between significant
and nonsignificant differences between any pair of treatment means. Once the
level of significance attached to the test is identified, Equation 7-4 can be
written as:

$$LSD_\alpha = (t_\alpha)(s_{\bar{d}}) \qquad\qquad\qquad (7\text{-}5)$$

where

$$t_\alpha = \text{the tabular } t \text{ value for a level of significance } (\alpha) \text{ with}$$
$$n = \text{error degrees of freedom}$$
$$s_{\bar{d}} = \sqrt{(s_1^2/r_1) + (s_2^2/r_2)}.$$

In computing the *LSD* at a prescribed level of significance (α), the 2
treatments are considered to be significantly different if their difference exceeds the computed *LSD* value. On the other hand, if the difference between
the treatments is less than the computed *LSD*, the results are not significantly
different. In using the *LSD* procedure, note should be taken that the appropriate
standard error of estimate ($s_{\bar{d}}$) is used. Depending on the design of the experiment, and the types of means compared, the standard error of estimate to be
used varies. The steps in computing the *LSD* are illustrated with the following
examples using different experimental design, and comparing different types
of treatment means.

Complete Block Design: In a complete block design where the experiment
may be a completely randomized design (CRD), a randomized complete block
(RCB), or a Latin square where only 1 error term is involved, the standard error
is computed as:

$$s_{\bar{d}} = \sqrt{\frac{2s^2}{r}} \qquad\qquad\qquad (7\text{-}6)$$

where

$$s^2 = \text{the error mean square } (MSE) \text{ in the analysis of variance}$$

$$r = \text{the number of replications common to both treatments}$$
$$\text{in the pair.}$$

On the other hand, if the treatments have different numbers of replications, then the standard error of estimate is computed as:

$$s_{\bar{d}} = \sqrt{\left(\frac{s^2}{r_1}\right) + \left(\frac{s^2}{r_2}\right)} \qquad (7\text{-}7)$$

We will illustrate the case with equal and unequal numbers of replications, using the data from previous examples.

Equal Replications: The data from a completely randomized experiment involving 7 treatments with a control (Example 4.2.1.2) are used to determine if 1 or more of the 7 vitamin supplementation treatments is better than the control treatment. The steps in performing the *LSD* test are given below.

STEP 1: Compute the mean difference between the control and each of the 7 treatments as shown in Table 7.2.1.1.

TABLE 7.2.1.1.

Mean Weight Gain Differences Between the 7 Treatments and the Control Treatment

Treatment	Treatment Mean (kg)	Difference from Control (kg)
A	4.94	0.64^{ns}
B	5.02	0.72^{ns}
C	6.58	2.28**
D	5.24	0.94^{ns}
E	4.44	0.14^{ns}
F	6.54	2.24**
G	5.18	0.88^{ns}
H (Control)	4.30	-

ns = nonsignificant, ** = significant at the 1% level.

STEP 2: Compute the *LSD* value at the 5% and 1% levels of significance. Since the error mean square for this example was 0.61 with 32 error degrees of freedom and 5 replications, the tabular t value from

Appendix I is 2.041 for the 5% level and 2.741 for the 1% level of significance. Thus, we have:

$$LSD_{.05} = (t_{.05}) \sqrt{\frac{2s^2}{r}}$$

$$= 2.041 \sqrt{\frac{2(0.61)}{5}}$$

$$= 1.00$$

$$LSD_{.01} = 2.741 \sqrt{\frac{2(0.61)}{5}}$$

$$= 1.34$$

STEP 3: Make a comparison of each of the mean differences computed in Step 1 with the computed *LSD* values in Step 2. It appears that only treatments C and F are significant, implying that these 2 treatments show increased weight gains while the other 5 treatments were not significantly different from the control treatment.

Unequal Replications: Researchers often are faced with situations where, for previously unseen reasons, elements of the experimental unit are destroyed or lost in the process. In such conditions, the researcher may not have equal numbers of replications in the experiment. When we have unequal numbers of replications in an experiment, the appropriate comparison is the planned-pair comparisons. Using the data from an unequal replication experiment (Example 4.2.1.3 in Chapter 4) where response to nitrogen fertilization on different maize hybrids was studied in a completely randomized design, the agronomist wishes to determine if any of the 5 hybrids (treatments) performed better than the control treatment.

STEP 1: Compute the mean difference between each of the 5 treatments and the control as shown in column 4 of Table 7.2.1.2.

STEP 2: Compute the *LSD* value at the 5% and 1% level of significance. As some treatments have 2 replications and others have 3, we must compute 2 sets of the *LSD* values. Since the error mean square for this example was 4,966.72 (see Table 4.2.1.6) with 9 error degrees of freedom, the tabular *t* value from Appendix I is 2.262 for the 5% level and 3.250 for the 1% level of significance. Thus, the *LSD* values are computed as:

A. Treatments with 3 replications compared with the control:

TABLE 7.2.1.2.

Mean Yield Differences Between the 5 Hybrids (Treatments) and the Control Treatment

Treatment	Number of Replications	Mean Yield bu/ac	Difference Control (bu/ac)	LSD Values 5%	1%
P3747	3	154.0	12.0^{ns}	130	187
P3732	3	145.7	3.7^{ns}	130	187
Mo17 × A634	2	149.0	7.0^{ns}	145.5	209
LH 74 × LH51	2	166.5	24.5^{ns}	145.5	209
CP18 × LH54	2	140.5	-1.5^{ns}	145.5	209
Control	3	142.0	–	–	–

Note: *ns* = non significant.

$$LSD_{.05} = 2.262 \sqrt{\frac{2(4,966.72)}{3}}$$

$$= 130 \text{ bu/ac}$$

$$LSD_{.01} = 3.250 \sqrt{\frac{2(4,966.72)}{3}}$$

$$= 187 \text{ bu/ac}$$

B. Treatments with 2 replications compared with control: as the number of replications of the control mean is 3 and each of the treatments has only 2 treatments, we use Equation 7-5 and substitute Equation 7-7 for the standard error of estimate in it. The *LSD* values are computed as follows:

$$LSD_\alpha = t_\alpha \sqrt{\left(\frac{s^2}{r_i}\right) + \left(\frac{s^2}{r_j}\right)} \qquad (7\text{-}8)$$

$$LSD_{.05} = 2.262 \sqrt{\left(\frac{(4,966.72)}{2}\right) + \left(\frac{(4,966.72)}{3}\right)}$$

$$= 145.5 \text{ bu/ac}$$

$$LSD_{.01} = 3.250 \sqrt{\left(\frac{(4,966.72)}{2}\right)\left(\frac{(4,966.72)}{3}\right)}$$

$$= 209 \text{ bu/ac}$$

STEP 3: Make a comparison of each of the mean differences computed in Step 1 with the computed *LSD* values in Step 2. Indicate by an appropriate asterisk or the "nonsignificant" sign (*ns*) whether the treatments are different from the control. From our analysis, it appears that none of the treatments are significantly better than the control treatment.

Balanced Lattice Design: Recall from Chapter 4 (Section 4.3.1) that in these designs every pair of treatments occurs once in the same incomplete block. Hence, the degree of precision in comparing all pairs of treatments is the same. In the analysis of variance, we computed the total, the treatment, and the replication sum of squares in the usual manner. However, the sum of squares for blocks within replications was adjusted for treatment effects. As we apply the *LSD* test to the data from a balanced lattice design, we must use the adjusted treatment means in computing the mean difference, and the effective error mean square to compute the standard error of the mean difference.

To illustrate the *LSD* test, we use the data from a 3 x 3 balanced lattice design experiment where each treatment is replicated 4 times. The researcher has gathered the following data on alfalfa forage yield. He wishes to know if there are any differences in yield between the 8 treatments and the control. In performing the analysis of variance, the researcher has found the effective error mean square with 16 degrees of freedom to be 2.1. The following adjusted mean data are reported and are presented in Table 7.2.1.3. The steps in performing the *LSD* test are given below.

TABLE 7.2.1.3.
The Adjusted Mean Yield of Alfalfa Forage from First-Harvest in Oklahoma

Treatment Number	Adjusted Mean ton/acre	Difference from Control ton/acre
1	3.69	2.19**
2	2.61	1.11^{ns}
3	2.42	0.92^{ns}
4	1.84	0.64^{ns}
5	1.97	0.47^{ns}
6	2.58	1.08^{ns}
7	1.79	0.29^{ns}
8	3.74	2.24**
9 (Control)	1.50	—

Note: ^{ns} = non significant. ** = significant at 5% level.

STEP 1: Compute the mean difference between the control and the 8 treatments, as shown in Column 3 of Table 7.2.1.3.

STEP 2: Calculate the *LSD* value at α level of significance as shown below:

$$LSD_{\alpha} = t_{\alpha} \sqrt{\frac{2(\text{effective error } MS)}{r}} \tag{7-9}$$

As the tabular *t* values from Appendix I with 16 degrees of freedom are 2.120 for the 5% and 2.921 for the 1% levels of significance, respectively, we have:

$$LSD_{.05} = 2.120 \sqrt{\frac{2(2.1)}{4}} = 2.17 \text{ ton/acre}$$

$$LSD_{.01} = 2.921 \sqrt{\frac{2(2.1)}{4}} = 2.99 \text{ ton/acre}$$

STEP 3: Compare the computed *LSD* values at the different levels of significance with the mean difference of each pair computed in Step 2. It appears that none of the treatments except treatments 1 and 8 are significantly better than the control treatment.

Partially Balanced Lattice Design: Performing the *LSD* test on the data from a partially balanced lattice design experiment is similar to the balanced lattice design, in that the computation of the mean difference requires the use of the adjusted mean, and the effective error mean square is used to compute the standard error of the mean difference. As we apply the *LSD* test to a partially balanced lattice experiment, we must remember there are 2 effective error mean squares to be considered. The first (MS_1) involved treatments that were tested in the same incomplete block, and the second (MS_2) dealt with the treatment pairs that never appeared in the same incomplete block. Thus, the *LSD* test for each would be:

$$LSD_{\alpha} = t_{\alpha} \sqrt{\frac{2(\text{effective error } MS_1)}{r}} \tag{7-10}$$

$$LSD_{\alpha} = t_{\alpha} \sqrt{\frac{2(\text{effective error } MS_2)}{r}} \tag{7-11}$$

For the purpose of illustrating the procedure for testing the *LSD*, we will take the data from Table 4.3.2.3 and assume that treatment number 15 is the control treatment, and that the researcher wishes to determine if any of the hybrid corns is significantly better than the control.

Before we start performing the *LSD* test, we have to adjust the data that were presented in Table 4.3.2.3. First, to compute the adjusted treatment means, we use the following formula:

$$\mu_a = \frac{T_a}{k+1} \tag{7-12}$$

where

$$\mu_a = \text{adjusted treatment mean}$$

$$T_a = \text{adjusted treatment total}$$

$$k+1 = \text{the number of replications } (r).$$

As Equation 7-11 calls for the adjusted treatment totals, we must convert the treatment totals given in Table 4.3.2.3 as follows:

$$T_a = T + \mu \sum C_b \tag{7-13}$$

where T is the treatment total, and C_b is the difference between the sum of treatment totals for all treatments appearing in a block and the product of the number of replications and block totals ($C_b = \Sigma T - nB_T$). The C_b value for our example was computed and presented in Table 4.3.2.4. The adjusted treatment totals are computed using Equation 7-13, and the results are presented in Table 7.2.1.4. With this information, we are ready to perform the *LSD* test. The step-by-step procedures are outline below.

STEP 1: Calculate the mean difference between the control treatment (treatment number 15) and each of the 24 hybrid corns. The mean difference is shown in column 5 of Table 7.2.1.4.

STEP 2: Calculate the *LSD* values, using the 2 effective error means MS_1 and MS_2. These values were computed in Chapter 5 as $MS_1 = 50.8$, and $MS_2 = 51.5$. The tabular t values from Appendix I with $n = 56$ degrees of freedom are 2.004 for 5% and 2.397 for 1% levels of significance, respectively. Thus, the *LSD* for comparing 2 corn hybrids tested in the same incomplete block at these 2 levels are:

TABLE 7.2.1.4.

Comparison of the Adjusted Treatment Means with the Control Treatment (Treatment 15)

Treatment Number[a]	Treatment Total (T)	Adjusted Treatment Total (T_a)	Adjusted Treatment Mean (μ_a)	Difference from Control
1	119	119.29	29.82	18.33**
2	72	72.29	17.82	6.33^{ns}
3	121	121.29	30.32	18.83**
4	66	66.29	16.57	5.08^{ns}
5	82	82.29	20.57	9.08^{ns}
6	71	70.62	17.66	6.17^{ns}
7	105	104.62	26.16	14.67**
8	70	69.62	17.41	5.92^{ns}
9	87	86.62	21.66	10.17*
10	92	91.62	22.91	11.42*
11	83	82.97	20.74	9.25^{ns}
12	112	111.97	27.99	16.50**
13	84	83.97	20.99	9.50^{ns}
14	60	59.77	14.99	3.50^{ns}
15	46	45.97	11.49	—
16	112	111.56	27.89	16.40**
17	83	82.56	20.64	9.15^{ns}
18	84	83.56	20.89	9.40^{ns}
19	90	89.56	22.39	10.90*
20	121	120.56	30.14	18.65**
21	80	80.56	20.14	8.65^{ns}
22	87	87.56	21.89	10.40*
23	104	104.56	26.14	14.65**
24	49	49.56	12.39	0.90^{ns}
25	106	106.56	26.64	15.15**

Note: ^{ns} = nonsignificant, * = significant at 1% level, ** = significant at 5% level. [a]Those treatments that were tested in the same incomplete block with the control are shown in bold.

$$LSD_1 = 2.004 \sqrt{\frac{2(50.8)}{4}}$$

$$= 10.10 \text{ bu/ac}$$

$$LSD_1 = 2.397 \sqrt{\frac{2(50.8)}{4}}$$

$$= 12.08 \text{ bu/ac}$$

For those corn hybrids that were not tested together in the same incomplete block, the *LSD* values are:

$$LSD_2 = 2.004 \sqrt{\frac{2(51.5)}{4}}$$

$$= 10.17 \text{ bu/ac}$$

$$LSD_2 = 2.397 \sqrt{\frac{2(51.5)}{4}}$$

$$= 12.16 \text{ bu/ac}$$

STEP 3: Compare each of the mean differences computed in Step 1 with the calculated *LSD* values. We use the LSD_1 for those pairs that were tested together in the same incomplete block, and the LSD_2 for those pairs of treatments that were not tested in the same incomplete block. The results of all pairs of comparison with the control treatment are shown with the appropriate signs to indicate their significance and nonsignificance in column 5 of Table 7.2.1.4.

Applying the *LSD* test to the next set of designs that involve 2 or more factors is as simple as those illustrated with the single factor. We will illustrate the application of the *LSD* test only for the split-plot design, as the procedures can be generalized to the strip-plot and the split-split-plot design easily. There are only 2 methodological adjustments that are necessary. One relates to the use of the appropriate standard error of the mean difference ($s_{\bar{d}}$), and the other is the use of the weighted *t* values in the calculation of the *LSD* values. Depending on the type of paired-comparison, the calculation of the standard error of the mean difference varies. When the standard error of the mean involves more than 1 error mean square, the appropriate *t* value to use will be the weighted *t* value. The condition which calls for the use of the weighted *t* values is explained in the next section when we discuss the split-plot design.

Split-Plot Design: Recall from Chapter 5 that in a split-plot design, the researcher is able to work with 2 variable factors in an experiment. The design allows the investigator to increase the precision in estimating certain effects while sacrificing certain other effects. Usually, the precision of the treatment assigned to the main plots is sacrificed for the increased precision of the average treatment effect assigned to the subplots. With this randomization of the treatments within a block, there are 2 error terms that the researcher will work with: the main-plot error and the subplot error. As one would expect, the main-plot error is usually larger as a result of the increased variability from widely spaced main plots, while the subplot error is smaller because of the reduced variability from closely spaced subplots within the main plot.

So in performing the *LSD* test on the data from a split-plot design, the researcher will work with 4 mean comparisons, as we have 2 factors and 2 error terms. The 4 mean comparisons and the standard error for the mean difference $s_{\bar{d}}$ are given in Table 7.2.1.5.

TABLE 7.2.1.5.

Standard Error of the Mean Difference for the 4 Types of Paired
Comparisons in a Split-plot Design

Paired Comparison	$s_{\bar{d}}$
A. 2 main-plot means averaged over all subplot treatments averaged over all vertical treatments.	$\sqrt{\dfrac{2E_a}{rb}}$
B. 2 subplot means averaged over all main-plot treatments, or 2 vertical means averaged over all horizontal treatments.	$\sqrt{\dfrac{2E_b}{ra}}$
C. 2 subplot means in the same main-plot treatment.	$\sqrt{\dfrac{2E_b}{r}}$
D. 2 main-plot means in the same or different subplot treatments.	$\sqrt{\dfrac{2\left[(b-1)E_b+E_a\right]}{rb}}$

In the above formulas, E_i or EMS_i refers to the error mean square of the ith factor, a is the number of main-plot treatments, and b is the number of subplot treatments. Whenever the standard error of the mean difference involves 2 error terms (as is the case in comparison D), we use the following equation to compute the weighted t values:

$$t_w = \frac{(b-1)EMS_b t_b + EMS_a t_a}{(b-1)EMS_b + EMS_a} \qquad (7\text{-}14)$$

To illustrate the *LSD* test for the split-plot design, let us use the data from Example 5.5.1, and compare the between-hybrid means with the same nitrogen rate. The steps are:

STEP 1: Calculate the standard error of the mean difference. As 2 subplot means within the same main plot are compared in this example, the formula to calculate the standard error of the mean difference is:

$$s_{\bar{d}} = \sqrt{\frac{2\,EMS_b}{r}} \qquad (7\text{-}15)$$

For our example, the error mean square (b), which is 29.99, is taken from Table 5.5.4 and is substituted in Equation 7-15 to get:

$$s_{\bar{d}} = \sqrt{\frac{2\,(29.99)}{2}}$$

$$= 5.48 \text{ bu/ac}$$

STEP 2: Calculate the *LSD* value at α level of significance, using Equation 7-5 as shown below:

$$LSD_{\alpha} = (t_{\alpha})(s_{\bar{d}}) \tag{7-5}$$

The tabular t values from Appendix I with $n = 16$ error b degrees of freedom are 2.120 for 5% and 2.921 for 1% levels of significance, respectively. Thus the *LSD* test is:

$$LSD_{.05} = (2.120)(5.48) = 11.62$$

$$LSD_{.01} = (2.921)(5.48) = 16.01$$

STEP 3: Compare the difference between 2 hybrid means at the same nitrogen fertilization rate with the *LSD* values. To do this, we need to construct a 2-way table of means that shows the mean yield of the hybrids over the 2 replications for each rate of fertilizer applied. This is shown in Table 7.2.1.6.

TABLE 7.2.1.6.

Mean Yields of 5 Corn Hybrids Tested with 4 Nitrogen Rates in a
Split-plot Design Experiment

Nitrogen Rate, lb/ac	Mean Yield from 2 Replications, bu/ac				
	P3747	P3732	Mo17 × A64	A632 × LH38	LH74 × LH51
0	132.5	137.5	122.5	122.5	130.0
70	160.0	155.0	147.5	145.0	175.0
140	180.0	170.0	160.0	167.5	177.5
210	175.0	182.5	162.5	155.0	185.0

We are now able to compare the difference between any 2 hybrid means at the same rate of nitrogen fertilization and the *LSD* value. The difference between the mean yield of hybrid P3747 and P3732 is 137.5 - 132.5 = 5.0. This yield difference is smaller than the *LSD* value of 11.62, implying the nonsignificance of the yield difference. On the other hand, the mean yield difference between hybrids P3732 and Mo17 × A64, as well as A632 × LH38, is 137.5 - 122.5 = 15. This yield difference is significant at the 5% level. The mean yield difference for the different hybrids at the same rate can be compared with the *LSD* values similarly, and a decision can be reached regarding the significance of the difference.

The procedure for performing the *LSD* test for the strip-plot and the split-split-plot is similar to the split-plot, except that the formulas for the standard error of the mean difference are different for the different comparisons. The reader is referred to Gomez and Gomez (1984), which provides a detailed list of the formulas for the standard error of the mean difference for the strip-plot and the split-split-plot designs.

7.2.2   Duncan's Multiple-Range Test (DMRT)

Among the several multiple-range tests, Duncan's test is the most widely used and misused in agriculture. The test is especially useful when the number of all possible pairs of treatment means compared is large and when several unrelated treatments are included in an experiment. However, its inappropriate use has been pointed out in the literature extensively (Maindonald and Cox, 1984; Bryan-Jones and Finney, 1983; and Morse and Thompson, 1981).

The principle behind the test is to take an ordered set of treatment means and test each group of 2, 3, or more successive means to determine whether the spread between the highest and the lowest of each group is significantly large in terms of the experimental error mean square, at the same time allowing for the number of treatments being tested in the group. Unlike other multiple comparison tests such as the Fisher's Significance Difference (*FSD*) test, the *DMRT* does not require that the *F* test be significant before making pairwise comparisons. Instead, the *DMRT* replaces the *F* test by a range test of the difference between the smallest mean in the set of *k* means. If this range is not significant, then none of the pairwise comparisons can be significant in terms of the *DMRT*.

As was stated in the beginning of Section 7.2.1, when several *t* tests have been performed and the *t* tests are independent, the probability that 1 of these tests is significant is reduced as the number of tests performed increases. This, of course, presents problems in reporting the results of a test. The same problem surfaces when several confidence intervals are computed. Duncan's multiple-range test tries to provide some level of protection against statements made in the summary of an experiment. Consider for a moment that a specified pair of true means are identical. As such, the combined *F*- and *t*-tests give a protection level of at least 95% that the means will be declared indistinguishable. Suppose now we have 3 identical true means. The protection level for these 3 means depends on the position of the fourth mean. If the fourth mean happens to be far removed from the 3 identical means in the array, the level of protection is about 87% (Cochran and Cox, 1957). Depending on how many means are compared and how far they are from each other, the level of protection varies. Duncan's multiple-range test specifies the level of protection that guards against erroneously finding significant results. The level of protection which has been specified by Duncan is 95% for sets of 2 means, 90.25% or $(0.95)^2$ for sets of 3 means, and 85.7% or $(0.95)^3$ for sets of 4 means, and so on. Even though there is a reduced level of protection in this test, the sensitivity of the test is greater than the Newman-Keuls test, which keeps all levels at 95% in detecting real differences.

To apply the Duncan's multiple-range test for comparing all pairs of means involving equal sample sizes, we arrange the k treatment averages in ascending (or descending) order, and compute the standard error of a mean as:

$$s_{\bar{d}} = \sqrt{\frac{2s^2}{r}} \tag{7-6}$$

In the case of unequal sample size (r), we replace the r in Equation 7-6 with the harmonic mean r_h of the (r_i), where

$$r_h = \frac{k}{\displaystyle\sum_{i=1}^{k} (1/r_i)} \tag{7-16}$$

In Equation 7-16 if $r_1 = r_2 = \ldots = r_k$, then $r_h = r$. Once we have computed the standard error of the mean, we then obtain from Appendix H the values of $r_\alpha(p, f)$, for $p = 2, 3, \ldots, k$, where α is the significance level, p is the number of means for range being tested, and f is the number of degrees of freedom for error. Convert these ranges into a set of $k - 1$ least significant ranges (R_p) for $p = 2, 3, \ldots, k$, by calculating

$$R_p = r_a(p, f)s_{\bar{d}} \qquad \text{for } p = 2, 3, \ldots, k \tag{7-17}$$

Now we are ready to test the observed differences between means by taking the difference between the largest and smallest mean and comparing it with the least significant range R_p. Next, the difference of the largest and the second smallest is computed and compared with the least significant range R_{k-1}. The comparisons are continued until all means have been compared with the largest mean. A pair of means is considered significantly different if the observed difference is greater than the corresponding least significant range. It should be kept in mind, however, that no differences between a pair of means is considered significant if the 2 means involved fall between 2 other means that do not differ significantly. What this means is that, for example, if among 6 means in an array (A B C D E F), we find A to be significantly different from B, and B not to be significantly different from E, then we draw a line joining B and E as shown below.

A   <u>B   C   D   E</u>   F

In the above situation it is not necessary to test C and D against E, as they fall within 2 other means that are not significantly different. Our next set of means to be compared will be D and F. If the difference is not significant, D and F will be joined by a line as shown below.

A   B   C   <u>D   E   F</u>

The steps in performing a Duncan's multiple-range test are given below, using the data from a previous example (Example 4.2.1.2), where the mean weight gain from vitamin supplementation of the 8 treatments with 5 replications were recorded as 4.94, 5.02, 6.58, 5.24, 4.44, 6.54, 5.18, and 4.30 kilograms. The error mean square was computed as 0.61.

STEP 1: Rank the treatment means in a descending order as shown in Table 7.2.2.1.

TABLE 7.2.2.1.

Rank Order of the Treatment Means from an Animal Weight-Gain Experiment Involving 8 Treatments and 5 Replications

Treatment	Treatment Mean (kg/animal)	Rank
C	6.58	1
F	6.54	2
D	5.24	3
G	5.18	4
B	5.02	5
A	4.94	6
E	4.44	7
H (Control)	4.30	8

STEP 2: Using Equation 7-6, compute the standard error of the mean as:

$$s_{\bar{d}} = \sqrt{\frac{2(0.61)}{5}} = 0.49 \, kg$$

STEP 3: From Appendix H, determine the $r_\alpha(p,f)$ for $\alpha = .05$ and 32 error degrees of freedom. Thus, we have:

$r_{.05}(2,32) = 2.88$

$r_{.05}(3,32) = 3.03$

$r_{.05}(4,32) = 3.11$

$r_{.05}(5,32) = 3.19$

$r_{.05}(6,32) = 3.24$

$r_{.05}(7,32) = 3.28$

STEP 4: Using Equation 7-17, determine the $(k - 1)$ values of the shortest significant range (R_p) as:

$R_2 = r_{.05}(2,32)s_{\bar{d}} = (2.88)(0.49) = 1.41$

$R_3 = r_{.05}(3,32)s_{\bar{d}} = (3.03)(0.49) = 1.48$

$R_4 = r_{.05}(4,32)s_{\bar{d}} = (3.11)(0.49) = 1.52$

$R_5 = r_{.05}(5,32)s_{\bar{d}} = (3.19)(0.49) = 1.56$

$R_6 = r_{.05}(6,32)s_{\bar{d}} = (3.24)(0.49) = 1.59$

$R_7 = r_{.05}(7,32)s_{\bar{d}} = (3.28)(0.49) = 1.61$

STEP 5: Compare the largest mean with the smallest, using the information provided in Table 7.2.2.1. If the difference between these means and the R_p value is equal to or greater, the means are significantly different. Repeat this process of comparison for the remaining treatment means as shown below.

C vs. H: 6.58 - 4.30 = 2.28 > 1.61 (R_7)

C vs. E: 6.58 - 4.44 = 2.14 > 1.59 (R_6)

C vs. A: 6.58 - 4.94 = 1.64 > 1.56 (R_5)

C vs. B: 6.58 - 5.02 = 1.56 > 1.52 (R_4)

C vs. G: 6.58 - 5.18 = 1.40 < 1.48 (R_3)

C vs. D: 6.58 - 5.24 = 1.34 < 1.41 (R_2)

F vs. H: 6.54 - 4.30 = 2.24 > 1.59 (R_6)

F vs. E: 6.54 - 4.44 = 2.10 > 1.56 (R_5)

F vs. A: 6.54 - 4.94 = 1.60 > 1.52 (R_4)

F vs. B: 6.54 - 5.02 = 1.52 > 1.48 (R_3)

F vs. G: 6.54 - 5.18 = 1.36 < 1.41 (R_2)

D vs. H: 5.24 - 4.30 = 0.90 < 1.56 (R_5)

STEP 6: Indicate statistical significance by line notation as shown below:

C	F	D	G	B	A	E	H
6.58	6.54	5.24	5.18	5.02	4.94	4.44	4.30

You will notice that in the above presentation there is no significant difference (at the 5% level) between treatments C and G, as well as between treatments D and H. Since treatments F and D are in a subset with nonsignificant range, it is not necessary to test F treatment against D and G treatments. Similarly, as there is no significant difference between the D and H treatments, it is not necessary to test all those treatments such as G, B, A, and E that are a subset of a nonsignificant range.

We can also use the alphabet notation to show the nonsignificance of the means compared. When using the alphabet notation to present the results, significant differences can be shown even if the means are not arrayed. Alphabet letters such as a , b , c, etc. are used to show that the means compared are not significantly different at the 5% level. For the present example we have:

C	F	D	G	B	A	E	H
6.58	6.54	5.24	5.18	5.02	4.94	4.44	4.30

a

b

7.3 COMPARISONS OF GROUPED MEANS

In the previous section we discussed 2 of the most widely used pairwise comparisons in agriculture. When an experiment involves $k \geq 3$ means and if we test all pairwise differences among means, we actually have $c = k (k - 1)/2$ comparisons or tests of significance to perform. In making these group comparisons, we are simply partitioning the treatment sum of squares into meaningful component comparisons that meet the experimental objectives. That is, the researcher must choose from a number of comparisons those sets of group comparisons that provide answers to the research questions. Once the desired group comparisons have been selected, the researcher then can compute the sum of squares for the components of each group and perform an F test, as was the case in the analysis of variance. The components may be orthogonal class comparisons or trend comparisons. In the following sections, we shall discuss these comparisons using the *orthogonal coefficients*.

7.3.1 Orthogonal Class Comparisons

A researcher, in making 2 comparisons such as d_i and d_j on the same set of k treatments, would have an *orthogonal* or *independent* comparison if the sum of the products of the corresponding coefficients for the d_i and d_j is equal to 0, that is, if

$$\sum_{i}^{k} a_{ki}a_{kj} = a_{1i}a_{1j} + a_{2i}a_{2j} + \ldots + a_{ki}a_{kj} = 0 \qquad (7\text{-}18)$$

Equation 7-18 states that i and j are distributed independently of each other. This independence is relevant to the partitioning of the treatment sum of squares. If the researcher wishes to divide the treatment sum of squares into the contribution from a comparison d_1 and the remainder, and wishes to further subdivide the remainder, he/she must choose comparisons that are orthogonal

251

Treatment Means Comparisons

to d_1. In this context, after removing the contribution of d_2, the next comparison d_3 must be orthogonal to both d_1 and d_2, and so on.

A group of comparisons are said to be *mutually orthogonal* if the sum of the products of the coefficients for all possible pairs of comparisons are equal to 0. In an experiment with k treatments, we can construct only $k - 1$ single degree of freedom mutually orthogonal contrasts. The sum of their sum of squares equals the treatment sum of squares as given below:

$$\text{Treatment}\,SS = SS\,(M_1) + SS\,(M_2) + \ldots + SS\,(M_{k-1})$$ (7-19)

where M_1, M_2, and M_{k-1} are mutually orthogonal single $d.f.$ contrasts. Each of the sum of squares on the right hand side of Equation 7-19 are also a mean square with 1 $d.f.$ Thus, in performing the F test, each of the treatment mean squares is divided by the error mean square.

To show how we construct a table of comparison coefficients, let us take 3 of the many comparisons that might be made on a set of $k = 4$ treatment means. Table 7.3.1.1 shows the comparisons and values of the coefficients and the notation for the coefficients.

TABLE 7.3.1.1.

Three Comparisons on $k = 4$ Treatment Means with Values of Coefficients for Each Comparison and Notation for the Coefficients

| | Values of Coefficients | | | | Notation for Coefficients | | | | |
Comparisons	$\bar{X}_1.$	$\bar{X}_2.$	$\bar{X}_3.$	$\bar{X}_4.$	$\bar{X}_1.$	$\bar{X}_2.$	$\bar{X}_3.$	$\bar{X}_4.$	$\sum a_i^2$
d_1	1	−1	0	0	a_{11}	a_{21}	a_{31}	a_{41}	2
d_2	0	0	−1	1	a_{12}	a_{22}	a_{32}	a_{42}	2
d_3	1/2	1/2	−1/2	−1/2	a_{13}	a_{23}	a_{33}	a_{43}	1

Comparisons such as those shown in Table 7.3.1.1 are linear functions of the treatment means. Any linear function of the treatment means such as the one in Equation 7-20 is called a *comparison* if at least 2 of the coefficients are not equal to 0 and if the sum of the coefficients is equal to 0. That is if $\Sigma a_i = 0$.

$$d_i = a_{1i}\bar{X}_1. + a_{2i}\bar{X}_2. + \ldots + a_{ki}\bar{X}_k.$$ (7-20)

In Table 7.3.1.1, the means of the treatments are given as $\bar{X}_1.$, $\bar{X}_2.$, $\bar{X}_3.$, and $\bar{X}_4.$. The numbers given in each row of the table are called the *coefficients* of the treatment means, and the value of a with the appropriate subscript shown in the right-hand side of the table represent these coefficients. For example, the first subscript refers to a particular treatment mean, which is to be multiplied by the coefficient, and the second subscript indicates a particular comparison. So by multiplying the treatment means by the coefficients in the first row, $\bar{X}_1.(1), \bar{X}_2.(-1), \bar{X}_3.(0)$ and $\bar{X}_4.(0)$ we obtain the following comparison:

$$d_1 = \bar{X}_1. - \bar{X}_2. \tag{7-21}$$

To obtain comparison d_2, we multiply the treatment means by the coefficients in the second row, and we have:

$$d_2 = \bar{X}_4. - \bar{X}_3. \tag{7-22}$$

As we multiply the treatment means by the coefficients in the last row, we have the comparison d_3, which is written as:

$$d_3 = \frac{1}{2}(\bar{X}_1. + \bar{X}_2.) - \frac{1}{2}(\bar{X}_3. + \bar{X}_4.) \tag{7-23}$$

Equation 7-23 states that comparison d_3 is the difference between the average of the means for treatments 1 and 2 and the average of the means for treatments 3 and 4. It should be noted that the first 2 comparisons are pairwise comparisons, but the third is not.

In constructing a table of comparison coefficients such as the one presented in Table 7.3.1.1, the following simple rules have been used:

1. When comparing groups of equal size, assign coefficients of +1 to one group and -1 to members of the other group. It does not matter which group is assigned a (+) or a (-).
2. When comparing groups that contain different numbers of treatments, assign to the first group coefficients equal to the number of treatments in the second group, and the number of treatments in the first to the second group, but with opposite signs. For example, if we have 5 treatments such that the first 3 treatments are to be compared with the last 2 treatments, then the coefficients would be +2, +2, +2, -3, -3. For simplicity, the coefficients can be reduced to the smallest possible integer. For example, if we are comparing a group of 4 treatments with a group of 2, the coefficients can be written as:

+2, +2, +2, +2, -4, -4

or in its reduced form:

+1, +1, +1, +1, -2, -2

3. To find the interaction coefficients, multiply the corresponding coefficients of the main effects.

Our discussion so far has dealt with mean comparisons. We can also use treatment sums for our comparisons. To do this, let D_i represent the difference obtained by multiplying each of the treatment sums by the coefficients; then we have

$$D_i = a_{1i} \sum X_{1.} + a_{2i} \sum X_{2.} + \ldots + a_{ki} \sum X_{k.} \tag{7-24}$$

To compute the sum of squares for a comparison coefficient, we use the following equation:

$$SS = \frac{D_i^2}{r \sum a_{.i}^2} \tag{7-25}$$

where

$$a_{.i} \quad = \text{comparison coefficient}$$

$$D_i \quad = \text{as defined above}$$

$$r \quad = \text{number of replications}$$

Equation 7-25 is a component of the treatment sum of squares with 1 degree of freedom. As such it is also a mean square and therefore can be written as:

$$MS_{D_i} = \frac{D_i^2}{r \sum a_{.i}^2} \tag{7-26}$$

To test the significance of a comparison on the treatment sum, we have

$$F = \frac{MS_{D_i}}{\text{Error } MS} \tag{7-27}$$

The d.f. associated with the numerator is 1, while the denominator d.f. is equal to those associated with the error mean square.

Since it is possible to analyze the treatment sum of squares into more than 1 set of orthogonal comparisons, the particular set of comparisons to be made should be determined by experimental interests. The testing of orthogonal comparisons for significance should be used only if the comparisons have been planned in advance.

The following example will show how a class comparison is made using the orthogonal contrasts.

Example 7.3.1.1

In determining the yield difference due to the application of P and K on kale, a horticulturist has conducted a randomized complete block design experiment with 4 replications and 4 treatments. She is particularly interested in the following questions: (1) Is there a difference in yield between the fields applied with P and those receiving K? (2) Are there yield differences between the fields receiving P and K separately and those receiving P and K together? (3) Is there a response to fertilization at all? In response to these questions, the horticulturist has designed the following set of treatments: (1) Control, (2) P added to the plots, (3) K added to the plots, and (4) both P and K added to the plots. The following yield data have been collected by the horticulturist and are shown in Table 7.3.1.2.

TABLE 7.3.1.2.

Yield Data, Treatment Total, and Treatment Mean from a Randomized Complete Block Design with 4 Replications and 4 Treatments

Treatment	Rep. I	Rep. II	Rep. III	Rep. IV	Treatment Total (T)	Treatment Mean
Control	21.2	20.5	19.8	19.6	81.1	20.28
P added	24.6	25.3	26.8	25.9	102.6	25.65
K added	23.2	22.5	25.4	21.4	92.5	23.13
P and K added	28.7	29.8	27.2	30.1	115.8	28.95
Rep. Total (R)	97.7	98.1	99.2	97.0		
Grand Total					392.0	
Grand Mean						24.5

Solution

To answer these questions, we first perform an analysis of variance for a randomized complete block design following the steps outlined in Section 4.2.2. The results are shown in Table 7.3.1.3.

TABLE 7.3.1.3.

Analysis of Variance of Kale Yield Data in a Randomized Complete Block Design with 4 Treatments and 4 Replications

Source of Variation	Degree of Freedom	Sum of Squares	Mean Square	F
Treatment	3	163.46	54.49	28.38**
Replication	3	0.64	0.21	
Error	9	17.28	1.92	
Total	15	181.38		

Note: ** = significant at 1% level.

In calculating the sum of squares for treatment components using the comparison coefficients, we follow these steps:

STEP 1: Construct a set of orthogonal contrasts among the treatments, following the rules stated earlier. This is shown in Table 7.3.1.4.

TABLE 7.3.1.4.

A Set of Orthogonal Contrasts Among the Treatments

	Treatment			
Contrast	Control	P applied	K applied	P and K applied
Response to fertilizer	3	−1	−1	−1
P vs. K	0	+1	−1	0
P and K vs. PK together	0	−1	−1	+2

STEP 2: Sum the squares for each comparison. Thus, we have:

$$\sum a_{.1}^2 = (3)^2 + (-1)^2 + (-1)^2 + (-1)^2 = 12$$

$$\sum a_{.2}^2 = (+1)^2 + (-1)^2 = 2$$

$$\sum a_{.3}^2 = (-1)^2 + (-1)^2 + (+2)^2 = 6$$

STEP 3: Compute the coefficients (D_i) for each of the 3 comparisons, using either the treatment means or the treatment sums. We have used the treatment sums in our computation, and the results are presented in Table 7.3.1.5. To compute D_1, for example, Equation 7-24 was used and we get:

$$81.1(3) + 102.6(-1) + 92.5(-1) + 115.8(-1) = -67.60$$

TABLE 7.3.1.5.

A Set of Orthogonal Contrasts Among the Treatments

Contrast	Control 81.1	P applied 102.6	K applied 92.5	P and K applied 115.8	Value of D_i
Response to fertilizer	3	−1	−1	−1	−67.60
P vs. K	0	+1	−1	0	10.10
P and K vs. PK together	0	−1	−1	+2	36.50

STEP 4: Compute the sum of squares for the treatment components, using Equation 7-25:

$$SS_{Fertilizer} = \frac{(-67.60)^2}{4(12)} = 95.20$$

$$SS_P = \frac{(10.10)^2}{4(2)} = 12.75$$

$$SS_K = \frac{(36.50)^2}{4(6)} = 55.51$$

As was pointed out earlier, the above sum of squares are also mean squares because we have only a single degree of freedom associated with each. Thus, in computing the F value, each of the above sum of squares are divided by the error mean square as shown in the next step. Note that the sum of all the mean squares is equal to the treatment sum of squares given in Table 7.3.1.3. That is,

$$MS_{D_1} + MS_{D_2} + MS_{D_3} = 95.20 + 12.75 + 55.51 = 163.46$$

STEP 5: Perform the F test using Equation 7-27 and the data from Table 7.3.1.3 as:

$$F_1 = \frac{95.20}{1.92} = 49.58$$

$$F_2 = \frac{12.75}{1.92} = 6.64$$

$$F_3 = \frac{55.51}{1.92} = 28.91$$

STEP 6: Compare the computed F values with the tabular F given in Appendix E with $\alpha = 0.05$, and $\alpha = 0.01$ for 1 and 9 degrees of freedom. The tabular F for 5% and 1% levels of significance are 5.12 and 10.56, respectively. Thus, we would conclude that D_1, D_2, and D_3 are significant. Based on these results we have an answer to each of the questions posed when the experiment was planned.

7.3.2 Trend Comparisons

Our discussions so far have dealt with mean comparisons that focus on specific treatments. In some experiments treatments may consist of different values of an ordered variable that covers the whole range of treatments. For example, we may test the response to different increments of a fertilizer or test different groups of animals after increasing the dosage of a drug. In such experimental situations a wider range of treatment responses is observed. Thus, it is not appropriate to use specific mean comparisons.

In experiments where the treatments consist of different values of an ordered variable, and if it can be assumed that the difference between the values are uniform (equal), then the researcher may be interested in determining whether the treatment means (or totals) are functionally related to the different values of the treatment variable: that is, what the nature of the response of the experimental unit is to the varying levels of a treatment. Specifically, the researcher may want to examine whether the treatment means are linearly related to the treatment variable, and if so, whether there is a significant curvature in the trend of the means.

To illustrate the computational procedures for a trend or a factorial comparison, we have used the data from Example 5.5.1 where corn hybrids received (0, 70, 140, and 210 lb/acre of nitrogen). Note that the treatments have an equal interval of 70 lb/acre of nitrogen fertilizer. The ordered treatment means were

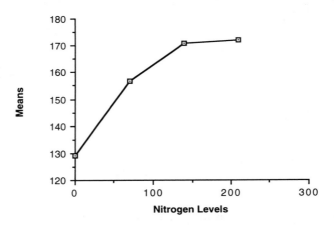

FIGURE 7.3.2.1.

Treatment means for each of the 4 levels of fertilizer applied.

129, 156.5, 171.2, and 172.0. Figure 7.3.2.1 shows the treatment means graphed against the level of nitrogen applied. The trend of the mean does not appear to be linear.

In order to determine whether the linear component of the trend of the sums is statistically significant and whether the treatment sums (or means) deviate significantly from linearity, use is made of the table of coefficient for orthogonal polynomials given in Appendix J. The steps in computing the sum of squares and performing the F test are given below.

STEP 1: Determine the coefficients for the linear, quadratic, and cubic component for $k = 4$ treatments from Appendix J. For the present example the coefficients are given in Table 7.3.2.1.

TABLE 7.3.2.1.

Coefficients for the Linear, Quadratic, and Cubic Components for $k = 4$ Treatments

	Treatment Totals			
Comparison	1,290	1,565	1,710	1,720
Linear	−3	−1	1	3
Quadratic	1	−1	−1	1
Cubic	−1	3	−3	1

STEP 2: Sum the squares for each comparison given in Table 7.3.2.1 as shown below.

$$\sum a_L^2 = (-3)^2 + (-1)^2 + (+1)^2 + (+3)^2 = 20$$

$$\sum a_{.Q}^2 = (+1)^2 + (-1)^2 + (-1)^2 + (+1)^2 = 4$$

$$\sum a_{.C}^2 = (-1)^2 + (+3)^2 + (-3)^2 + (+1)^2 = 20$$

STEP 3: Compute the coefficients for each of the three comparisons, linear (L), quadratic (Q), and cubic (C), by multiplying the treatment sums by the appropriate coefficient as given below:

$$L = (-3)(1,290) + (-1)(1,565) + (1)(1,710) + (3)(1,720) = 1,435$$
$$Q = (1)(1,290) + (-1)(1,565) + (-1)(1,710) + (1)(1,720) = -265$$
$$C = (-1)(1,290) + (3)(1,565) + (-3)(1,710) + (1)(1,720) = -5$$

STEP 4: Compute the sum of squares (mean squares) for the treatment components, using Equation 7-25:

$$MS_L = SS_L = \frac{(1,435)^2}{10(20)} = 10,296.13$$

$$MS_Q = SS_Q = \frac{(-265)^2}{10(4)} = 1,755.63$$

$$MS_C = SS_C = \frac{(-5)^2}{10(20)} = 0.13$$

As a check, note that the sum of squares for the treatment components add up to the treatment sum of squares given in Table 5.5.4. Since the components are an orthogonal set, they must equal the sum of squares partitioned.

STEP 5: Perform the test of significance for the linear, quadratic, and cubic comparisons as shown below:

$$F = \frac{MS_L}{\text{Error } MS} = \frac{10,296.13}{93.96} = 109.58$$

$$F = \frac{MS_Q}{\text{Error } MS} = \frac{1,755.63}{93.96} = 18.68$$

$$F = \frac{MS_C}{\text{Error } MS} = \frac{0.13}{93.96} = {<}1$$

STEP 6: Compare the F values with the tabular F from Appendix E at the prescribed level of significance. This is shown in Table 7.3.2.2.

TABLE 7.3.2.2.

Analysis of Variance of Hybrid × N Response Experiment Partitioned into Linear, Quadratic, and Cubic Components of Response Curve

Source of Variation	Degree of Freedom	Sum of Squares	Mean Square	F	Required F 5%	1%
Replication	1	4,100.63	4,100.63			
Nitrogen (A)	3	12,051.88	4,017.29	42.76**	9.28	29.46
Linear	(1)	10,296.13	10,296.13	109.58**	10.13	34.12
Quadratic	(1)	1,755.63	1,755.63	18.68*		
Cubic	(1)	0.13	0.13	<1		
Error (a)	3	281.87	93.96			
Hybrid (B)	4	2,466.25	822.10	27.41**		
Nitrogen × Hybrid (A × B)	12	863.75	71.98	2.40^{ns}		
Error (b)	16	479.97	29.99			
Total	39	20,244.38				

The results of Table 7.3.2.2 show that there is a significant linear trend in the treatment sums. This means that the relationship between yield and nitrogen rate of application is linear within the range of rates tested. The results also indicate that there is significant curvature in the trend of the sums. From Table 7.3.2.2 we note that beyond the quadratic component, the results are nonsignificant. Thus, we conclude that the yield is a quadratic function of the nitrogen rates. The interpretation of the above results should be done with caution. Recall from earlier chapters that when the number of error degrees of freedom is not adequate, as is the case in the present example, the test of significance is not valid.

The same procedures applied in partitioning the treatment components can also be applied to make between-group comparison on the means of the 3 hybrids.

REFERENCES AND SUGGESTED READINGS

Bryon-Jones, J. and Finney, D. J. 1983. "On an error in 'Instruction to Authors'," *Hort. Sci.* 18:279-82.

Cochran, W. G. and Cox, G. M. 1957. *Experimental Designs*, 2nd ed. New York: John Wiley & Sons, chap. 6A.

Cramer, S. G. and Swanson, M. R. 1973. "An evaluation of ten pairwise multiple comparison procedures by Monte Carlo methods," *J. Am. Stat. Assoc.* 68:66-74.

Chew, V. 1976. "Comparing treatment means: a compendium," *Hort. Sci.* 11:348-357.

Duncan, D. B. 1955. "Multiple range and multiple F tests." *Biometrics* 11:1-42.

Gomez, K. A. and Gomez A. A. 1984. *Statistical Procedures for Agricultural Research.* 2nd ed. New York: John Wiley & Sons, chap. 5.

Maindonald, J. H. and Cox, N. R. 1984. "Use of statistical evidence in some recent issues of DSIR agricultural journals," *N. Z. J. Agric.*, 27:597-610.

Mead, R. 1988. *The Design of Experiments: Statistical Principles for Practical Applications.* Cambridge: Cambridge University Press, chap. 4 and 12.

Miller, R. G., Jr. 1977. "Developments in multiple comparisons, 1966-1976." *J. Am. Stat. Assoc.* 72:779-788.

Morse, P. M. and Thompson B.K. 1981. "Presentation of experimental results," *Can. J. Plant Sci.* 61:799-802.

Peterson, R. G. 1977. "Use and misuse of multiple comparison procedures," *Agron. J.* 69:205-208.

Wyman, J. A., Chapman, R. K., and Longridge, J. L. 1982. *Project Report: Vegetable Crops Entomology Field Research.* Madison: University of Wisconsin.

EXERCISES

1. In a completely randomized field experiment, measurements were made to study the response of field-grown cassava (*Manihot esculenta* Crantz) to changes in the application of a fertilizer. The objective of the research was to find out if there were any differences in the amount of the dry matter produced under 5 different fertilizer regimes. The mean yield from the experiment with 4 replications were:

Mean yield of dry matter production (tons/ha) of cassava as a result of 5 different fertilizer applications.

Treatment	Treatment Mean (tons/ha)
Control	2.14
50 kg/ha	2.54
100 kg/ha	2.68
150 kg/ha	3.81
200 kg/ha	4.42

(a) Compute the *LSD* value at 5% and 1% levels of significance.
(b) What conclusions can you draw from the comparisons?

2. Suppose that in the previous example the treatment means were computed from the 4 replications where some data were lost as a result of a natural disaster, as shown below:

Mean yield of dry matter production (tons/ha) of cassava as a result of 5 different fertilizer applications.

Treatment	Rep. I	Rep. II	Rep. III	Rep. IV	Treatment Mean
Control	2.20	—	2.25	2.01	2.15
50 kg/ha	2.40	2.56	2.66	2.52	2.54
100 kg/ha	—	2.68	7.79	—	2.74
150 kg/ha	3.00	3.56	4.00	4.66	3.81
200 kg/ha	3.50	4.98	5.00	4.20	4.42

(a) What type of a comparison of means would be appropriate?
(b) Compute the *LSD* at the 5% and 1% levels of significance.
(c) Compare each of the mean differences.

3. In a balanced lattice design experiment where each treatment was replicated 4 times, a natural scientist studied the impact of chlorpyrifos on ash borer infestation. The ash borer represents serious threat to ash and olive trees. The treatments were applied during the moth flight season. The scientist was interested in knowing if there are differences in the number of attacks on the ash tree as a result of the treatments. In performing the analysis of variance, the scientist found the effective error mean square with 16 degrees of freedom to be 1.8.

Treatment Number	Adjusted Mean No. of attack sites
1	3
2	4
3	2
4	3
5	1
6	4
7	2
8	5
9 (Control)	8

(a) Compute the *LSD* for the 5% and 1% levels of significance.
(b) What conclusions can be drawn about the effectiveness of the treatments as compared with the control?

4. To make a mean comparison on the data from a 4 x 4 partially balanced triple lattice design experiment, an animal scientist recorded the following data on the role of progestron in stimulating sexual receptivity in estrogen-treated gilts. The experiment involved 16 ovariectomized gilts who were treated with estradil benzoate (EB), and a control group which did not receive estradil benzoate (treatment number 16). After EB treatment, gilts were moved to an evaluation pen where bores were brought in. Gilts remained in the evaluation pen for 5 minutes, during which time the number of mounts attempted by the bore were recorded. The treatment numbers appear in parentheses.

Incomplete Block Number	Mounts, Number/5 min.			
	Replication I			
1	7(01)	5(02)	4(03)	4(04)
2	5(05)	2(06)	1(07)	3(08)
3	4(09)	3(10)	4(11)	5(12)
4	3(13)	3(14)	2(15)	1(16)
	Replication II			
1	6(01)	6(05)	6(09)	2(13)
2	4(02)	2(06)	3(10)	3(14)
3	3(03)	3(07)	1(11)	5(15)
4	1(04)	4(08)	3(12)	2(16)
	Replication III			
1	7(01)	5(06)	4(11)	2(16)
2	4(05)	4(02)	3(15)	6(12)
3	1(09)	2(14)	2(03)	4(08)
4	5(13)	3(10)	4(07)	2(04)

(a) How many effective error mean squares are involved in this experiment?

(b) Before performing the *LSD* test, do you need to compute the adjusted treatment means?

(c) Compute the mean difference between the control treatment and each of the 15 treatments.

5. To study the impact of 4 insecticides on the omnivorous looper, a split-plot design experiment was conducted. The insecticide treatments were assigned to the subplots, whereas the concentration rate of active ingredients were assigned to the main plots. The results of the experiment are shown below.

The average number of larvae found per tree 14 days posttreatment from an experiment replicated 3 times

Active Ingredient	Dylox 80 SP	Kryocide 8 F	Lannate L	Orthene 75 SP	Control
1 lb/ac	9.66	10.33	3.00	2.33	17.33
2 lb/ac	7.33	5.00	2.66	2.33	20.00
4 lb/ac	3.66	5.33	2.00	1.33	16.33

(a) Perform a *LSD* test on the data.
(b) What conclusions can be drawn from the analysis?

6. Research has shown that aldicarb is the most preferred pesticide used by potato farmers. In a study conducted in Wisconsin, researchers used a variety of pest management strategies including 6 different chemicals: aldicarb (Temik), phorate (Thimet), disulfoton (Disyston), carbofuran (Furadan), oxamyl (Vydate), and terbufos (Counter) on Atlantic variety potatoes. The data on the yield per acre of potato are given below.

Mean yield of Atlantic variety potatoes treated soil-applied systemic insecticide[a]	
Treatment (lb a.i./acre)	Mean yield(cwt per acre)
Temik 15G (3) with fertilizer at planting	327.4
Temik 15G (3) topdress at emergence	369.8
Temik 15G (2) topdress at emergence	360.5
Disyston 15G (3) with fertilizer at planting	318.4
Disyston 15G (3) topdress at emergence	307.4
Disyston 15G (3+3) with fertilizer at planting and topdress at emergence	322.0
Thimet 20G (3) with fertilizer at planting	319.7
Thimet 20G (3) with fertilizer at planting, foliar spray	317.5
Furadan 15G (3) with fertilizer at planting	329.3
Vydate 10G (3) with fertilizer at planting	296.9
Vydate 10G (3) topdress at emergence	315.7
Counter 15G (3) with fertilizer at planting	317.8
Untreated	239.6

[a]Data adapted from Wyman, J. A., Chapman, R. K. and Longridge, J. L. 1982. *Project Report: Vegetable Crop Entomology Field Research.* Madison: University of Wisconsin.

Perform a Duncan's Multiple-Range Test on the data.

7. In an attempt to provide adequate moisture for germination, cotton farmers in California irrigate their fields before planting. This preseason irrigation has been identified as a major contributor to drainage problems. To remedy this problem a study was conducted to determine if plastic mulch could be used effectively to replace preseason irrigation of cotton fields. Three different plastic mulches (white, black) were used in this experiment on Acala cotton. The experiment was replicated 4 times. Plant height was

used as a criterion in determining the differences between the mulches and the control. Perform a grouped mean comparison on the data.

Height (inches) of Acala cotton under different treatment conditions				
Treatment	Rep. I	Rep. II	Rep. III	Rep. IV
Control	10.0	8.5	9.2	11.0
White plastic	20.3	19.5	21.0	19.8
Black Plastic	24.5	23.8	23.6	22.4

(a) Is there a difference in the height of cotton plants from fields with white mulch and black mulch?
(b) Is there a difference between the height of plants receiving white mulch and the control?
(c) Is there a difference between the height of plants receiving black mulch and the control?

8. To determine whether application of fertilizers produce more range forage in drought than normal years, a study was conducted by a scientist in Colorado. The results of the experiment produced the following oven dry weight of range forage.

Range forage yield according to rate of nitrogen application in normal and 5 drought years						
		Fertilizer Treatment, lb/ac				
Replication	Precipitation(in)	0	30	60	90	120
I.	Normal (16.5)	3520	3600	3750	4150	4390
	Drought (8.5)	2650	3211	3476	4485	4863
	Drought (10.5)	2652	3356	4291	6310	6050
	Drought (6.0)	1432	2397	2513	3200	3252
	Drought (8.0)	1451	2498	2568	2734	3682
II.	Normal (16.5)	3319	3560	3552	4230	4410
	Drought (8.5)	2578	3276	3399	4345	4674
	Drought (10.5)	2602	3416	4300	5810	5980
	Drought (6.0)	1392	2278	2567	3660	3000
	Drought (8.0)	1400	2399	2510	2932	3862

Replication	Precipitation(in)	Fertilizer Treatment, lb/ac				
		0	30	60	90	120
III.	Normal (16.5)	3423	3545	3759	4252	4270
	Drought (8.5)	2760	3200	3498	4375	4542
	Drought (10.5)	2752	3452	4312	6110	6000
	Drought (6.0)	1487	2377	2542	3189	3342
	Drought (8.0)	1493	2456	2575	2831	3732

Perform a trend or a factorial comparison on the data.

Chapter 8

SAMPLE DESIGNS OVER TIME

8.1 TERMINOLOGY AND CONCEPTS

Our discussions so far have centered around those experimental designs that are conducted in one period in time and on a single site. Care must be taken in the generalization of the results of such experiments. No matter how well such experiments are conducted, their results differ from those obtained from the same experiments conducted over a longer period of time (several seasons or several years) and at several sites, or both. In this chapter we will address how such experiments are conducted and the unique conditions associated with these experiments.

Agricultural researchers are often interested in the applicability of their findings over a wide range of physical and environmental conditions. For example, they would be interested in the applicability of their recommendations from a corn experiment conducted in Nebraska not only to the neighboring states, but also to other corn-producing regions around the world. To accomplish such an objective, the experiment on a single-site would have to emulate all the physical and climatic conditions. This is very difficult, as there are certain factors that are not easily controlled. To modify such factors as rainfall, sunshine, or the inherent fertility of the soil is outside the control of the experimenter. Given this dilemma, researchers would have to rely on experiments conducted at several sites and over a number of periods of time, or both, so that they can generalize the results of their experiments over a wide range of climatic and physical conditions.

To analyze the results of a series of experiments conducted at different sites and times requires great care. Such experiments and their analyses are usually carried out in stages. The preliminary stage entails conducting individual experiments at different sites. The data obtained from such experiments are then combined to see if there are differences among the treatments from the different experiments. At this stage the experimenter is able to determine if a particular treatment consistently performs well in comparison to others. Furthermore, it is possible to determine whether certain treatments do well in certain circumstances. The second stage of the analysis can be used for estimation and comparison of the average effects of treatments over the whole series of experiments.

Multisite experiments, while helpful in overcoming the shortfalls of the single-site experiments, present difficulties that must be recognized at the outset. One such problem is that treatments × places interactions may not be homogeneous. The other is the inequality of error variance at different sites. When the treatments × places interactions are not homogeneous, it implies that some sites will give a better estimate of the difference than will others. Under such conditions the difference is estimated overall by weighing the difference at each site by the reciprocal of its variance. The reason for such weighing is to give greater weight to the sites that provide a better determination, on account of either higher replication or lower error variance.

In cases where the interactions are large, interpreting the results of the combined analysis is not too difficult. A large F ratio for interactions permits the researcher to be confident in the statistical significance of the experiment. Additionally, when interactions are large, the F test of the treatment mean square against the interaction mean square is little affected by the inequality in the error variance.

When the researcher suspects that there is interaction between treatments and places, as well as the presence of unequal error at different sites, it is best to work with specified contrasts between treatments. Under such circumstances it is advisable to test any component of the treatment sum of squares against its own interaction with places (Snedecor and Cochran, 1967). This means that the treatments mean square has to be subdivided into sets of comparisons; then, the interactions mean squares for each set are computed and tested separately (Snedecor and Cochran, 1967; Pearce, 1983).

To check for the inequality of the error variance, Bartlett's test could be used. If the test shows that variances are heterogeneous, the validity of using the F test of treatments × places interactions is in question. When comparisons are made over a subset of places, the pooled error for these places should be used instead of the overall pooled error (Yates and Cochran, 1938; Snedecor and Cochran, 1967).

In summary, experiments that are conducted at several sites, with equal care, design and structure, are easily combined for analysis. The results of such an analysis are efficient and theoretically sound. However, care should be taken in the generalization of the results, especially from the ordinary test of significance. As was pointed out, experiments conducted at several sites are prone to heterogeneous error and interaction variances.

8.1.1 Representing Time Spans

Agricultural field experiments are often repeated over a number of seasons and for a number of years. This is done to determine if the subject under investigation performs differently at different time periods. In conducting a time-span experiment, the aim of the researcher is to obtain data (from successive years) that are regarded as independent. To accomplish this objective, researchers should select a new site and use a new randomization for the experiment. The experimental arrangement, however, need not be uniform throughout the whole series of experiments, in order to obtain data considered independent. In such experiments, the interaction between treatment and seasons are more important than the treatment and year.

The statistical analysis that we should use with the data from a series of experiments depends on the objectives of the study. However, it is safe to say that the analysis in the preliminary stages tends to be the same. In this chapter we will present an introductory account of these procedures. In the next 2 sections, procedural steps for the combined analysis of experiments over the years and over the seasons are presented.

8.2 ANALYSIS OF EXPERIMENTS OVER YEARS

Researchers, in conducting experiments over the years, are interested in determining how treatments respond to the different environmental conditions that normally occur from year to year or over a number of years. Because researchers are not able to predict what environmental conditions may prevail, years are generally considered as a random variable in such experiments. We will illustrate the combined analysis for experiments that are performed over several years, by using only the data from an experiment that was conducted over a 2-year period. In the following example, we will illustrate the step-by-step procedure in analyzing the data from such an experiment.

Example 8.2.1

Field experiments to evaluate potential differences in yield response of 5 corn hybrids widely grown in the northern Corn Belt were conducted during 1990 and 1991. A randomized complete block design with 4 replications was used for the study. The experiments were conducted on separate but adjacent sites, where corn had been grown since 1985. The 1991 experimental plots

	1990 Yield, bu/acre					1991 Yield, bu/acre				
Hybrid	Rep. I	Rep. II	Rep. III	Rep. IV	Total	Rep. I	Rep. II	Rep. III	Rep. IV	Total
P3747	110	135	120	120	485	115	120	113	117	465
P3732	150	150	148	160	608	140	155	138	140	573
A630 × LH2	113	120	135	116	484	118	129	136	138	521
LH70 × LH3	120	130	138	120	508	130	142	150	149	571
P3742	122	126	132	130	510	128	143	110	139	520
Total	615	661	673	646	2595	631	689	647	683	2650

were randomized differently than those in 1990. Soil fertility conditions, planting date, plant density were all similar in both years. The above results in bushels/acre were obtained from the experiments. Perform the combined analysis of variance on the data.

Solution

To perform the combined analysis, we first need to perform a separate analysis of variance for each year in the experiment. The procedure for computing the sums of squares is shown below, using the 1990 data. The sum of squares for the year 1991 is computed similarly. The results of the analysis are shown in Table 8.2.1. The steps in performing the individual analysis of variance and the combined analysis are given below.

STEP 1: Compute the correction factor and the various sums of squares as:

$$C = \frac{G^2}{rt} \qquad (8\text{-}1)$$

$$= \frac{(2,595)^2}{4 \times 5} = \frac{6,734,025}{20}$$

$$= 336,701.25$$

$$\text{Total } SS = \sum_{j=1}^{t} \sum_{i=1}^{r} X_{ij}^2 - C \qquad (8\text{-}2)$$

$$\text{Total } SS = \left[(110)^2 + (150)^2 + \dots + (130)^2 \right] - 336,701.25$$

$$= 340,307 - 336,701.25$$

$$= 3,605.75$$

$$\text{Replications}\,SS = \frac{\sum R_j^2}{t} - C \qquad (8\text{-}3)$$

$$= \frac{(615)^2 + (661)^2 + (673)^2 + (646)^2}{5} - 336{,}701.25$$

$$= 337{,}078.20 - 336{,}701.25$$

$$= 376.95$$

$$\text{Treatment}\,SS = \frac{\sum T_i^2}{r} - C \qquad (8\text{-}4)$$

$$= \frac{(485)^2 + (608)^2 + (484)^2 + (508)^2 + (510)^2}{4} - 336{,}701.25$$

$$= 339{,}327.25 - 336{,}701.25$$

$$= 2{,}626$$

$$\text{Error}\,SS = \text{Total}\,SS - \text{Rep.}\,SS - \text{Treatment}\,SS \qquad (8\text{-}5)$$

$$= 3{,}605.75 - 376.95 - 2{,}626$$

$$= 602.8$$

STEP 2: Compute the mean square for each source of variation as follows:

$$\text{Replication}\,MS = \frac{\text{Replication}\,SS}{r - 1} \qquad (8\text{-}6)$$

$$= \frac{376.3}{3} = 125.65$$

$$\text{Treatment}\,MS = \frac{\text{Treatment}\,SS}{t - 1} \qquad (8\text{-}7)$$

$$= \frac{2{,}626}{4} = 656.50$$

$$\text{Error } MS = \frac{\text{Error } SS}{(r-1)(t-1)} \tag{8-8}$$

$$= \frac{602.80}{12} = 50.23$$

STEP 3: Compute the F value for testing the treatment difference as:

$$F = \frac{\text{Treatment } MS}{\text{Error } MS} \tag{8-9}$$

$$= \frac{656.50}{50.23} = 13.07$$

To perform the combined analysis of variance, we use the data given in Table 8.2.1 and proceed as follows:

TABLE 8.2.1.

Results of the Analyses of Variance for the Individual Years Using an RCB Design with 4 Replications and 5 Corn Hybrids

Source of Variation	Degree of Freedom	Sum of Squares	Mean Square	F
		1990		
Replication	3	376.95	125.65	
Hybrid	4	2,626.00	656.50	13.07**
Error	12	602.80	50.23	
		1991		
Replication	3	471.00	157.00	
Hybrid	4	1,984.00	496.00	6.64**
Error	12	896.00	74.67	

** = significant at 1% level.

STEP 4: Determine the degrees of freedom associated with each source of variation. Since in the present case year (Y) is considered a random variable, the error term is represented by the year $\times$ hybrid interaction MS. Thus, we have:

Year (Y) $d.f.$ $= y - 1 = 1$

Reps. within year $= y (r - 1) = 6$

Hybrid	$= t - 1 = 4$
Year × hybrid	$= (y - 1)(t - 1) = 4$
Pooled error	$= y(r - 1)(t - 1) = 24$
Total $d.f.$	$= rt - 1 = 39$

STEP 5: Compute the sum of squares for the replications within the year and the pooled error as follows:

$$\text{Rep. within year } SS = \sum_{i=1}^{y} (\text{Rep. } SS)_i \qquad (8\text{-}10)$$

$$\text{Pooled error } SS = \sum_{i=1}^{y} (\text{Error } SS)_i \qquad (8\text{-}11)$$

Thus, we have:

$$\text{Reps. within year } SS = 376.95 + 471.00 = 847.95$$

$$\text{Pooled error } SS = 602.80 + 896.00 = 1,498.80$$

STEP 6: Calculate the correction factor as:

$$C = \frac{\left(\sum_{i=1}^{y} G_i\right)^2}{yrt} \qquad (8\text{-}12)$$

$$C = \frac{(2,595 + 2,650)^2}{40} = 687,750.63$$

STEP 7: Compute the sum of squares for the year, treatment, and the year × treatment interaction as follows:

$$\text{Year } SS = \sum_{i=1}^{y} \frac{G_i^2}{rt} - C \qquad (8\text{-}13)$$

$$\text{Year } SS = \frac{(2,595)^2 + (2,650)^2}{4(5)} - 687,750.63$$

$$= 687,826.25 - 687,750.63$$

$$= 75.62$$

$$\text{Treatment (Hybrid) } SS = \sum_{j=1}^{t} \frac{T_j^2}{yr} - C \tag{8-14}$$

$$= \frac{\left[(950)^2 + (1,181)^2 + (1,005)^2 + (1,079)^2 + (1,030)^2 \right]}{2(4)}$$

$$-687,750.63$$

$$= 5,532,427/8 - 687,750.63$$

$$= 3,802.75$$

$$\text{Year} \times \text{Treatment } SS = \sum_{i=1}^{y} \sum_{j=1}^{t} \frac{(YT)_{ij}^2}{r} - (C + \text{Year } SS + \text{Treatment } SS) \tag{8-15}$$

where

$(YT)_{ij}$ =the total of the jth treatment in the ith year.

$$\text{Year} \times \text{Treatment } SS = \frac{\left[(485)^2 + (608)^2 + \dots + (1,030)^2 \right]}{4}$$

$$-(687,750.63 + 75.62 + 3,802.75)$$

$$= 2,769,745/4 - 691,629$$

$$= 692,436.25 - 691,629$$

$$= 807.25$$

STEP 8: Compute the mean square for each source of variation as shown below:

$$\text{Year } MS = \frac{\text{Year } SS}{y - 1} = \frac{75.62}{1} = 75.62$$

$$\text{Reps. within year } MS = \frac{\text{Reps. within year } SS}{y(r - 1)} = \frac{847.95}{6} = 141.33$$

$$\text{Treatment (hybrid)} \ MS = \frac{\text{Treatment} SS}{t - 1} = \frac{3,802.75}{4} = 950.69$$

$$\text{Year} \times \text{Treatment} \ MS = \frac{\text{Year} \times \text{Treatment} SS}{(y - 1)(t - 1)} = \frac{807.25}{1(4)} = 201.81$$

$$\text{Pooled error} \ MS = \frac{\text{Pooled error} SS}{y(r - 1)(t - 1)} = \frac{1,498.80}{2(3)(4)} = 62.45$$

STEP 9: Perform the test of homogeneity of error variance, using the data from the individual analysis of variance given in Table 8.2.1. This test could be performed either through the application of the chi-square or the F test. We have used the F test to determine the homogeneity as follows:

$$F = \frac{\text{Larger error} \ MS}{\text{Smaller error} \ MS} \tag{8-16}$$

$$F = \frac{74.67}{50.23} = 1.49$$

In comparing the computed F value with the corresponding F value (2.69 at the 5% level) found in Appendix E for 12 degrees of freedom, we cannot reject the hypothesis of homogeneous error variances. Had we rejected the homogeneity of the error variances, then we would need to partition the pooled error sum of squares into components corresponding to the Year × Treatment sum of squares. To compute the F value for each component of the Year × Treatment interaction, we use the corresponding component in the pooled error as its error term. Since in the present case we cannot reject the hypothesis of homogeneous error variance, we proceed in computing the F value for testing the significance of the various effects.

STEP 10: Compute the F value for each source of variation. As was mentioned earlier, when data are combined over years, the error term is the Year × Treatment interaction mean square. Thus the F value for the replications within year, treatment, and the Year × Treatment interaction is computed below and shown in Table 8.2.2.

TABLE 8.2.2.

Combined Analysis of Variance of a Randomized Complete Block Experiment over 2 years

Source of Variation	Degree of Freedom	Sum of Squares	Mean Square	F
Year	1	75.62	75.62	a
Reps. within year	6	847.95	141.33	
Treatment (hybrid)	4	3,802.75	950.69	4.71^{ns}
Year × Treatment	4	807.25	201.81	3.23*
Pooled error	24	1,498.80	62.45	
Total	39	7,032.37		

Note: ^{a} Since the d.f. is not adequate, a valid test of significance cannot be performed.
^{ns} = nonsignificant, * = significant at 5% level.

$$F \text{ (Treatment)} = \frac{\text{Treatment}\,MS}{\text{Year} \times \text{Treatment}\,MS} \tag{8-17}$$
$$= \frac{950.69}{201.81} = 4.71$$

$$F(\text{Year} \times \text{Treatment}) = \frac{\text{Year} \times \text{Treatment}\,MS}{\text{Pooled error}\,MS} \tag{8-18}$$
$$= \frac{201.81}{62.45} = 3.23$$

The results of the F tests show that the hybrid effect is nonsignificant, while the interaction effect between the hybrid and year is significant. When the interaction between year and treatment (hybrid) is significant, it is important to determine the relative size of such interaction with the average effect of the treatments.

If the interaction is large relative to the average effect, and the ranking of treatments (hybrids) changes over years—one treatment performs better than another some years and does not do well in other years—then we need to examine the nature of the interaction. However, if the interaction effect is small and ranking of the treatments (hybrids) over the years is stable—hybrid A will do better than hybrid B in all years—then the interaction could be ignored.

As can be seen in Table 8.2.2, the interaction for the present example is large, and the average effect of the hybrid is not significant. We need to further examine the mean difference between years for the 5 hybrids. Table 8.2.3 shows the mean yield of the 5 corn hybrids tested in 1990 and 1991.

In examining the mean difference between years for each of the 5 corn hybrids, we make the following observations:

TABLE 8.2.3.

Mean Yield of 5 Corn Hybrids Tested in 1990 and 1991

| Hybrid | Mean Yield, bu/acre | | | |
	1990	1991	Average	Difference
P3747	121.25	116.25	118.75	−5.00
P3732	152.00	143.25	147.63	−8.75
A630 × LH2	121.00	130.25	125.63	9.25
LH70 × LH3	127.00	142.75	134.88	15.75
P3742	127.50	130.00	128.75	2.50
Ave.	129.75	132.50	156.25	2.75

1. With the exception of the first 2 hybrids P3747, and P3732 which had lower yield in the second year of the experiment, all the other hybrids had higher yields in the second year than in the first.
2. Our analysis shows that hybrid P3732 gave the highest average yield (147.63 bu/acre) over the 2-year period; however, its performance was superior only in the first year of the experiment.
3. Hybrid P3742 shows some consistency in its performance over the 2 years.

Based on these results, the experimenter is well advised to conduct further studies on the hybrids with special attention to P3742 and P3732.

8.3 ANALYSIS OF EXPERIMENTS OVER SEASONS

The object of the combined analysis of experiments over seasons is to provide the experimenter with information on the nature of the interaction between treatment and seasons. Treatment may be any factor of production. Once it is determined that the interaction between the treatment and season is significant, then such information is used to make recommendations on the use of treatments for different seasons. For example, if data from a study on a new hybrid (treatment) shows that it performs better in spring planting rather than winter, then the recommendation for its use will undoubtedly favor spring planting.

In the previous section, we mentioned that because researchers cannot predict environmental conditions such as rainfall or sunshine from one year to the next, year was used as an independent variable in the analysis. In conducting seasonal analysis, planting seasons within a year are distinct periods and therefore can be considered as a fixed variable. This fact makes it possible to determine the superiority of a treatment for a specific season.

The following example will illustrate the step-by-step procedures for analyzing the data from an experiment conducted over 2 seasons.

Example 8.3.1

Field studies were conducted to determine the influence of planting season using 5 nitrogen rates on wheat. A randomized complete block design with 4 replications was used in this experiment. The data in Table 8.3.1 show the yield in bushels per acre for winter and spring planting. The step-by-step procedures are outlined below.

TABLE 8.3.1.
Yield Data from Spring and Winter Planting of Wheat

Nitrogen	Yield (bu/acre)					
Rate, lb/ac	Rep. I	Rep. II	Rep. III	Rep. IV	Total	Mean
Spring Planting						
0	27.8	24.6	28.2	26.9	107.5	26.88
50	30.0	29.2	30.1	28.9	118.2	29.55
100	29.9	28.3	29.7	30.0	117.9	29.48
150	31.4	32.0	31.7	31.8	126.9	31.73
200	30.8	31.3	29.9	32.0	124.0	31.00
250	30.5	31.2	33.0	31.8	126.5	31.63
Total	180.4	176.6	182.6	181.4	721.0	
Winter Planting						
0	25.1	24.9	26.2	24.2	100.4	25.10
50	24.4	29.2	28.1	26.9	108.6	27.15
100	30.4	26.8	28.2	29.5	114.9	28.73
150	30.3	34.3	32.1	36.2	132.9	33.23
200	31.5	33.6	35.8	32.9	133.8	33.45
250	34.2	35.4	33.6	31.2	134.4	33.60
Total	175.9	184.2	184.0	180.9	725.0	

Solution

To perform a combined analysis, we first need to conduct a separate analysis of variance for each planting season in the experiment. The procedure for performing the individual analysis of variance is similar to that of Example 8.2.1 and need not be repeated. Table 8.3.2 shows the result of the analysis, which is based on a randomized complete block design.

STEP 1: Determine the degrees of freedom associated with each source of variation based on the RCB design used in the experiment. The following sources of variation are identified:

TABLE 8.3.2.

Results of the Analyses of Variance for the Individual Seasons Using an RCB Design with 5 Nitrogen Application Rates and 4 Replications

Source of Variation	Degree of Freedom	Sum of Squares	Mean Square	F
		Spring Planting Season		
Replication	3	3.37	1.12	
Nitrogen	5	67.40	13.48	15.32**
Error	15	13.25	0.88	
		Winter Planting Season		
Replication	3	7.50	2.50	
Nitrogen	5	275.05	55.01	15.41**
Error	15	53.51	3.57	

Note: ** = significant at 1% level.

Season (S) $d.f.$	$= s - 1 = 1$
Reps. within season	$= s(r - 1) = 6$
Treatment (Nitrogen)	$= t - 1 = 5$
Season $\times$ Nitrogen	$= (s - 1)(t - 1) = 5$
Pooled error	$= s(r - 1)(t - 1) = 30$
Total $d.f.$	$= srt - 1 = 47$

STEP 2: Compute the sum of squares for the replications within the season and the pooled error as follows:

$$\text{Rep. within season } SS = \sum_{i=1}^{s} (\text{Rep. } SS)_i \qquad (8\text{-}19)$$

$$\text{Pooled error } SS = \sum_{i=1}^{s} (\text{Error } SS)_i \qquad (8\text{-}20)$$

Thus we have:

$$\text{Reps. within season } SS = 3.37 + 7.50 = 10.87$$

Pooled error $SS = 13.25 + 53.51 = 66.76$

STEP 3: Calculate the correction factor as:

$$C = \frac{\left(\sum_{i=1}^{s} G_i\right)^2}{srt} \qquad (8\text{-}21)$$

$$C = \frac{(721.0 + 725.0)^2}{48} = 43{,}560.75$$

STEP 4: Compute the sum of squares for the season, the nitrogen, and the Season × Nitrogen interaction as follows:

$$\text{Season } SS = \sum_{i=1}^{s} \frac{G_i^2}{rt} - C \qquad (8\text{-}22)$$

$$\text{Season } SS = \frac{(721)^2 + (725)^2}{4(6)} - 43{,}560.75$$

$$= 43{,}561.08 - 43{,}560.75$$

$$= 0.33$$

$$\text{Nitrogen } SS = \sum_{j=1}^{t} \frac{T_j^2}{sr} - C \qquad (8\text{-}23)$$

$$= \frac{\left[(207.9)^2 + (226.8)^2 + (232.8)^2 + (259.8)^2 + (257.8)^2 + (260.9)^2\right]}{2(4)}$$

$$- 43{,}560.75$$

$$= 43{,}860.27 - 43{,}560.75$$

$$= 299.52$$

$$\text{Season} \times \text{Treatment } SS = \sum_{i=1}^{s}\sum_{j=1}^{t} \frac{(st)_{ij}^2}{r} - (C + \text{Season } SS + \text{Treatment } SS) \quad (8\text{-}24)$$

where

$(st)_{ij}$ = the total of the jth treatment in the ith season.

$$\text{Season} \times \text{Treatment}\, SS = \frac{\left[(107.5)^2 + (118.2)^2 + \dots + (134.4)^2\right]}{4}$$

$$- (43,560.75 + 0.33 + 299.52)$$

$$= 43,903.53 - 43,860.60$$

$$= 42.93$$

STEP 5: Compute the mean square for each source of variation as shown below:

$$\text{Season}\, MS = \frac{\text{Season}\, SS}{s - 1} = \frac{0.33}{1} = 0.33$$

$$\text{Reps. within season}\, MS = \frac{\text{Reps. within season}\, SS}{s(r-1)} = \frac{10.87}{6} = 1.81$$

$$\text{Nitrogen}\, MS = \frac{\text{Nitrogen}\, SS}{t - 1} = \frac{299.52}{5} = 59.90$$

$$\text{Season} \times \text{Nitrogen}\, MS = \frac{\text{Season} \times \text{Nitrogen}\, SS}{(s-1)(t-1)} = \frac{42.93}{1(5)} = 8.59$$

$$\text{Pooled error}\, MS = \frac{\text{Pooled error}\, SS}{s(r-1)(t-1)} = \frac{66.76}{2(3)(5)} = 2.23$$

STEP 6: Compare the computed F value for each source of variation shown in Table 8.3.3 with the table F value in Appendix E.

The tabular F value for 5 and 30 degrees of freedom at the 1% level of significance is 3.70. The computed F value for the nitrogen treatment and the Season × Nitrogen interaction is significant. This implies that wheat yield is responsive to the application of nitrogen, but there is a difference in response between the spring and winter planting seasons.

TABLE 8.3.3.

Combined Analysis of Variance of a Randomized Complete Block Experiment over 2 Planting Seasons

Source of Variation	Degree of Freedom	Sum of Squares	Mean Square	F
Season	1	0.33	0.33	a
Reps. within season	6	10.87	1.81	<1
Nitrogen	5	299.52	59.90	26.86**
Season × Nitrogen	5	42.93	8.59	3.85**
Pooled error	30	66.76	2.23	
Total	47	420.41		

Note: ^{a}Since the *d.f.* is not adequate, a valid test of significance cannot be performed.
** = significant at 1% level.

STEP 7: Perform the test of homogeneity of error variance using the data from the individual analysis of variance given in Table 8.3.2. As was pointed out in Section 8.2, this test could be performed either through the application of the chi-square or the F test. We have used the F test to determine the homogeneity as follows:

$$F = \frac{\text{Larger error } MS}{\text{Smaller error } MS} \qquad (8\text{-}25)$$

$$F = \frac{3.57}{0.88} = 4.06$$

In comparing the computed F value with the corresponding F value (3.52 at the 1% level) found in Appendix E for 15 degrees of freedom, we reject the hypothesis of homogeneous error variances over seasons. Thus, we need to partition the Season × Nitrogen interaction into a set of orthogonal contrasts. Such a partition will shed light on why the relative performance of nitrogen applications differed over the spring and winter planting seasons.

STEP 8: Construct a set of mutually orthogonal contrasts on one of the factors. Since nitrogen is highly significant, the most natural set of contrasts that can explain the nature of the interaction between nitrogen and season will be the orthogonal polynomials on nitrogen. This means that the Season × Nitrogen sum of squares needs to be partitioned into Season × Nitrogen$_{linear}$, and Season × Nitrogen$_{quadratic}$, and so on. However, for simplicity of analysis we will consider only the linear and quadratic orthogonal polynomial contrasts, since the nitrogen rates tested in this experiment have equal intervals and $t = 6$. The single-degree of freedom contrast coefficients are obtained directly from Appendix J and are given below.

Nitrogen Rates, lb/acre	Orthogonal Polynomial Coefficient	
	Linear	Quadratic
0	-5	+5
50	-3	-1
100	-1	-4
150	+1	-4
200	+3	-1
250	+5	+5

If the treatment intervals were not equal, the polynomial coefficients have to be derived. Gomez and Gomez (1984) provide the derivation of the coefficients when treatment intervals are not equal.

STEP 9: Calculate the sum of squares for each single $d.f.$ contrast, using the following equations:

$$\text{Nitrogen}_{\text{Linear}}\ SS = \frac{\left[(-5)(N_1)+(-3)(N_2)+(-1)(N_3)+(1)(N_4)+(3)(N_5)+(5)(N_6)\right]^2}{(r)(a)\left[(-5)^2+(-3)^2+(-1)^2+(1)^2+(3)^2+(5)^2\right]} \quad (8\text{-}26)$$

$$\text{Nitrogen}_{\text{Quadratic}}\ SS = \frac{\left[(5)(N_1)+(-1)(N_2)+(-4)(N_3)+(-4)(N_4)+(-1)(N_5)+(5)(N_6)\right]^2}{(r)(a)\left[(5)^2+(-1)^2+(-4)^2+(-4)^2+(-1)^2+(5)^2\right]} \quad (8\text{-}27)$$

where

$N_1 \ldots N_6$ = Nitrogen totals

r = number of replications

a = number of seasons.

To compute the sum of squares, we need to construct the Season × Nitrogen table of totals, as shown in Table 8.3.4.

TABLE 8.3.4.

The Season × Nitrogen Table of Yield Totals Computed from Data in Table 8.3.1.

Nitrogen	Rep. I	Rep. II	Rep. III	Rep. IV	Nitrogen Total
0	52.9	49.5	54.4	51.1	207.9
50	54.4	58.4	58.2	55.8	226.8
100	60.3	55.1	57.9	59.5	232.8
150	61.7	66.3	63.8	68.0	259.8
200	62.3	64.9	65.7	64.9	257.8
250	64.7	66.6	66.6	63.0	260.9

The sum of squares are computed as follows:

$Nitrogen_{Linear}SS =$

$$\frac{\left[(-5)(207.9)+(-3)(226.8)+(-1)(232.8)+(1)(259.8)+(3)(257.8)+(5)(260.9)\right]^2}{(4)(2)\left[25+9+1+1+9+25\right]}$$

$$= \frac{(385)^2}{560} = 264.69$$

$Nitrogen_{Quadratic}SS =$

$$\frac{\left[(5)(207.9)+(-1)(226.8)+(-4)(232.8)+(-4)(259.8)+(-1)(257.8)+(5)(260.9)\right]^2}{(4)(2)\left[(5)^2+(-1)^2+(-4)^2+(-4)^2+(-1)^2+(5)^2\right]}$$

$$= 18.33$$

STEP 10: Following the procedures of Step 9, compute the sum of squares for each contrast based on the nitrogen totals at each season. This is shown below, and the results are presented in Table 8.3.5.

TABLE 8.3.5.

Treatment Total and the Linear and Quadratic Sum of Squares
Computed from Data in Table 8.3.1

Season	Treatment Total						Sum of Squares	
	0	50	100	150	200	250	Linear	Quadratic
Spring	107.5	118.2	117.9	126.9	124.0	126.5	52.64	9.44
Winter	100.4	108.6	114.9	132.9	133.8	134.4	248.16	12.69

Nitrogen$_{Linear}SS = $ (8-28)

$$\frac{\left[(-5)(N_1) + (-3)(N_2) + (-1)(N_3) + (1)(N_4) + (3)(N_5) + (5)(N_6)\right]^2}{r\left[(-5)^2 + (-3)^2 + (-1)^2 + (1)^2 + (3)^2 + (5)^2\right]}$$

$$= \frac{\left[(-5)(107.5) + (-3)(118.2) + (-1)(117.9) + (1)(126.9) + (3)(124.0) + (5)(126.5)\right]^2}{(4)\left[25 + 9 + 1 + 1 + 9 + 25\right]}$$

$= 52.64$

and the quadratic sum of square is:

(8-29)

Nitrogen$_{Quadratic}SS = $

$$\frac{\left[(5)(N_1) + (-1)(N_2) + (-4)(N_3) + (-4)(N_4) + (-1)(N_5) + (5)(N_6)\right]^2}{r\left[(5)^2 + (-1)^2 + (-4)^2 + (-4)^2 + (-1)^2 + (5)^2\right]}$$

$$= \frac{\left[(5)(107.5) + (-1)(118.2) + (-4)(117.9) + (-4)(126.9) + (-1)(124.0) + (5)(126.5)\right]^2}{(4)\left[25 + 1 + 16 + 16 + 1 + 25\right]}$$

$= 9.44$

STEP 11: Calculate the component of the Nitrogen × Season interaction corresponding to the set of mutually orthogonal contrasts as follows:

TABLE 8.3.6.

Analysis of Variance of a Partitioned Treatment Sum of Squares of a Randomized Complete Block Experiment

Source of Variation	Degree of Freedom	Sum of Squares	Mean Square	F
Season	1	0.33	0.33	a
Reps. within season	6	10.87	1.81	<1
Nitrogen	(5)	(299.52)	59.90	26.86**
Nitrogen$_L$	1	264.69	264.69	59.34**
Nitrogen$_Q$	1	18.33	18.33	8.21**
Nitrogen$_{Res.}$	3	16.50	5.50	2.47^{ns}
Nitrogen × Season	(5)	(42.93)	8.59	3.85**
Nitrogen$_L$ × Season	1	36.11	36.11	16.19**
Nitrogen$_Q$ × Season	1	3.80	3.80	1.70^{ns}
Nitrogen$_{Res.}$ × Season	3	3.02	1.01	1.00^{ns}
Pooled error	30	66.76	2.23	
Total	47			

Note: ^{a}Since the *d.f.* is not adequate, a valid test of significance cannot be performed.
^{ns} = nonsignificant, ** = significant at 1% level.

$$\text{Nitrogen}_i \times \text{Season} SS = \sum_{j=1}^{s} (\text{Nitrogen}_i\, SS)_j - \text{Nitrogen}_i\, SS \qquad (8\text{-}30)$$

The subscript i is used to distinguish between the linear and quadratic contrasts, and subscript j corresponds to the 2 seasons. Thus we have:

$$\text{Nitrogen}_{\text{Linear}} \times \text{Season} SS = (52.64 + 248.16) - 264.69$$
$$= 36.11$$
$$\text{Nitrogen}_{\text{Quadratic}} \times \text{Season} SS = (9.44 + 12.69) - 18.33$$
$$= 3.80$$

STEP 12: Enter all the values of the analysis in an ANOVA table as shown in Table 8.3.6. The analysis shows that only the linear component of the sum of squares varied significantly with season. This implies that the Season × Nitrogen interaction is mainly due to the difference in the linear part of the yield responses to nitrogen rates of the different seasons. We also note that both the linear and the quadratic component of the Season × Nitrogen interaction are significant. This implies that yield initially increased at lower rates, tapered off at a maximum, and finally decreased at higher rates.

The results indicate that different nitrogen rates need to be used for the spring and winter seasons. The average yield response to nitrogen fertilizer between seasons is important in making accurate economic fertilizer recommendations that are environmentally sound. Experiments such as this serve as the basis for making recommendations on the actual rates of nitrogen fertilizer to be used. Predictions based on multiyear and multilocation studies provide the researcher with information on changes in nitrogen fertilizer requirements for different seasons.

If the test of homogeneity is not significant, the combined analysis of the experiment would have been completed at this stage and the results reported as shown in Table 8.3.6. However, if the test of homogeneity is significant, then the appropriate test of significance requires that the error term be the component of the pooled error and not the pooled error itself. The following steps show how this is accomplished.

STEP 13: Partition the pooled error sum of squares into components corresponding to those of the Season × Nitrogen sum of squares as:

$$\text{Rep. within Season} \times \text{Nitrogen}_{Lin} = \tag{8-31}$$

$$\sum_{i=1}^{s} \left[\sum_{j=1}^{r} (\text{Nitrogen}_{Lin} \ SS)_{ji} - (\text{Nitrogen}_{Lin} \ SS)_{i} \right]$$

$$\text{Rep. within Season} \times \text{Nitrogen}_{Quad} = \tag{8-32}$$

$$\sum_{i=1}^{s} \left[\sum_{j=1}^{r} (\text{Nitrogen}_{Quad} \ SS)_{ji} - (\text{Nitrogen}_{Quad} \ SS)_{i} \right]$$

To determine what the values are for the left-hand side of the above equations, we first need to compute the sum of squares for the right-hand side. To do this, we need to compute the $(\text{Nitrogen}_{Lin} \ SS)_{ij}$, which is the linear component of the Nitrogen SS computed from the jth replication in the ith season. The $\text{Nitrogen}_{Lin} \ SS_i$ is the corresponding component that is computed from the totals over all replications, as shown in Table 8.3.7.

TABLE 8.3.7.

Treatment Totals and the Linear and Quadratic Sum of Squares for the Pooled Error SS

Replication	Treatment Total						Sum of Squares	
Number	0	50	100	150	200	250	Linear	Quadratic
				Spring				
I	27.8	30.0	29.9	31.4	30.8	30.5	4.33	2.50
II	24.6	29.2	28.3	32.0	31.3	31.2	26.41	6.13
III	28.2	30.1	29.7	31.7	29.9	33.0	9.22	0.002
IV	26.9	28.9	30.0	31.8	32.0	31.8	18.11	2.54
Total	107.5	118.2	117.9	126.9	124.0	126.5	52.64	9.44
				Winter				
I	25.1	24.4	30.4	30.3	31.5	34.2	63.56	0.06
II	24.9	29.2	26.8	34.3	33.6	35.4	76.55	0.39
III	26.2	28.1	28.2	32.1	35.8	33.6	58.51	0.44
IV	24.2	26.9	29.5	36.2	32.9	31.2	50.92	24.75
Total	100.4	108.6	114.9	132.9	133.8	134.4	248.16	12.69

$$\text{Nitrogen}_{Lin}SS = \tag{8-33}$$

$$\frac{[(-5)(N_1) + (-3)(N_2) + (-1)(N_3) + (1)(N_4) + (3)(N_5) + (5)(N_6)]^2}{[(-5)^2 + (-3)^2 + (-1)^2 + (1)^2 + (3)^2 + (5)^2]}$$

$$\text{Nitrogen}_{Lin}SS =$$

$$\frac{[(-5)(27.8) + (-3)(30.0) + (-1)(29.9) + (1)(31.4) + (3)(30.8) + (5)(30.5)]^2}{[(-5)^2 + (-3)^2 + (-1)^2 + (1)^2 + (3)^2 + (5)^2]}$$

$$= \frac{[-258.9 + 276.3]^2}{70} = \frac{302.76}{70}$$

$$= 4.33$$

Similarly, the quadratic sum of squares for nitrogen is computed as:

$$\text{Nitrogen}_{Quad}SS =$$

$$\frac{[(5)(27.8) + (-1)(30.0) + (-4)(29.9) + (-4)(31.4) + (-1)(30.8) + (5)(30.5)]^2}{[(5)^2 + (-1)^2 + (-4)^2 + (-4)^2 + (-4)^2 + (5)^2]}$$

$$= \frac{[-306 + 291.5]^2}{84} = \frac{210.25}{84}$$

$$= 2.50$$

Now we are ready to compute the components of the pooled error sum of squares as:

$$\text{Rep. within Season} \times \text{Nitrogen}_{Lin} = [(4.33 + 26.41 + 9.22 + 18.11) - 52.64]$$

$$+ [(63.56 + 76.55 + 58.51 + 50.92) - 248.16]$$

$$= 6.81$$

$$\text{Rep. within Season} \times \text{Nitrogen}_{Quad} = [(2.50 + 6.13 + 0.002 + 2.54) - 9.44]$$

$$+ [(0.06 + 0.39 + 0.44 + 24.75) - 12.69]$$

$$= 14.68$$

$$\text{Rep. within Season} \times \text{Nitrogen}_{Res} = \text{Pooled Error} - \text{Reps. within Season} \times \text{Nitrogen}_{Lin}SS$$

$$- \text{Reps within Season} \times \text{Nitrogen}_{Quad}SS$$

$$= 66.76 - (6.81 + 14.68)$$

$$= 66.76 - 21.49$$

$$= 45.27$$

The results summarized in Table 8.3.8 show each source of variation of the pooled error.

TABLE 8.3.8.
Components of the Pooled Error Sum of Squares

Source of Variation	Degree of Freedom	Sum of Squares	Mean Square
Pooled Error	(30)	(66.76)	2.23
Reps. within Season × Nitrogen$_L$	6	6.81	1.14
Reps. within Season × Nitrogen$_Q$	6	14.68	2.45
Reps. within Season × Nitrogen$_{Res.}$	18	45.27	2.52

STEP 14: Test the significance of the Season × Nitrogen interaction, using the 3 components of the pooled error, as follows:

$$F_{\text{Nitrogen}_{Lin} \times \text{Season}} = \frac{\text{Nitrogen}_{Lin} \times \text{Season}\,MS}{\text{Reps. within Season} \times \text{Nitrogen}_{Lin}MS} \tag{8-34}$$

$$F_{\text{Nitrogen}_{Quad} \times \text{Season}} = \frac{\text{Nitrogen}_{Quad} \times \text{Season}\,MS}{\text{Reps. within Season} \times \text{Nitrogen}_{Quad}MS} \tag{8-35}$$

$$F_{\text{Nitrogen}_{Res.} \times \text{Season}} = \frac{\text{Nitrogen}_{Res.} \times \text{Season} \, MS}{\text{Reps. within Season} \times \text{Nitrogen}_{Res.} MS} \qquad (8\text{-}36)$$

Thus, we have:

$$F_{\text{Nitrogen}_{Res.} \times \text{Season}} = \frac{36.11}{1.14} = 31.68$$

$$F_{\text{Nitrogen}_{Res.} \times \text{Season}} = \frac{3.80}{2.45} = 1.55$$

$$F_{\text{Nitrogen}_{Res.} \times \text{Season}} = \frac{3.02}{2.52} = 1.20$$

Again we note that only the linear component of the Nitrogen × Season is significant (F values from Appendix E for the 5% and 1% levels of significance are 5.99 and 13.74, respectively). The interpretation of the results remain the same as given in Step 12. The only difference in the analysis is that now we have used the appropriate error term in the F test of significance.

REFERENCES AND SUGGESTED READINGS

Cochran, G. W. and Cox, G. M. 1957. *Experimental Designs*. New York: John Wiley & Sons.

Cochran, G. W. 1954. "The combination of estimates from different experiments." *Biometrics* 10:101-129.

Fisher, R. A. 1971. *The Design of Experiments*. 9th ed. New York: Hafner Press.

Gomez, K. A., and Gomez, A. A. 1984. *Statistal Procedures for Agricultural Research*. 2nd ed. New York: John Wiley & Sons, Chap. 5.

Pearce, S. E. 1983. *The Agricultural Field Experiment: A Statistical Examination of Theory and Practice*. Chichester, England: John Wiley & Sons, chap 9.

Snedecor, G. W. and Cochran, W. G. 1967. *Statistical Methods*. 6th ed. Ames: Iowa State University Press, chap. 12.

Yates, F. and Cochran, W.G. 1938. "The analysis of groups of experiments."
J. Agric. Sci. 28:556-580.

EXERCISES

1. Agricultural engineers were interested in evaluating the impact of limited
capacity sprinkler irrigation systems on soybean yields. A 3-year field
study was conducted using a randomized complete block experimental
design. The irrigation treatments included a nonirrigated check (NI), irrigation scheduled by soil moisture depletion (SCH), irrigation which began
no earlier than the flowering stage (FL), and irrigation which began no
earlier than pod elongation (POD). Soybean yields as influenced by irrigation treatments were recorded as follows.

Soybean yield (bu/acre) as influenced by irrigation treatments

Year	Irrigation Treatment	Rep. I	Rep. II	Rep. III	Rep. IV
1990	NI	37	34	35	35
	POD	50	51	50	49
	FL	53	48	50	51
	SCH	46	48	47	47
1991	NI	32	29	25	30
	POD	34	33	30	35
	FL	43	42	40	38
	SCH	49	48	45	43
1992	NI	30	31	33	32
	POD	40	41	40	38
	FL	43	47	50	45
	SCH	43	47	48	46

(a) Perform a combined analysis over the years.
(b) Perform a test of homogeneity of variance. Can the hypothesis of
homogeneous variances be accepted?
(c) What can be said about the nature of the interaction effect
between the irrigation treatments and years?

2. As more attention is being paid to the issue of soil conservation, a number of practices are suggested as viable methods. Conventional and no-tillage systems are 2 being tested growing different crops. In a study in Nebraska, the performances of commercial corn hybrids were investigated using the conventional (CT) and no-till (NT) systems. The other objective of the study was to determine if hybrid × tillage system interactions exits for grain yield. The experimental design was a randomized complete block in a split-plot arrangement with 4 replications. The data collected are shown below.

Grain yield (bu/acre) under conventional and no-tillage systems for 10 corn hybrids

		CT Replications				NT Replications			
Year	Hybrid	I	II	III	IV	I	II	III	IV
1992	Pioneer 3737	135.4	132.0	133.9	136.1	125.2	122.4	126.0	122.8
	Pioneer 3744	100.9	101.8	102.0	105.3	110.0	109.6	106.4	108.5
	Funk G-4312	105.0	104.9	102.1	106.3	111.3	110.2	110.9	111.0
	Funk G-4342	122.5	121.3	122.9	122.0	110.9	110.2	111.2	110.5
	DeKalb 484	128.2	127.6	125.9	126.3	123.9	124.0	124.1	124.2
	DeKalb 524	126.8	127.2	126.9	128.0	133.1	132.5	131.9	132.3
	Cargil 842	138.1	137.6	137.2	135.7	120.0	123.3	124.7	120.5
1993	Pioneer 3737	133.4	132.0	133.9	131.1	118.2	126.4	123.0	123.7
	Pioneer 3744	117.9	115.8	112.0	115.5	120.0	129.7	126.3	128.4
	Funk G-4312	109.0	108.9	105.3	108.7	108.3	106.4	109.9	107.0
	Funk G-4342	104.5	101.3	106.9	107.4	120.9	122.2	121.2	120.8
	DeKalb 484	124.8	126.6	125.3	124.9	113.8	114.0	114.1	114.2
	DeKalb 524	136.9	137.2	136.9	140.0	132.1	136.5	131.3	135.8
	Cargil 842	108.9	107.6	107.2	105.8	118.3	119.3	120.7	121.4

(a) Perform a combined analysis over the years.
(b) Perform a test of homogeneity of variance. Can the hypothesis of homogeneous variances be accepted?
(c) What can be said about the nature of the interaction effect between the hybrid and tillage interaction?

3. In an attempt to determine the effect of harvest management on forage yield from "Nitro" a cultivar of alfalfa (*Medicago sativa* L.), a randomized complete block experiment with 4 replicates was conducted. The treatments consisted of the following 4 harvest management systems:

H_1: No harvest during the growing season;
H_2: Two harvests at bud and herbage regrowth harvested in the fall;
H_3: Three harvests at bud and herbage regrowth harvested in the fall;
H_4: Two harvests at first flower and herbage regrowth harvested in the fall.

The data on forage yield collected for the Summer and Fall are given below.

Effect of harvest management on forage yield (ton/acre)

Harvest Management	Rep. I	Rep. II	Rep. III	Rep. IV
		Summer		
H_1	—	—	—	—
H_2	1.9	2.0	2.1	2.0
H_3	3.0	3.1	3.0	2.8
H_4	2.5	2.4	2.8	2.3
		Fall		
H_1	0.5	2.3	1.4	2.1
H_2	0.7	0.9	0.8	0.7
H_3	0.4	0.6	0.2	0.1
H_4	0.3	0.5	0.3	0.4

(a) Perform a combined analysis over the seasons.
(b) Perform a test of homogeneity of variance. Can the hypothesis of homogeneous variances be accepted?
(c) What can be said about the nature of interaction effect between the seasons and harvest management practices?

4. Since grazing of hay fields during some part of the year is used as an alternative to harvested hay for beef cattle, an animal scientist compared 4 management systems to determine the yield of forage produced during the spring and fall. The data gathered from the randomized complete block experiment with 5 replications are shown below.

Total seasonal herbage dry matter harvested as influenced by management (lb/acre)

Management	Rep. I	Rep. II	Rep. III	Rep. IV	Rep. V
			Spring		
Hay only	4015	4020	4000	3995	4005
Spring and fall grazing	3756	3776	3789	3886	3554
Spring grazing	3300	3312	3310	3345	3305
Fall grazing	4583	4539	4601	4520	4495
			Fall		
Hay only	2334	2486	2300	2319	2398
Spring and fall grazing	3300	3198	3264	3310	3315
Spring grazing	2404	2400	2445	2487	2404
Fall grazing	3360	3354	3329	3352	3390

(a) Perform a combined analysis over the seasons.
(b) Perform a test of homogeneity of variance. Can the hypothesis of homogeneous variances be accepted?
(c) What can be said about the nature of the interaction effect between the seasons and the management systems?

Chapter **9**

REGRESSION AND CORRELATION ANALYSIS

9.1 BIVARIATE RELATIONSHIPS

In many agricultural research problems you will be faced with a situation where you are interested in the relationship that exists between 2 different random variables X and Y. Such a relationship is known as a bivariate relationship. For example, an animal scientist may be interested in the relationship between the amount of TDN (total digestible nutrient) and the average daily weight gain of the animal. An agricultural economist may be interested in the bivariate relationship between the appraised value of farm real estate, X, and its sale price, Y; or an agronomist may use a crop/weather model to analyze the effects of weather (X) on the yield of a crop (Y).

To determine if one variable is a predictor of another variable, we use the bivariate modeling technique. The simplest model for relating a variable Y to a single variable X is a straight line. This is referred to as a *linear* relationship. *Simple linear regression* is used as a technique to judge whether a relationship exists between Y and X. Furthermore, the technique is used to estimate the mean value of Y, and to predict a future value of Y for a given value of X.

In simple regression analysis, we are interested in describing the pattern of the functional nature of the relationship that exists between 2 variables. This is accomplished by estimating an equation called the regression equation. The variable to be estimated in the regression equation is called the dependent variable and is plotted on the vertical (or Y) axis. The variable used as the predictor of Y, which exerts influence in explaining the variation in the dependent variable, is called the independent variable. This variable is plotted on the horizontal (or X) axis.

The linear relationship between the 2 variables Y and X is expressed by the general equation for a straight line as:

$$Y = a + bX \tag{9-1}$$

where

Y = dependent variable

a = regression constant, or the Y intercept

b = regression coefficient, or the slope of the regression line

X = independent variable.

Other linear equations where the parameters occur in a linear fashion, although the relationship between X and Y is definitely not linear but quadratic and cubic, are:

$$Y = a + bX + cX^2 \tag{9-2}$$

$$Y = a + bX + cX^2 + dX^3 \tag{9-3}$$

The graphical presentations of the 3 models are shown in Figure 9.1.1.

In correlation analysis, on the other hand, we are simply interested in the magnitude or closeness of the relationship between 2 variables. Regression and correlation analyses work together. One asks if there is any relationship between 2 variables, and the other seeks to provide an answer to how close this relationship is.

In the following sections of this chapter we will discuss the techniques for estimating a regression equation, the standard error, and coefficients of determination and correlation. The concepts developed in Section 9.1 are directly applicable to our discussion of multiple regression in Section 9.4.

9.2   REGRESSION ANALYSIS

As was pointed out in the previous section, simple linear regression analysis is concerned with the relationship that exists between 2 variables. Researchers use prior knowledge and past research as a basis for selecting the

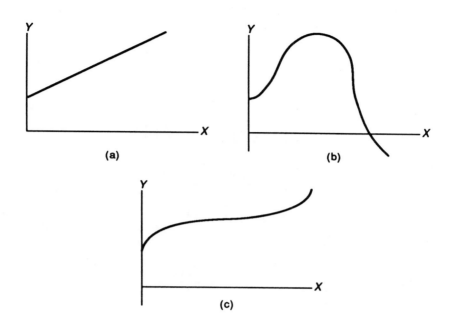

FIGURE 9.1.1

Polynomial relationships: (a) linear, (b) quadratic, and (c) cubic.

independent variables that are helpful in predicting the values of the dependent variable.

Once we have determined that there is a logical relationship between 2 variables, we can portray the relationship between the variables through a scatter diagram. A scatter diagram is a graph of the plotted points, each of which represents an observed pair of values of the dependent and independent variables. The scatter diagram serves 2 purposes: (1) it provides for a visual presentation of the relationship between 2 variables, and (2) it aids in choosing the appropriate type of model for estimation.

Example 9.2.1 presents a set of data that is used to illustrate how a scatter diagram is helpful in determining the presence or lack of linear relationship between the dependent and the independent variables.

Example 9.2.1 _____

An animal scientist interested in the milk yield of a lactating ewe has measured the milk yield (kg/day) by weighing the lamb before and after suckling. The following observations were obtained at different time intervals. The scatter diagram is constructed as follows.

Day	Yield
10	1.78
14	1.66
18	1.62
22	1.59
26	1.55
30	1.60
34	1.58
38	1.54
42	1.50
46	1.48
50	1.43
54	1.40
58	1.37
62	1.35
66	1.32

Solution

Following the standard convention of plotting the dependent variable along the Y axis and the independent variable along the X axis, we have the milk yield plotted along the Y axis and the day-intervals along the X axis. Figure 9.2.1 shows the scatter diagram for this problem.

An examination of Figure 9.2.1 shows that there is an *inverse*, or negative, relationship between the 2 variables. That is, as the X variable increases, the Y tends to decrease. If, on the other hand, Y increases as X increases, then there is a *direct*, or positive, relationship between variables.

The scatter diagram is also used to determine if there is a *linear* or *curvilinear* relationship between variables. If a straight line can be used to describe the relationship between variables X and Y, there exists a linear relationship. If the observed points in the scatter diagram fall along a curved line, there exists a *curvilinear relationship* between variables.

Figure 9.2.2 illustrates a number of different scatter diagrams depicting different relationships between variables. You will notice that scatter diagrams *a* and *b* illustrate a positive and negative linear relationship between 2 variables, respectively. Diagrams *c* and *d* show the positive and negative curvilinear relationship between variables X and Y.

Another curvilinear relationship is illustrated in *e* where X and Y rise at first, and then as X increases, Y decreases. Such a relationship is observed in agricul-

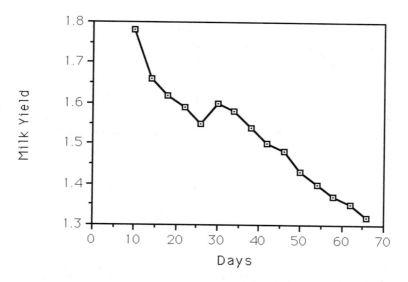

FIGURE 9.2.1.

A scatter diagram of the milk yield of a lactating ewe at 4-day intervals.

tural economics and business, for instance, where a farmer's earned income tends to rise with the age of the farmer, then decline after the farmer retires.

Figure 9.2.2(*f*) shows no relationship between the variables.

The Linear Regression Equation: The mathematical equation of a line such as the one in the scatter diagram in Figure 9.2.1 that describes the relationship between 2 variables is called the *regression* or *estimating equation*. The regression equation has its origins in the pioneering work of Sir Frances Galton (1877), who fitted lines to scatter diagrams of data on the heights of fathers and sons. Galton found that the heights of children of tall parents tended to regress toward the average height of the population. Galton referred to his equation as the regression equation.

The regression equation is determined by the use of a mathematical method referred to as the *least squares*. This method simply minimizes the sum of the squares of the vertical deviations about the line. Thus, the least squares method is a best fit in the sense that the $\sum (Y - \hat{Y})^2$ is less than it would be for any other possible straight line. Additionally, the least squares regression line has the following property:

$$\sum (Y - \hat{Y}) = 0$$

(9-4)

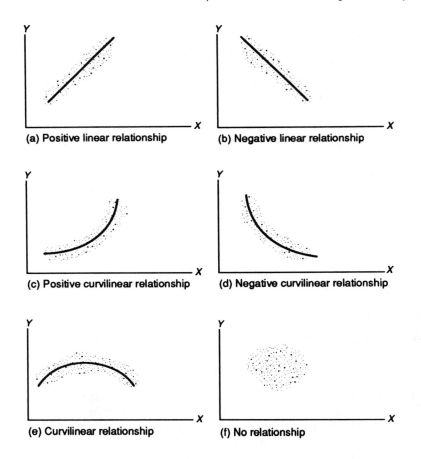

FIGURE 9.2.2.

Examples of linear and curviliner relationships found for scatter diagrams.

This characteristic makes the total of positive and negative deviations equal to 0.

You should note that the linear regression equation, Equation 9-1, is just an estimate of the relationship between the 2 variables in the population given in Equation 9-5 below:

$$\mu_{y.x} = A + BX \qquad (9\text{-}5)$$

where

$$\mu_{y.x} = \text{the mean of the } Y \text{ variable for a given } X \text{ value}$$

A and B = population parameters that must be estimated from sample data.

The regression equation can be calculated by 2 methods. The first involves solving simultaneously 2 equations called the *normal equations*. They are:

$$\sum Y = na + b \sum X \tag{9-6}$$

$$\sum XY = a \sum X + b \sum X^2 \tag{9-7}$$

We use Equations 9-6 and 9-7 to solve for a and b, and obtain the estimating or regression equation.

The second method of arriving at a least squares regression equation is using a computationally more convenient equation given below:

$$b = \frac{n \left(\sum XY\right) - \left(\sum X\right)\left(\sum Y\right)}{n \left(\sum X^2\right) - \left(\sum X\right)^2} \tag{9-8}$$

$$a = \frac{\sum Y}{n} - b \left(\frac{\sum X}{n}\right) = \bar{Y} - b\bar{X} \tag{9-9}$$

The following example will illustrate the computation of the regression equation, using either the normal equations or the shortcut formula given in Equations 9-8 and 9-9.

Example 9.2.2

An agronomist is interested in the relationship between maize yield, Y, and the amount of fertilizer applied to maize, X. In order to determine whether there is a relationship, the agronomist has divided a field into 9 plots of equal size in different localities in Iowa and has applied different amounts of fertilizer to each plot. The following yield data (in bushels) were recorded for the different amounts of fertilizer (in pounds) applied.

Yield Y	Fertilizer X
50	5
57	10
60	12
62	18
63	25
65	30
68	36
70	40
69	45
66	48

Solution

To compute the regression equation, the data in Table 9.2.1 are used.

TABLE 9.2.1.

Computation of Intermediate Values Needed for Calculating the Regression Equation

Yield Y	Fertilizer X	Y^2	XY	X^2
50	5	2,500	250	25
57	10	3,249	570	100
60	12	3,600	720	144
62	18	3,844	1,116	324
63	25	3,969	1,575	625
65	30	4,225	1,950	900
68	36	4,624	2,448	1,296
70	40	4,900	2,800	1,600
69	45	4,761	3,105	2,025
66	48	4,356	3,168	2,304
630	269	40,028	17,702	9,343

We will substitute the appropriate values from Table 9.2.1 into Equations 9-6 and 9-7 as follows:

$$\sum Y = na + b\sum X$$

$$\sum XY = a\sum X + b\sum X^2$$

$$630 = 10a + 269b$$

$$17,702 = 269a + 9,343b$$

To solve the above 2 equations for either of the unknowns a and b, we must eliminate one of the unknown coefficients. For example, to eliminate the unknown coefficient a in the above 2 equations, we multiply the first equation by 26.9. By doing so, the value of the a coefficient in the first equation is now equal to the value of a in the second equation. We then subtract the second from the first to obtain the value of b as shown below:

$$16,947 = 269a + 7,236.1b$$

$$-17,702 = 269a - 9.343b$$

$$\overline{}$$

$$-755 = -2,106.9b$$

$$b = 0.358$$

The value of b can be substituted in either Equation 9-6 or 9-7 to solve for the value of a.

$$630 = 10a + 269(0.358)$$

$$630 = 10a - 96.302$$

$$533.69 = 10a$$

$$a = 53.369$$

The least squares regression equation is:

$$\hat{Y} = 53.369 + 0.358X$$

To use the shortcut formula, you will obtain the following:

$$b = \frac{n\left(\sum XY\right) - \left(\sum X\right)\left(\sum Y\right)}{\left(\sum X^2\right)\left(\sum X\right)^2}$$

$$= \frac{10(17,702) - (269)(630)}{10(9,343) - (269)^2}$$

$$= \frac{7,550}{21,069}$$

$$= 0.358$$

We will now substitute the value of b into Equation 9-9 to obtain the intercept of the line, or a, as follows:

$$a = \frac{630}{10} - 0.358\left(\frac{269}{10}\right)$$

$$= 63.0 - 0.358(26.9)$$

$$= 63.0 - 9.630$$

$$= 53.369$$

Hence, the least squares regression line is:

$$\hat{Y} = 53.369 + 0.358X$$

In this estimated equation, $a = 53.369$ and is an estimate of the Y intercept. The value of a means that the yield of maize from a plot of land with no fertilizer added is equal to 53.369 bu. The b value indicates that the slope of the line is positive. This means that as fertilizer usage increases, so does the yield. The value of b implies that for each additional pound of fertilizer applied, the yield will increase by 0.358 bu within the range of values observed.

The above regression equation can also be used to estimate values of the dependent variable for given values of the independent variable. For example, if the agronomist wishes to estimate the yield from 42 pounds of fertilizer, the estimated yield will be:

$$\hat{Y} = 53.369 + 0.358(42)$$
$$= 53.369 + 15.036$$
$$= 68.41 \text{ bu. of maize}$$

You must keep in mind that the sample regression equation should not be used for prediction outside the range of values of the independent variable given in a sample.

In order to graph the regression line, we need 2 points. Since we have determined only one $(X = 42, Y = 68.41)$, we need one other point for graphing the regression line. The second point $(X = 10, Y = 56.95)$ is shown, along with the original data in Figure 9.2.3.

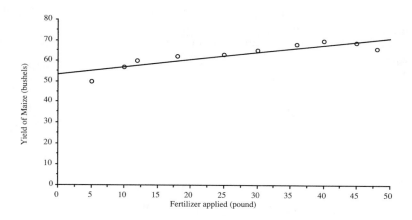

FIGURE 9.2.3.

Regression line of the maize yield.

The estimated yield values of 56.95 and 68.41 should be treated as average values. This means that in the future, the average yield will vary from sample to sample due to the fertility of the soil, the temperature, the amount of water used during the growing season, and a host of other factors.

In the next section, we will examine a measure that helps us determine whether the estimate made from the regression equation is dependable.

The Standard Error of Estimate: The regression equation is primarily used for estimation of the dependent variable, given values of the independent variable. Once we have estimated a regression equation, it is important to determine whether the estimate is dependable or not. Dependability is measured by the closeness of the relationship between the variables. If in a scatter diagram the points are scattered close to the regression line, there exists a close relationship between the variables. If, on the other hand, there is a great deal of dispersion between the points and the regression line, the estimate made from the regression equation is less reliable.

The standard error of estimate is used as a measure of scatter or dispersion of the points about the regression line, just as one uses the standard deviation to measure the deviation of the individual observations about the mean of those values. The smaller the standard error of estimate, the closer the estimate is likely to be to the ultimate value of the dependent variable. In the extreme case where every point falls on the regression line, the vertical deviations are all 0; that is, $S_{y.x} = 0$. In such a situation, the regression line provides perfect predictions. On the other hand, when the scatter is highly dispersed, making the vertical deviation large ($S_{y.x}$ is large), the predictions of Y made from the regression line are subject to sampling error.

The standard error of estimate ($S_{y.x}$) is computed by solving the following equation:

$$S_{y.x} = \sqrt{\frac{\sum (Y - \hat{Y})^2}{n - 2}}$$

(9-10)

where

Y = the dependent variable

$\hat{Y}$ = the estimated value of the dependent variable

n = the sample size.

The $n - 2$ value in the denominator represents the number of degrees of freedom around the fitted regression line. Generally, the denominator is $n - k$ where k represents the number of constants in the regression equation. In the case of a simple linear regression, we lose 2 degrees of freedom when a and b are used as estimates of the constants in the population regression line. Notice that Equation 9-10 requires a value of Y for each value of X. We must therefore compute the difference between each $\hat{Y}$ and the observed value of Y as shown in Table 9.2.2.

TABLE 9.2.2.

Computation of Intermediate Values Needed for Calculating the Standard Error of Estimate

Yield Y	Fertilizer X	$\hat{Y}$	$Y - \hat{Y}$	$(Y - \hat{Y})^2$
50	5	55.159	−5.159	26.615
57	10	56.949	0.051	0.003
60	12	57.665	2.335	5.452
62	18	59.813	2.187	4.783
63	25	62.319	0.681	0.464
65	30	64.109	0.891	0.794
68	36	66.257	1.743	3.038
70	40	67.689	2.311	5.341
69	45	69.479	−0.479	0.229
66	48	70.553	−4.553	20.729
630	269	630.000	0.0	67.488

The standard error of estimate for Example 9.2.2 is calculated as follows:

$$S_{y.x} = \sqrt{\frac{\sum (Y - \hat{Y})^2}{n - 2}}$$ (9-10)

$$= \sqrt{\frac{67.448}{8}}$$

$$= 2.90 \text{ bu. of maize}$$

The above computational method requires a great deal of arithmetic especially when large numbers of observations are involved. To minimize cumbersome arithmetic, the following shortcut formula is used in computing the standard error of estimate.

$$S_{y.x} = \sqrt{\frac{\sum Y^2 - a\left(\sum Y\right) - b\left(\sum XY\right)}{n - 2}}$$ (9-11)

All the values needed to compute the standard error of estimate are available from Table 9.2.1 and the previously obtained values of a and b.

$$S_{y.x} = \sqrt{\frac{40,028 - 53.369(630) - 0.358(17,702)}{8}}$$

$$= \sqrt{\frac{68.214}{8}}$$

$$= 2.92 \text{ bu. of maize}$$

The answers from the 2 approaches are similar, as expected. The minute difference in the 2 values of $S_{y.x}$ is due to rounding.

Since the standard error of estimate is theoretically similar to the standard deviation, there is also similarity in the interpretation of the standard error of estimate. If the scatter about the regression line is normally distributed, and we have a large sample, approximately 68% of the points in the scatter diagram will fall within 1 standard error of estimate above and below the regression line; 95.4% of the points will fall within 2 standard errors of estimate above and below the regression line, and virtually all points above

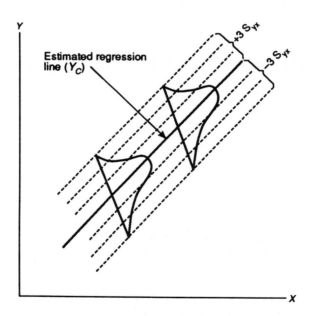

FIGURE 9.2.4.

Illustration of the standard error of estimate about the estimated regression line.

and below the regression line will fall within 3 standard errors of estimate, as shown in Figure 9.2.4.

Confidence Interval Estimate: We used the regression equation to estimate the value of Y given a value of X. An estimate of Y_1 was obtained by simply inserting a value for X_1 into the regression equation $Y_1 = a + bX_1$. The estimate of Y_1 is nothing more than a *point estimate*. To attach some confidence to this point estimate, we use the standard error of estimate to compute an interval estimate where probability value may be assigned to it. For a small sample, the interval estimate for an individual Y given value of X, say X_o, is computed by the following equation:

$$Y_i = \hat{Y} \pm t\,(S_{y.x}) \sqrt{1 + \frac{1}{n} + \frac{(X_o - \bar{X})^2}{\sum X^2 - \frac{(\sum X)^2}{n}}} \tag{9-12}$$

Earlier, the agronomist estimated the yield of maize to be 68.41 bushels when he applied 42 pounds of fertilizer. How much confidence he has in this estimate depends on probability value attached to such estimate. To construct a 95% prediction interval, we use Equation 9-12 and the data from Table 9.2.2. The critical value of t for the 95% level of confidence is given as 2.306 from Appendix I. The computation of the prediction interval is as follows:

$$Y_i = 68.41 \pm 7.35$$

or

$$61.72 < Y_i < 75.77 \text{ bushels}$$

Hence, the prediction interval is from 61.72 bushels to 75.77 bushels. To make a probabilistic interpretation of this number, we would say that we are 95% confident that the single prediction interval constructed includes the true yield.

The prediction interval for an individual value of Y when using large samples is determined by the following expression:

$$\hat{Y} \pm z\,S_{y.x} \tag{9-13}$$

9.3   CORRELATIONAL ANALYSIS

In regression analysis we emphasized estimation of an equation that describes the relationship between 2 variables. In this section, we are interested in those measures that verify the degree of closeness or association between 2 variables, and the strength of the relationship between them. We will examine 2 correlation measures: the *coefficient of determination* and the *coefficient of correlation*.

Before using these correlation measures, it is important to have an understanding of the assumptions of the 2-variable correlation models. In correlation analysis, we make the assumption that both X and Y are random variables; furthermore, they are normally distributed. Also, the standard deviations of the Ys are equal for all values of X, and vice versa.

Coefficient of Determination: As a measure of closeness between variables, the coefficient of determination (r^2) can provide an answer to how well the least-squares line fits the observed data. The relative variation of the Y values around the regression line and the corresponding variation around the mean of the Y variable can be used to explain the correlation that may exist between X and Y. Conceptually, Figure 9.3.1 illustrates 3 different deviations—namely, the total deviation, the explained deviation, and the unexplained deviation—that exist between a single point Y and the mean and the regression line.

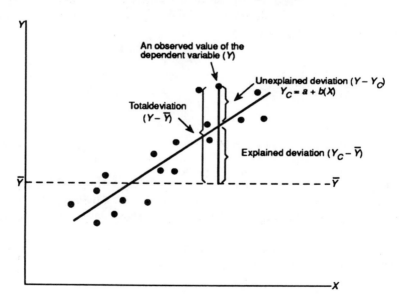

FIGURE 9.3.1.

Total deviation, explained deviation, and unexplained deviation for one observed value of Y.

The vertical distance between the regression line and the line is the explained deviation, and the vertical distance of the observed Y from the regression line is the unexplained deviation. The unexplained deviation represents that portion of total deviation that was not explained by the regression line. The distance between Y and $\bar{Y}$ is called the total deviation. Stated differently, the total deviation is the sum of the explained and the unexplained deviation, as given below:

Total deviation = explained deviation + unexplained deviation

$$Y - \bar{Y} = (\hat{Y} - \bar{Y}) + (Y - \hat{Y}) \tag{9-14}$$

To transform the above equation into a measure of variability, we simply square each of the deviations in Equation 9-14 and sum for all observations to obtain the squared deviations.

Total sum of squares = explained sum of squares + unexplained sum of squares

$$\sum (Y - \bar{Y})^2 = \sum (\hat{Y} - \bar{Y})^2 + \sum (Y - \hat{Y})^2 \tag{9-15}$$

The term on the left-hand side of Equation 9-15 is now the *total sum of squares,* which measures the dispersion of the observed value of the Y about their mean $\bar{Y}$. Similarly, on the right-hand side of Equation 9-15 we have the explained and unexplained sum of squares, respectively.

Given the above relationships, we can define the sample coefficient of determination as:

$$r^2 = 1 - \frac{\text{Unexplained sum of squares}}{\text{Total sum of squares}}$$

or

$$r^2 = \frac{\text{Explained sum of squares}}{\text{Total sum of squares}}$$

or

$$r^2 = \frac{\sum (\hat{Y} - \bar{Y})^2}{\sum (Y - \bar{Y})^2} \tag{9-16}$$

To compute r^2 using Equation 9-16, we need the value of the explained and total sum of squares, as computed in Table 9.3.1.

TABLE 9.3.1.
Calculation of the Sum of Squares when $\overline{Y}$ = 63

Yield Y	Fertilizer X	$\widehat{Y}$	$Y - \overline{Y}$	$(\widehat{Y} - \overline{Y})^2$
50	5	55.159	169	61.48
57	10	56.949	36	36.61
60	12	57.665	9	28.46
62	18	59.813	1	10.15
63	25	62.319	0	0.46
65	30	64.109	4	1.22
68	36	66.257	25	10.60
70	40	67.689	49	21.98
69	45	69.479	36	41.97
66	48	70.553	9	57.04
630	269	630.000	338	269.97

Substituting the value of the explained and the total sum of squares into Equation 9-15 we get

$$r^2 = \frac{269.97}{338} = 0.798$$

As is apparent, computation of r^2 using Equation 9-16 is tedious, particularly with a large sample. To remedy the situation, we use a shortcut formula that utilizes the estimated regression coefficients and the intermediate values used to compute the regression equation. Thus, the shortcut formula for computing the sample coefficient of determination is:

$$r^2 = \frac{a \sum Y + b \sum XY - n\,\overline{Y}^2}{\sum Y^2 - n\,\overline{Y}^2} \tag{9-17}$$

To calculate the r^2 for the agronomist in Example 9.2.2, we have:

$$r^2 = \frac{53.369(630) + 0.358(17,702) - 10(63)^2}{40,028 - 10(63)^2}$$

$$= \frac{33,622.47 + 6,337.32 - 39,690}{40,028 - 39,690}$$

$$= \frac{269.79}{338}$$

$$= 0.798$$

The calculated $r^2 = 0.798$ signifies that 79.8% of the total variation in yield of maize Y can be explained by the relationship between the yield and the amount of fertilizer X applied. Since the greatest possible value that r can have is 1, the calculated r in this example implies a strong linear relationship between X and Y. Similar to the sample coefficient of determination, the population coefficient of determination ρ^2 is equal to the ratio of the explained sum of squares to the total sum of squares.

Sample Correlation Coefficient Without Regression Analysis: Another parameter that measures the strength of the linear relationship between 2 variables X and Y is the *coefficient of correlation*. The sample coefficient of correlation (r) is defined as the square root of the coefficient of determination. The population correlation coefficient measures the strength of the relationship between 2 variables in the population. Thus, the sample and population correlation coefficients are

$$r = \sqrt{r^2} \tag{9-18}$$

$$\rho = \sqrt{\rho^2} \tag{9-19}$$

The correlation coefficient can be any value between -1 and +1 inclusive. When r equals -1, there is a perfect inverse linear correlation between the variables of interest. When r equals 1, there is a perfect direct linear correlation between X and Y. When r equals 0, the variables X and Y are not linearly correlated. Figure 9.3.2 shows different scatter diagrams representing various simple correlation coefficients.

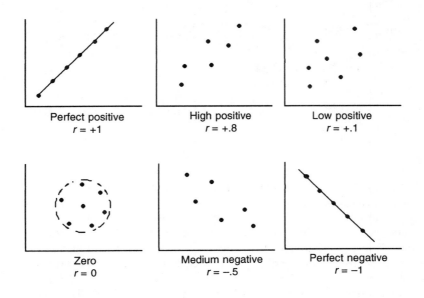

FIGURE 9.3.2.

Various scatter diagrams with different correlation coefficients.

The algebraic sign of r is the same as that of b in the regression equation. Thus, if the regression coefficient b is positive, then r will also have a positive value. Similarly, if the regression coefficient is negative, we will expect r to have a negative value.

The sample coefficient of correlation for Example 9.2.2 is :

$$r = \sqrt{0.798}$$
$$= 0.89$$

Note that the sign of this sample correlation coefficient is positive, as was the b coefficient.

A word of caution is needed with respect to the interpretation of the correlation coefficient. In the case of the coefficient of determination r^2, we could interpret that parameter as a proportion or as a percentage. However, when the square root of a percentage is taken, as is the case with the correlation coefficient, the specific meaning attached to it becomes obscure. Therefore, we could merely conclude that the closer the r value is to 1, the better the correlation is between X and Y. Since r^2 is a decimal value, its square root r is a larger number. This may give a false impression that a high degree of correlation exists between X and Y. For example, consider a situation where r = 0.70 which indicates a relatively high degree of association. However, since $r^2 = 0.49$, the reduction in total variation is only 49%, or less than half. Despite the common use of the correlation coefficient, it is best to use the coefficient of determination to explain the degree of association between X and Y when regression analysis is employed.

The correlation coefficient that is computed without performing regression analysis does provide more meaningful results. In the following section, you will be shown how to compute the correlation coefficient without performing a regression analysis.

There are many instances where a researcher is not interested in making a prediction, but is interested in whether there is a relationship between X and Y. For example, an animal scientist may be interested in whether increasing the concentration of a drug's dosage reduces the symptoms. In such cases the following formula, referred to as the *Pearson sample correlation coefficient*, can be used:

$$r = \frac{\sum XY - n\bar{X}\,\bar{Y}}{\sqrt{(\sum X^2 - n\bar{X}^2)(\sum Y^2 - n\bar{Y}^2)}} \tag{9-20}$$

We may use the intermediate calculations found earlier in Example 9.2.2 to compute the sample correlation coefficient as shown below:

$$n = 10 \qquad\qquad \bar{X} = 26.9 \qquad\qquad \bar{Y} = 63$$

$$\sum XY = 17{,}702 \qquad\qquad \sum X^2 = 9{,}343 \qquad\qquad \sum Y^2 = 40{,}028$$

$$r = \frac{17{,}702 - 10(26.9)(63)}{\sqrt{[(9{,}343 - 10(26.9)^2][40{,}028 - 10(63)^2]}}$$

$$r = \frac{17{,}702 - 16{,}947}{\sqrt{712{,}132.2}}$$

$$= \frac{755}{843.881}$$

$$= 0.89$$

The correlation coefficient 0.89 computed with Equation 9-22 is the same as the value we found earlier by taking the positive square root of r^2.

Inferences Regarding Regression and Correlation Coefficients: So far, our discussions have centered around the computation and interpretation of the regression and correlation coefficients. The topic of this section is our degree of confidence that these coefficients do not contain sampling error and that they correspond to the population parameters. Hypothesis testing, or confidence interval estimation, is often used to determine whether the sample data provide sufficient evidence to indicate that the estimated regression coefficient differs from 0. If we can reject the null hypothesis that b is not equal to 0, we can conclude that X and Y are linearly related.

To illustrate the hypothesis-testing procedure for a regression coefficient, let us use the data from Example 9.2.2. The null and alternative hypotheses regarding the regression coefficient may be stated as:

$$H_0 = \beta = 0$$

$$H_1 = \beta \neq 0$$

To test this hypothesis, the agronomist wishes to use a 0.05 level of significance. The procedure involves a 2-tailed test in which the test statistic is:

$$t = \frac{b - \beta}{S_b} \tag{9-21}$$

where S_b is the estimated standard error of the regression coefficient and is computed as follows:

$$S_b = \frac{S_{y.x}}{\sqrt{\sum X^2 - n\bar{X}^2}} \tag{9-22}$$

We substitute the appropriate values into Equations 9-22 and 9-21 to perform the test.

$$S_b = \frac{2.90}{\sqrt{9,343 - 10(26.9)^2}}$$

$$= 0.06$$

The calculated test statistic is

$$t = \frac{0.358}{0.06}$$

$$= 5.966$$

Given an $\alpha = 0.05$ and 8 degrees of freedom $(10 - 2 = 8)$, the critical value of t from Appendix I is 2.306. Since the computed t exceeds the critical value of 2.306, the null hypothesis is rejected at the 0.05 level of significance, and we conclude that the slope of the regression line is not 0.

Similar to the test of significance of b, the slope of the regression equation, we can perform a test for the significance of a linear relationship between X and Y. In this test, we are basically interested in knowing whether there is correlation in the population from which the sample was selected. We may state the null and alternative hypotheses as:

$$H_0 = \rho = 0$$

$$H_1 = \rho \neq 0$$

The test statistic for samples of small size is:

$$t = \frac{r\sqrt{n-2}}{\sqrt{1-r^2}} \qquad (9\text{-}23)$$

Again, using a 0.05 level of significance and $n - 2$ degrees of freedom, the critical t from Appendix I is 2.306. The decision rule states that if the computed t falls within ± 2.306, we should accept the null hypothesis; otherwise, reject it.

$$t = \frac{0.89\sqrt{10-2}}{\sqrt{1-(0.89)^2}}$$

$$= \frac{0.89\,(2.828)}{\sqrt{1-0.792}}$$

$$= 5.52$$

Since 5.52 exceeds the critical t value of 2.36, the null hypothesis is rejected, and we conclude that X and Y are linearly related.

For large samples, the test of significance of correlation can be performed using the following equation:

$$z = \frac{r}{\frac{1}{\sqrt{n-1}}} \tag{9-24}$$

F Test: An Illustration: Instead of using the t distribution to test a coefficient of correlation for significance, we may use an analysis of variance or the F ratio. In computing the r^2 and the $S_{y.x}$, we partitioned the total sum of squares into explained and unexplained sums of squares. In order to perform the F test, we first set up the variance table, determine the degrees of freedom, and then compute F as a test of our hypothesis. The null and alternative hypotheses are:

$$H_0 = \rho = 0$$

$$H_1 = \rho \neq 0$$

We will test the hypothesis at the 0.05 level of significance. Table 9.3.2 shows the computation of the F ratio. The numerical values of the explained and total sums of squares used in Table 9.3.2 were presented earlier in Table 9.3.1. The degrees of freedom associated with the explained variation is always equal to the number of independent variables used to explain variations in the dependent variable ($n - k$, where k refers to the number of independent variables). Thus, for the present problem we have only 1 degree of freedom for the explained variation. The degrees of freedom associated with the unexplained variation is found simply by subtracting the degrees of freedom of the explained sum of squares from the degrees of freedom of the total sum of squares. Hence, the unexplained sum of squares is equal to 8. Computing the F ratio, as has been shown before, requires the variance estimate, which is found by dividing the explained and unexplained sums of squares by their respective degrees of freedom. The variance estimates are shown in column 4 of Table 9.3.2.

TABLE 9.3.2.
Variance Table for Testing Significance of Correlation by F Ratio

(1) Source of Variation	(2) Sum of Squares	(3) Degree of Freedom	(4) Variance Estimate	(5) F
Explained	269.97	1	269.97	31.76
Unexplained	68.03	8	8.50	
Total	338.00	9		

The F test is the ratio between the variance explained by the regression and the variance that is not explained by the regression. The F ratio is shown in Table 9.3.2.

The computed F ratio is compared with the critical value of F given in Appendix E. The critical F for the 0.05 level of significance and 1 and 8 degrees of freedom is 5.32. Our tested $F = 31.76$ is much greater than 5.32. Therefore, we reject the null hypothesis of no correlation, and conclude that the relationship between amount of fertilizer applied and yield is significant.

9.4 CURVILINEAR REGRESSION ANALYSIS

In our discussions so far, we have only considered the simplest form of a relationship between 2 variables, namely the linear relationship. While the simple linear regression may be considered a powerful tool in analyzing the relationship that may exist between 2 variables, the assumption of linearity has serious limitations. There are many instances when the straight-line does not provide an adequate explanation of the relationship between 2 variables. When theory suggests that the underlying bivariate relationship is nonlinear, or a visual check of the scatter diagram indicates a curvilinear relationship, it is best to perform a nonlinear regression analysis. For example, the crop yield may increase with added application of fertilizer up to a point, beyond which it will decrease. This type of a relationship suggests a curvilinear (in this case a parabola) model.

How we decide on which curve to use depends on the natural relationship that exists between the variables and our own knowledge and experience of the relationship. In dealing with agricultural data, there may be instances when the relation between 2 variables is so complex that we are unable to use a simple equation for such a relationship. Under such conditions it is best to find an equation that provides a good fit to the data, without making any claims that the equation expresses any natural relation.

There are many nonlinear models that can be used with agricultural data. The following are some examples of the nonlinear equations that can easily be transformed to their linear counterparts and analyzed as linear equations.

1.
$$Y = \alpha \, e^{\beta X} \tag{9-25}$$

2.
$$Y = \alpha \beta^X \tag{9-26}$$

3.
$$\frac{1}{Y} = \alpha + \beta X \tag{9-27}$$

4.
$$Y = \alpha + \frac{\beta}{X} \tag{9-28}$$

5.
$$Y = \left(\alpha + \frac{\beta}{X}\right)^{-1}$$
(9-29)

For illustrative purposes we have selected 2 of the nonlinear models that are extensively used with agricultural and economic data to perform the analysis.

The Exponential Growth Model: In its simplest form, the model is used for the decay or growth of some variable with time. The general equation for this curve is:

$$Y = \alpha e^{\beta X}$$

In this equation, X (time) appears as an exponent, and the coefficient β describes the rate of growth or decay, and $e \cong 2.718$ is the Euler's constant, which appears in the formula for the normal curve and is also the base for natural logs. The advantage of using this base is that $\beta \cong$ the growth rate.

The assumptions of the model are that the rate of decay is proportional to the current value of Y, and the error term is multiplicative rather than additive, as it is reasonable to assume that large errors are associated with large values of the dependent variable Y. Thus, the statistical model is:

$$Y = \alpha e^{\beta X}.u$$
(9-30)

The nonlinear models can be transformed (as was discussed in Chapter 2, Section 2.4) to a linear form by taking the logarithm of the equation. For example, transforming Equation 9-30 into its logarithm we get:

$$\log Y = \log \alpha + \beta X + \log u$$
(9-31)

where

$$Y' \equiv \log Y$$

$$\alpha' \equiv \log \alpha$$

$$e \equiv \log u$$

Then, we can rewrite (9-31) in the standard linear form as:

$$Y' = \alpha' + \beta X + e$$

Exponential models have been extensively used in laboratory experiments where the studies of insects treated with chemicals are conducted. In agricultural field experiments that study the decay of chemicals in the soil or in animals, the exponential model is used. Other situations in which the exponential models have been useful are growth studies where the response variable either increases with time, t, or as a result of an increasing level of a stimulus variable, X. Figure 9.4.1 shows the simple exponential model with growth and decay curves.

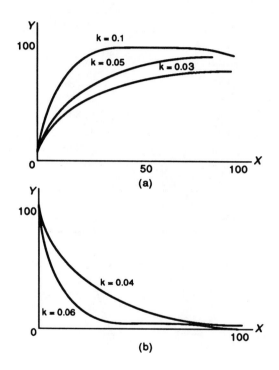

FIGURE 9.4.1.

Simple exponential models: (a) gradual approach of yield to an upper limit, (b) decay curve with time.

When considering different forms of writing a particular relationship, it must be kept in mind that in fitting the model to the data, the choice of correct error structure assumption is critically important.

Example 9.4.1 illustrates the use of an exponential model and how it can be fitted.

Example 9.4.1

From the early 1900s, there have been major resettlements of the population in many parts of California. Population data for the period 1900 to 1980 show the growth in a southern California community. Figure 9.4.2 shows the growth in population with a continual, though varying, percentage increase. This type of growth is similar to compound interest or unrestrained biological growth. Therefore, the appropriate model to use in analyzing the data is the exponential model.

Year	Population
1900	1,520
1910	2,800
1920	3,200
1930	18,600
1940	19,800
1950	32,560
1960	65,490
1970	150,450
1980	245,900

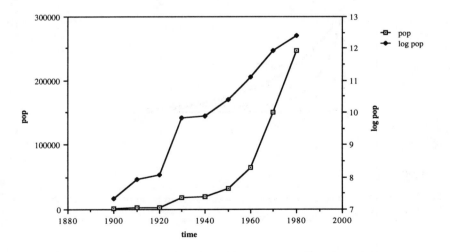

FIGURE 9.4.2.

Population growth, 1900-1980, and its exponential fit.

Solution _____

STEP 1: Transform the population data into the natural logs and compute the totals as shown in Table 9.4.1.

TABLE 9.4.1.

Population of a Southern California Community, 1900-1980

Year	Time (X)	Population (Y)	Log Y
1900	0	1,520	7.326
1910	1	2,806	7.940
1920	2	3,210	8.074
1930	3	18,630	9.833
1940	4	19,800	9.893
1950	5	32,560	10.391
1960	6	65,495	11.090
1970	7	150,457	11.921
1980	8	245,904	12.413
Total	36		88.881
Sum of Squares	204		903.627
Sum of X log Y			394.405

STEP 2: Compute the regression coefficients as shown below:

$$\sum x^2 = \sum X^2 - \frac{(\sum X)^2}{n} \tag{9-33}$$

$$= 204 - \frac{(36)^2}{9} = 60$$

$$\sum y'^2 = \sum Y'^2 - \frac{(\sum Y'^2)}{n} \tag{9-34}$$

$$= 903.627 - \frac{(88.881)^2}{9} = 25.867$$

$$\sum xy' = \sum XY' - \frac{(\sum X \sum Y')}{n} \tag{9-35}$$

$$= 394.405 - \frac{36(88.881)}{9} = 38.881$$

$$b' = \frac{\sum xy'}{\sum x^2} \tag{9-36}$$

$$= \frac{38.881}{60} = 0.648$$

$$a' = \frac{\sum Y'^2}{n} - b' \frac{\sum X}{n} \tag{9-37}$$

$$= \frac{88.881}{9} - 0.648 \left(\frac{36}{9} \right) = 7.284$$

The resultant regression equation is:

$$\hat{Y}' = 7.284 + 0.648x$$

STEP 3: Transform the above equation back into the exponential model, as follows:

$$\log Y = 7.284 + 0.648x$$

Taking antilogs (exponential)

$$\hat{Y} = e^{7.284} e^{.648x}$$

That is,

$$\hat{Y} = 1,457 e^{.648x}$$

For convenience we have left time (x) in deviation form. Thus, we can interpret the coefficient $\hat{\alpha}$ = 1,457 as the estimate of the population in 1900 (when $x = 0$). The coefficient $\hat{\beta}$ = .648 = 64.80% is the appropriate population growth rate every 10 years.

STEP 4: Compute the coefficient of determination as follows:

$$r^2 = \frac{\left(\sum xy'\right)^2}{\sum x^2 \sum y'^2} \tag{9-38}$$

$$= \frac{(38.881)^2}{(60)(25.867)}$$

$$= 0.97$$

From the scatter diagram in Figure 9.4.2 and the analysis performed, it appears that the exponential curve fits the data to past population growth better than any straight line. However, it is important to keep in mind that using it for any short-term prediction of the population is unwarranted. The concern mostly stems from the fact that in this simple growth model, the error u is likely to be serially correlated and thus has to be accounted for in any prediction.

Another example of the exponential model is the Cobb-Douglas production function, which is given as:

$$Q = \alpha K^\beta L^\gamma u \tag{9-39}$$

where

Q = quantity produced

K = capital

L = labor

u = multiplicative error term

α, β, and γ = parameters to be estimated

As before, by taking the logs of Equation 9-39, we obtain

$$\log Q = (\log \alpha) + \beta (\log K) + \gamma (\log L) + (\log u) \tag{9-40}$$

which is of the standard form:

$$Y = \alpha' + \beta X + \gamma Z + e \qquad (9\text{-}41)$$

The demand function is also an exponential model shown below:

$$Q = \alpha P^{\beta} u \qquad (9\text{-}42)$$

where

$$Q \quad = \text{quantity demanded}$$

$$P \quad = \text{price.}$$

Equation 9-42 can be linearized by taking its logarithms as shown:

$$\log Q = (\log \alpha) + \beta \log P + \log u \qquad (9\text{-}43)$$

which in standard form is:

$$Y = \alpha' + \beta X + e \qquad (9\text{-}44)$$

The Polynomial Curve: The polynomial is by far the most widely used equation to describe the relation between 2 variables. Snedecor and Cochran (1980) pointed out that when faced by nonlinear regression, and where one has no knowledge of a theoretical equation to use, the second-degree polynomial in many instances provides a satisfactory fit.

The general form of a polynomial equation is:

$$Y = a + \beta_1 X + \beta_2 X^2 + \ldots + \beta_k X^k \qquad (9\text{-}45)$$

In its simplest form, the equation will have the first 2 terms on the right-hand side of the equation, and is known as the equation for a straight line. Adding another term $(\beta_2 X^2)$ to the straight-line equation gives us the second-degree or a quadratic equation, the graph of which is a parabola. As more terms are added to the equation, the degree or power of X increases. A polynomial equation with a third degree (X^3) is called a cubic, and those with a fourth and fifth degrees are referred to as the quartic and quintic, respectively. Figure 9.4.3 shows examples of the polynomial curves with different degrees.

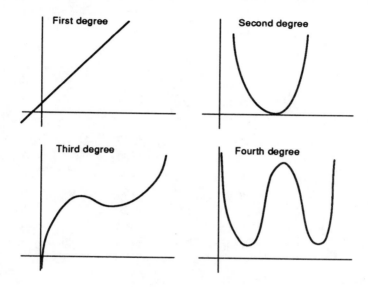

FIGURE 9.4.3.

Typical shapes of polynomial curves.

To illustrate the method and some of its applications, the following example is used.

Example 9.4.2

In order to determine the relationship between yield and the protein content of soybean, an agronomist gathered the following data. The agronomist is particularly interested in estimating the protein content for the different yields.

Furthermore, she wishes to test if there is any departure from linearity in the data. Thus, the regression equation is assumed to have the following functional form:

$$Y = \alpha + \beta X + \gamma X^2$$

Soybean Yield (bu/ac)	Protein (%)	Soybean Yield (bu/ac)	Protein (%)
29.50	40.80	31.20	37.60
28.60	42.10	34.50	35.10
28.10	41.30	38.10	35.80
31.20	40.30	33.20	36.90
32.40	39.60	36.10	35.80
30.20	43.40	33.60	36.80
38.30	35.30	33.00	35.20
30.40	41.50	32.30	36.60
30.00	39.70	37.30	34.40
31.30	39.20	35.50	34.00
29.80	41.90	34.30	37.20
34.50	36.05	33.90	37.60
30.20	39.00	39.20	37.80
27.00	43.10	39.15	35.60
32.00	37.30	38.40	36.25

Solution

STEP 1: Since it is assumed by the agronomist that the functional form of the regression equation is quadratic, and the data gathered are only on 2 variables (yield and protein content), we need to add another variable to the regression equation. The added variable is the square of the yield of soybean (column 2 of Table 9.4.2) and is treated like a third variable in the equation. Because of this added variable, we now have 2 independent variables, and thus a multiple regression equation. Since the step-by-step analysis of the multiple regression is given in Section 9.5, we will not elaborate on the calculation procedures now. For this particular example, we will simply compare the estimated regression equations and their fit to the data.

TABLE 9.4.2.
Soybean Yield (X), and Protein Content (Y) from 30 Plots

Soybean Yield (bu/ac) X	X^2	Protein (%) Y	Soybean Yield (bu/ac) X	X^2	Protein (%) Y
29.50	870.25	40.80	31.20	973.44	37.60
28.60	817.96	42.10	34.50	1190.25	35.10
28.10	789.61	41.30	38.10	1451.61	35.80
31.20	973.44	40.30	33.20	1102.24	36.90
32.40	1049.76	39.60	36.10	1303.21	35.80
30.20	912.04	43.40	33.60	1128.96	36.80
38.30	1466.89	35.30	33.00	1089.00	35.20
30.40	924.16	41.50	32.30	1043.29	36.60
30.00	900.00	39.70	37.30	1391.29	34.40
31.30	979.69	39.20	35.50	1260.25	34.00
29.80	888.04	41.90	34.30	1176.49	37.20
34.50	1190.25	36.05	33.90	1149.21	37.60
30.20	912.04	39.00	39.20	1536.64	37.80
27.00	729.00	43.10	39.15	1532.72	35.60
32.00	1024.00	37.30	38.40	1474.56	36.25

STEP 2: Calculate a simple regression using the yield and the protein data to determine the fit. The estimated regression equation and the coefficient of determination are given below:

$$\hat{Y} = 58.923 - 0.629X$$
$$R^2 = 0.64$$

The coefficient of determination shows that the straight line accounts for 64% of the variability in protein content. Figure 9.4.4 shows how the estimated simple linear equation fits the data.

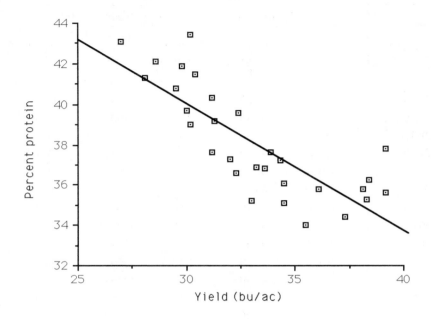

FIGURE 9.4.4.

Graph of the soybean data fitted to a linear equation.

STEP 3: To determine whether the relationship between the yield and the protein content is quadratic, as hypothesized by the agronomist, we will estimate a quadratic equation with the square of the yield data serving as the third variable. The estimated regression equation and the multiple coefficient of determination are given below:

$$\hat{Y} = 148.832 - 6.039X + 0.081X^2$$

$$R^2 = 0.77$$

Figure 9.4.5 shows that the quadratic equation accounts for 77% of the variability in the protein content, and is a much better fit than the straight line.

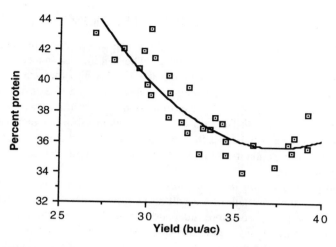

FIGURE 9.4.5.

Graph of the soybean data fitted to a quadratic equation.

STEP 4: Compare the results from the simple and quadratic equations as summarized in Table 9.4.3.

From Table 9.4.3 we observe that the reduction in the sum of squares tested against the mean square remaining after curvilinear regression is highly significant, thus confirming the agronomist's hypothesis that the relationship between yield and protein content is curvilinear.

TABLE 9.4.3.

Test of Significance of Departure from Linear Regression

Source of Variation	Degree of Freedom	Sum of Squares	Mean Square	F
Deviations from linear regression	28	75.357		
Deviations from quadratic regression	27	49.217	1.823	14.34**
Reduction in sum of squares	1	26.140	26.140	

Note: **= highly significant.

Experimental Research Design and Analysis

9.5 MULTIPLE REGRESSION AND CORRELATION

In the previous sections, it was shown how regression and correlation analysis are used in agricultural experiments. The techniques and concepts presented were used as a tool in analyzing the relationship that may exist between 2 variables. A single independent variable was used to estimate the value of the dependent variable. A brief introduction to the multiple regression was made in connection with fitting a curvilinear relationship between variables. In this section we will discuss the concepts of regression and correlation where 2 or more independent variables are used to estimate the dependent variable in more detail.

Since *multiple regression and correlation* is simply an extension of the simple regression and correlation, we will show how to derive the multiple regression equation using 2 or more independent variables. Second, attention will be given to calculating the standard error of estimate and related measures. Finally, the computation of multiple coefficient of determination and correlation will be explained.

The advantage of multiple regression over simple regression analysis is in enhancing our ability to use more available information in estimating the dependent variable. To describe the relationship between a single variable Y and several variables X, we may write the multiple regression equation as:

$$Y = \alpha + \beta_1 X_1 + \beta_2 X_2 + \ldots + \beta_k X_k + \varepsilon \qquad (9\text{-}46)$$

where

$$Y \;=\; \text{the dependent variable}$$

$$X_1 \ldots X_k \;=\; \text{the independent variables}$$

$$\varepsilon \;=\; \text{the error term, which is a random variable with a}$$
$$\text{mean of 0 and a standard deviation of } \sigma.$$

The numerical constants, α, and β_1 to β_k must be determined from the data, and are referred to as the partial regression coefficients. The underlying assumptions of the multiple regression model are:

1. the explanatory variables ($X_1 \ldots X_k$) may be either random or nonrandom (fixed) variables,
2. the value of Y selected for one value of X is probabilistically independent,
3. the random error has a normal distribution with mean equal to 0 and variance equal to σ^2.

These assumptions imply that the mean, or expected value $E(Y)$, for a given set of values of $X_1 \ldots .X_k$ is equal to

$$E(Y) = \alpha + \beta_1 X_1 + \beta_2 X_2 + \ldots + \beta_k X_k \qquad (9\text{-}47)$$

The coefficient α is the Y intercept when the expected value of all independent variables is 0. Equation 9-47 is called a *linear* statistical model. The scatter diagram for a 2-independent-variables case is a regression plane, as shown in Figure 9.5.1.

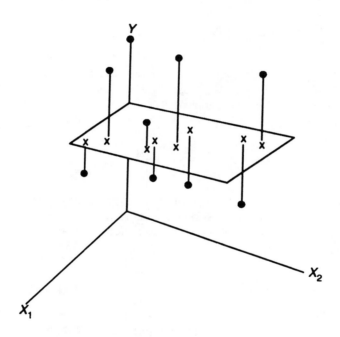

FIGURE 9.5.1.

Scatter diagram for multiple regression analysis involving 2 independent variables.

Estimating the Multiple Regression Equation—The Least Squares Method: In Section 9.2 we mentioned that if a straight line is fitted to a set of data using the least squares method, that line is the *best fit* in the sense that the sum of squared deviations is less than it would be for any other possible line.

The least-square formula provides a best-fitting *plane* to the data. The *normal equations* for k variables are as follows:

$$\sum x_1 y = b_1 \sum x_1^2 + b_2 \sum x_1 x_2 + \ldots + b_k \sum x_1 x_k \qquad (9\text{-}48)$$

$$\sum x_2 y = b_1 \sum x_1 x_2 + b_2 \sum x_2^2 + \ldots + b_k \sum x_2 x_k$$

$$\ldots \qquad \ldots\ldots\ldots\ldots\ldots\ldots$$

$$\ldots \qquad \ldots\ldots\ldots\ldots\ldots\ldots$$

$$\sum x_k y = b_1 \sum x_1 x_k + b_2 \sum x_2 x_k + \ldots + b_k \sum x_k^2$$

In the above equation, b_1, b_2, ...b_k are estimates of β_1, β_2, ...β_k, and the number of equations is equal to the number of parameters to be estimated. Given a sample of n observations on (Y, X_1, X_2), the sample regression or prediction equation is:

$$\hat{Y} = a + b_1 X_1 + b_2 X_2 \qquad (9\text{-}49)$$

With the least-squares model, the resulting estimates a, b_1, b_2 and $\hat{Y}$ are unbiased and have the smallest standard errors of any unbiased estimates that are linear expressions in the Y's. To estimate the value of a, we use the following equation:

$$a = \bar{Y} - b_1 \bar{X}_1 - b_2 \bar{X}_2 \qquad (9\text{-}50)$$

You should note that computing the coefficients, although not theoretically complex, is difficult and tedious even for k as small as 4 or 5. We generally use a computer program to solve for the coefficients.

To illustrate the concept and how the multiple regression coefficients are computed, let us take an example with 2 independent variables.

Example 9.5.1

In Section 9.2, an agronomist had taken 10 observations on the yield of maize as different amounts of fertilizer were applied. In this example, he wishes to consider an additional explanatory variable (the amount of rainfall).

He believes that the amount of fertilizer and rainfall are better predictors of yield. Using the information given below, determine for the agronomist the least-square multiple regression equation.

Yield of Maize bu/ac Y	Fertilizer lb N/ac X_1	Rainfall (in.) X_2
50	5	5
57	10	10
60	12	15
62	18	20
63	25	25
65	30	25
68	36	30
70	40	30
69	45	25
66	48	30

Solution

STEP 1: Compute the intermediate values needed in the *normal equations* from the data given in Table 9.5.1.

The normal equations for the 2 independent variables, as is the case with our example, are written as follows:

$$\sum x_1 y = b_1 \sum x_1^2 + b_2 \sum x_1 x_2 \tag{9-51}$$

$$\sum x_2 y = b_1 \sum x_1 x_2 + b_2 \sum x_2^2 \tag{9-52}$$

STEP 2: Solve for b_1 and b_2, using the following formulas:

$$b_1 = \frac{(\sum x_2^2)(\sum x_1 y) - (\sum x_1 x_2)(\sum x_2 y)}{(\sum x_1^2)(\sum x_2^2) - (\sum x_1 x_2)^2} \tag{9-53}$$

338   Experimental Research Design and Analysis

TABLE 9.5.1.

Calculation of Coefficients for Normal Equations

Yield of Maize bu/ac Y	Fertilizer lb N/ac X_1	Rainfall (in.) X_2
50	5	5
57	10	10
60	12	15
62	18	20
63	25	25
65	30	25
68	36	30
70	40	30
69	45	25
66	48	30
$\bar{Y} = 63$	$\bar{X}_1 = 26.9$	$\bar{X}_2 = 21.5$
$\sum y^2 = 338$	$\sum x_1^2 = 2,106.9$	$\sum x_2^2 = 702.5$
$\sum x_1 y = 755$	$\sum x_2 y = 460$	$\sum x_1 x_2 = 1,110.5$

and

$$b_2 = \frac{(\sum x_1^2)(\sum x_2 y) - (\sum x_1 x_2)(\sum x_1 y)}{(\sum x_1^2)(\sum x_2^2) - (\sum x_1 x_2)^2} \tag{9-54}$$

Substituting the appropriate values into Equations 9-53 and 9-54, we get:

$$b_1 = \frac{(702.5)(755) - (1,101.5)(460)}{(2,106.9)(702.5) - (1,101.5)^2}$$

$$= 0.088$$

and

$$b_2 = \frac{(2,106.9)(460) - (1,101.5)(755)}{(2,106.9)(702.5) - (1,101.5)^2}$$

$$= 0.516$$

STEP 3: Solve for a, using the following formula:

$$a = \bar{Y} - b_1\bar{X}_1 - b_2\bar{X}_2$$
$$= 63 - (0.088)(26.9) - (0.516)(21.5)$$
$$= 63 - 2.637 - 11.094$$
$$= 49.539$$

The estimated multiple linear regression is:

$$\widehat{Y} = 49.539 + 0.088X_1 + 0.516X_2$$

The interpretation of the coefficient a, b_1, and b_2 is analogous to the simple linear regression. The constant a is the intercept of the regression line. However, we interpret it as the value of $\widehat{Y}$ when both X_1 and X_2 are 0. The value b_1 and b_2 are called the *partial regression coefficients*. Coefficient b_1 simply measures the change in $\widehat{Y}$ per unit change in X_1 when X_2 is held constant. Likewise, coefficient b_2 measures the change in $\widehat{Y}$ per unit change in X_2 when X_1 is held constant. Thus, we may say that the b coefficients measure the net influence of each independent variable on the estimate of the dependent variable.

In the present example, the b_1 value of 0.088 indicates that for each increase of 1 lb of fertilizer, the yield increases by 0.088 bushels, regardless of the rainfall, (i.e., the amount of rainfall is held constant). The b_2 coefficient indicates that for each increase of 1 in. of rainfall, the yield increases by 0.516 bushels, regardless of the amount of fertilizer used.

Before we can compute the standard error of estimate and the coefficient of multiple determination, we need to partition the sum of squares for the dependent variable.

The total sum of squares (SST) has already been computed before as:

$$SST = \sum y^2 = 338$$

The explained or regression sum of squares (SSR) is computed as follows:

$$SSR = b_1 \sum x_1 y + b_2 \sum x_2 y \qquad (9\text{-}55)$$

Substituting the appropriate values into Equation 9-55 we get:

$$SSR = 0.088(755) + 0.516(460)$$
$$= 303.80$$

The unexplained or error sum of squares (*SSE*) is the difference between the *SST* and *SSR*:

$$SSE = SST - SSR \qquad\qquad (9\text{-}56)$$

Therefore, the error sum of squares is

$$SSE = 338 - 303.80 = 34.20$$

STEP 4: Compute the *standard error of estimate*, which measures the standard deviation of the residuals about the regression plane and thus specifies the amount of error incurred when the least-squares regression equation is used to predict values of the dependent variable. The smaller the standard error of estimate is, the closer the fit of the regression equation is to the scatter of observations.

The standard error of estimate is computed by using the following equation:

$$S_{y.12} = \sqrt{\frac{SSE}{n-k}} \qquad\qquad (9\text{-}57)$$

where

$$SSE \;=\; \text{the error sum of squares}$$

$$n \;=\; \text{the number of observations}$$

$$k \;=\; \text{the number of parameters.}$$

Hence, in a multiple regression analysis involving 2 independent variables and a dependent variable, the divisor will be $n - 3$. Having computed the error sum of squares earlier, we substitute its value into Equation 9-57 to compute the standard error of estimate as follows:

$$S_{y.12} = \sqrt{\frac{34.20}{7}}$$
$$= 2.21 \text{ bushels}$$

The standard error of estimate about the regression plane may be compared with the standard error of estimate of the simple regression.

In Section 9.2, when only fertilizer was used to explain the variation in yield, we computed a standard error of estimate of 2.92. Including an additional variable (rainfall) to explain the variation in yield has given us a standard error of estimate of 2.21. As was mentioned before, the standard error expresses the amount of variation in the dependent variable that is left unexplained by regression analysis. Since the standard error of the regression plane is smaller than the standard error of the regression line, inclusion of this additional variable will provide for a better prediction.

STEP 5: Compute the multiple coefficient of determination, using the following equation:

$$R^2 = \frac{SSR}{SST} \tag{9-58}$$

where

SSR = the regression sum of squares

$SST = \sum y^2$ = the total sum of squares.

Substituting the values of the regression and the total sum of squares into Equation 9-58, we get:

$$R^2 = \frac{303.80}{338} = 0.90$$

The coefficient of determination measures the contribution of the k independent variables to the variation in Y. This means that 90% of the variation in yield of maize is explained by the amount of fertilizer applied and the amount of rainfall in a locality. The above R^2 value is not adjusted for degrees of freedom. Hence, we may overestimate the impact of adding another independent variable in explaining the amount of variability in the dependent variable. Thus, it is recommended that an adjusted R^2 be used in interpreting the results.

The adjusted coefficient of multiple determination is computed as follows:

$$R_a^2 = 1 - (1 - R^2)\frac{n - 1}{n - k} \tag{9-59}$$

where

$$R_a^2 = \text{adjusted coefficient of multiple determination}$$

$$n = \text{number of observations}$$

$$k = \text{total number of parameters.}$$

For the present example, the adjusted R^2 is

$$R_a^2 = 1 - (1 - .90)\frac{10 - 1}{10 - 3}$$

$$= 0.87$$

We may wish to compare the coefficient of determination of the simple regression model where 1 independent variable, namely the impact of application of fertilizer, was analyzed with the coefficient of multiple determination where, in addition to fertilizer use, the impact of rainfall on yield was observed. The adjusted coefficient of determination for the simple regression was $r^2 = 0.76$, whereas the adjusted coefficient of multiple determination was 0.87. The difference of 0.11 indicates that an additional 11% of the variation in the yield of maize is explained by the amount of rainfall, beyond that already explained by the amount of fertilizer applied.

STEP 6: Test the significance of R^2 by computing the F value as:

$$F = \frac{SSR/k}{SSE/(n - k - 1)} \tag{9-60}$$

Substituting the appropriate values into Equation 9-60, we get:

$$F = \frac{303.80/2}{34.20/(10 - 2 - 1)}$$

$$= 31.09$$

STEP 7: Compare the computed F with the tabular F value given in Appendix E. For this example, the tabular F (for 2 and 7 degrees of freedom) is 8.65 at the 1% level of significance. Because the computed F value is greater than the tabular F value, the estimated multiple linear regression is highly significant.

Assumptions and Problems in Multiple Linear Regression: As with the simple regression, a number of assumptions apply to the case of the multiple regression. These assumptions are:

1. The regression model is linear and of the form

$$E\,(Y) = a + b_1 X_1 + b_2 X_2 + \ldots + b_k X_k \qquad\qquad (9\text{-}61)$$

2. The values of Y are independent of each other.
3. The values of Y are normally distributed.
4. The variance of Y values is the same for all values of $X_1, X_2 \ldots X_k$.

Violation of the above assumptions leads to a number of problems such as serial or autocorrelation, heteroscedasticity, and multicollinearity, which are explained as follows.

Serial or Autocorrelation. This problem arises when the assumption of the independence of Y values is not met. That is, there is dependence between successive values. This problem is often observed when time series data are employed in the analysis. To be sure that there is no autocorrelation, plotting the residuals against time is helpful. Figure 9.5.2 suggests the presence of autocorrelated terms. Methods for measuring serial correlation, such as the Durbin-Watson test, should be employed.

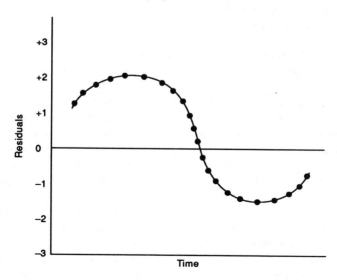

FIGURE 9.5.2.

Positive and negative autocorrelation in the residuals.

Remedying the serial correlation problems requires proper transformation of the dependent variable. Such a transformation is discussed elsewhere, and the reader should consult the references given at the end of this chapter.

Equal Variances. One of the assumptions of the regression model was that the error terms all have equal variances. This condition of equal variance is known as *homoscedasticity*. When this assumption is violated, the problem of *hetero-scedasticity* arises. As an example, in studying the yield of milk with different high-energy rations, we might find that yield rises with different high-energy rations. In such case as this, the yield function is probably heteroscedastic. When heteroscedasticity exists, it is difficult to make statistical inferences regarding the regression equation. Kelijian and Oats (1974) suggested a test for heterosce-dasticity, and gave a procedure to overcome problems arising from this error.

Multicollinearity. The problem of multicollinearity arises when 2 or more independent variables are highly correlated with each other. This implies that the regression model specified is unable to separate out the effect of each individual variable on the dependent variable. When multicollinearity exists between the independent variables, estimates of the parameters have larger standard errors, and the regression coefficients tend to be unreliable.

How do we know whether we have a problem of multicollinearity? When a researcher observes a large coefficient of determination (R^2) accompanied by statistically insignificant estimates of the regression coefficients, the chances are that there is *imperfect multicollinearity*. When 1 (or more) independent variable(s) is an exact linear combination of the others, we have *perfect multicollinearity*.

Once it is determined that multicollinearity exists between the independent variables, a number of possible steps can be taken to remedy this problem.

1. Drop the correlated variable from the equation. Which independent variable to drop from the equation depends on the test of significance of the regression coefficient, and the judgment of the researcher. If the *t* test indicates that the regression coefficient of an independent variable is statistically insignificant, that variable may be dropped from the equation. Dropping a highly correlated independent variable from the equation will not affect the value of R^2 very much.
2. Change the form of 1 or more independent variables. For example, an agricultural economist, in a demand equation for beef (Y), finds that income (X_1) and another independent variable (X_2) are highly correlated. In such a situation, dividing the income by the variable of population yields per capita income, which may result in less correlated independent variables. Other approaches are suggested in Kelijian and Oats (1974) and Hamburg (1983).

REFERENCES AND SUGGESTED READINGS

Galton, F. 1908 *Memories of My Life*. New York: E. P. Dutton.

Hamburg, M. 1983. *Statistical Analysis for Decision Making*. 3rd ed. New York: Harcourt Brace Jovanovich, chap. 12.

Hoshmand, A. R. 1988. *Statistical Methods for Agricultural Sciences*. Portland, OR: Timber Press, chap. 11 and 12.

Kelejian, H. H. and Oats, W. E. 1974. *Introduction to Econometrics*. New York: Harper & Row.

Mead, R. and Curnow, R. N. 1983. *Statistical Methods in Agriculture and Experimental Biology*. New York: Chapman & Hall.

Netter, J. and Kutner, M. H. 1983. *Applied Linear Regression Models*. Homewood, IL: Irwin.

Snedecor, G. W. and Cochran, W. G. 1980. *Statistical Methods*. Ames, IA: Iowa State University Press.

EXERCISES

1. A researcher interested in the relationship between the rate of germination of warm-season forage grasses and temperature has postulated a linear regression model where the number of seeds germinated per day is dependent on the average daily temperature. The following data were collected by the researcher.

Germinated Seed (no./day)	Temperature (°C)	Germinated Seed (no./day)	Temperature (°C)
5	10	28	22
7	11	31	23
9	13	35	24
10	15	38	26
14	16	49	27
20	18	55	29
24	20	61	30
25	21	73	32

(a) Use the least-square technique in estimating the equation.
(b) Compute the standard error of estimate. What is the interpretation of the standard error of estimate?
(c) Is there a significant correlation between the 2 variables? Use a 5% level of significance.

2. In a 2-factor experiment, an environmental horticulturist wishes to determine the impact of nitrogen fertilization and hourly exposure to sunlight on leaf thickness in the rubberplant. After 2 months of experimentation, the following average observations of leaf thickness, amount of light in the greenhouse, and fertilization were recorded from an experiment with 4 replications. Plants of the same age were grown in the same environment, varying only the 2 factors of interest.

Treatment Number	Leaf Thickness (mm)	Hours of Sunlight	Nitrogen Fertilization (mg/pot)
1	2.15	5.0	.20
2	2.28	5.0	.30
3	4.56	5.0	.40
4	7.68	5.0	.50
5	8.76	5.0	.60
6	12.80	5.0	.70
7	15.62	5.0	.80
8	16.54	5.0	.90
9	3.24	6.0	.20
10	6.32	6.0	.30
11	8.76	6.0	.40
12	7.48	6.0	.50
13	12.69	6.0	.60
14	16.87	6.0	.70
15	20.92	6.0	.80
16	18.66	6.0	.90
17	5.33	7.0	.20
18	6.38	7.0	.30
19	6.87	7.0	.40
20	8.88	7.0	.50
21	9.89	7.0	.60
22	20.87	7.0	.70
23	21.92	7.0	.80
24	22.96	7.0	.90

(a) Estimate the regression equation.
(b) Compute the coefficient of multiple determination.
(c) Test the significance of R^2.
(d) Compute the coefficient of correlation.

3. An animal researcher is interested in the relationship between urea nitrogen and the energy balance and the phosphorous balance. The researcher has randomly selected 15 cows that have been fed rations that contain high and low levels of energy and phosphorous. The data for the experiment are given below:

Treatment Number	Y Urea N (mg/100 ml)	X_1 Energy Balance (k cal/day)	X_2 P Balance (g/day)
1	9.5	8.1	21.3
2	10.2	8.4	25.9
3	11.0	8.5	24.8
4	11.2	10.9	28.4
5	12.3	7.5	-2.7
6	10.8	7.8	22.2
7	10.8	11.3	19.0
8	11.1	-5.4	8.5
9	11.7	7.6	17.9
10	11.8	8.3	-9.9
11	10.9	9.6	-8.4
12	12.4	9.5	-10.1
13	12.6	10.5	12.4
14	10.9	-5.4	7.8
15	11.3	6.4	8.9
16	10.8	5.7	9.4

(a) Estimate the least-square regression equation.
(b) Compute the standard error of estimate.
(c) Test the significance of the regression coefficients.
(d) Compute the correlation coefficient.

4. An agricultural engineer wishes to determine the relationship between the monthly electrical usage in a greenhouse and the size of the greenhouse. He believes that the relationship is best exemplified by the following model:

$$Y = a + b_1 X_1 + b_2 X_2^2 + \varepsilon$$

Given the following data

Monthly electrical usage (kw/h)	Greenhouse size (ft²)
2000	2800
2225	2900
2540	3100
2678	3200
2700	3250
2890	3300
2980	3400
3000	3500
3220	3550
3454	3600
3590	3650
3760	3850
4000	3900
4325	4000
4550	5050

(a) Estimate the multiple regression equation. (Use a computer to solve this problem.)
(b) Is the overall model useful in this investigation?
(c) Compute the correlation coefficient.
(d) Test the overall significance of the computed coefficients at $\alpha = .01$.

5. Suppose that in the previous example, the agricultural engineer hypothesized that there is a strong relationship between the monthly electrical usage in the greenhouse and the size of the greenhouse, as well as the average monthly temperature. Data gathered are shown below.

Monthly electrical usage (kw/h)	Greenhouse size (ft²)	Average monthly temperature (°F)
2000	2800	68
2225	2900	70
2540	3100	78
2678	3200	80
2700	3250	82
2890	3300	84
2980	3400	86
3000	3500	75
3220	3550	75
3454	3600	78
3590	3650	72
3760	3850	70
4000	3900	69
4325	4000	65
4550	5050	62

(a) Compute the estimated regression equation. (Use a computer to solve this problem.)

(b) Test the overall significance of the regression model when $\alpha = .05$.

6. An equine researcher has postulated that there is a nonlinear relationship between the life span of Arabian horses and the gestation period. The researcher believes that the following model best exemplifies the relationship:

$$Y = a + b_1 X_1 + b_2 X_2^2 + \varepsilon$$

Life Span (years)	Gestation Period (days)
15.2	210
17.8	230
18.2	240
20.0	265
22.1	285
23.4	370
20.5	345
22.2	340
21.5	385
19.6	295
18.8	279
20.4	305
22.3	315
20.0	395
18.4	290
15.5	300

(a) Estimate the multiple regression equation. (Use a computer to solve this problem.)
(b) Interpret the meaning of the coefficient of multiple determination.
(c) Are the regression coefficients significant at the $\alpha = .01$?
(d) Test the overall significance of the model at $\alpha = .01$.

APPENDICES

Appendix A

Chi-Square Distribution

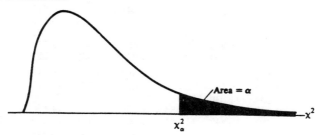

The following table provides the values of χ_α^2 that correspond to a given upper-tail area α and a specified number of degrees of freedom.

Degrees of Freedom	Upper-Tail Area α						
	.99	.98	.95	.90	.80	.70	.50
1	$.0^3157$	$.0^3628$	.00393	.0158	.0642	.148	.455
2	.0201	.0404	.103	.211	.446	.713	1.386
3	.115	.185	.352	.584	1.005	1.424	2.366
4	.297	.429	.711	1.064	1.649	2.195	3.357
5	.554	.752	1.145	1.610	2.343	3.000	4.351
6	.872	1.134	1.635	2.204	3.070	3.828	5.348
7	1.239	1.564	2.167	2.833	3.822	4.671	6.346
8	1.646	2.032	2.733	3.490	4.594	5.527	7.344
9	2.088	2.532	3.325	4.168	5.380	6.393	8.343
10	2.558	3.059	3.940	4.865	6.179	7.267	9.342
11	3.053	3.609	4.575	5.578	6.989	8.148	10.341
12	3.571	4.178	5.226	6.304	7.807	9.034	11.340
13	4.107	4.765	5.892	7.042	8.634	9.926	12.340
14	4.660	5.368	6.571	7.790	9.467	10.821	13.339
15	5.229	5.985	7.261	8.547	10.307	11.721	14.339
16	5.812	6.614	7.962	9.312	11.152	12.624	15.338
17	6.408	7.255	8.672	10.085	12.002	13.531	16.338
18	7.015	7.906	9.390	10.865	12.857	14.440	17.338
19	7.633	8.567	10.117	11.651	13.716	15.352	18.338
20	8.260	9.237	10.851	12.443	14.578	16.266	19.337
21	8.897	9.915	11.591	13.240	15.445	17.182	20.337
22	9.542	10.600	12.338	14.041	16.314	18.101	21.337
23	10.196	11.293	13.091	14.848	17.187	19.021	22.337
24	10.856	11.992	13.848	15.659	18.062	19.943	23.337
25	11.524	12.697	14.611	16.473	18.940	20.867	24.337
26	12.198	13.409	15.379	17.292	19.820	21.792	25.336
27	12.879	14.125	16.151	18.114	20.703	22.719	26.336
28	13.565	14.847	16.928	18.939	21.588	23.647	27.336
29	14.256	15.574	17.708	19.768	22.475	24.577	28.336
30	14.953	16.306	18.493	20.599	23.364	25.508	29.336

Degrees of Freedom	Upper-Tail Area α						
	.30	.20	.10	.05	.02	.01	.001
1	1.074	1.642	2.706	3.841	5.412	6.635	10.827
2	2.408	3.219	4.605	5.991	7.824	9.210	13.815
3	3.665	4.642	6.251	7.815	9.837	11.345	16.268
4	4.878	5.989	7.779	9.488	11.668	13.277	18.465
5	6.064	7.289	9.236	11.070	13.388	15.086	20.517
6	7.231	8.558	10.645	12.592	15.033	16.812	22.457
7	8.383	9.803	12.017	14.067	16.622	18.475	24.322
8	9.524	11.030	13.362	15.507	18.168	20.090	26.125
9	10.656	12.242	14.684	16.919	19.679	21.666	27.877
10	11.781	13.442	15.987	18.307	21.161	23.209	29.588
11	12.899	14.631	17.275	19.675	22.618	24.725	31.264
12	14.011	15.812	18.549	21.026	24.054	26.217	32.909
13	15.119	16.985	19.812	22.362	25.472	27.688	34.528
14	16.222	18.151	21.064	23.685	26.873	29.141	36.123
15	17.322	19.311	22.307	24.996	28.259	30.578	37.697
16	18.418	20.465	23.542	26.296	29.633	32.000	39.252
17	19.511	21.615	24.769	27.587	30.995	33.409	40.790
18	20.601	22.760	25.989	28.869	32.346	34.805	42.312
19	21.689	23.900	27.204	30.144	33.687	36.191	43.820
20	22.775	25.038	28.412	31.410	35.020	37.566	45.315
21	23.858	26.171	29.615	32.671	36.343	38.932	46.797
22	24.939	27.301	30.813	33.924	37.659	40.289	48.268
23	26.018	28.429	32.007	35.172	38.968	41.638	49.728
24	27.096	29.553	33.196	36.415	40.270	42.980	51.179
25	28.172	30.675	34.382	37.652	41.566	44.314	52.620
26	29.246	31.795	35.563	38.885	42.856	45.642	54.052
27	30.319	32.912	36.741	40.113	44.140	46.963	55.476
28	31.391	34.027	37.916	41.337	45.419	48.278	56.893
29	32.461	35.139	39.087	42.557	46.693	49.588	58.302
30	33.530	36.250	40.256	43.773	47.962	50.892	59.703

Source: From Table IV of Fisher and Yates, *Statistical Tables for Biological, Agricultural and Medical Research*, published by Longman Group Ltd., London (previously published by Oliver & Boyd, Edinburgh, 1963). Reproduced with permission of the authors and publishers.

Appendix B

The Arc Sine $\sqrt{\text{Percentage}}$ Transformation.

(Transformation of binomial percentages, in the margins, to angles of equal
information in degrees. The + or − signs following angles ending in 5 are for
guidance in rounding to one decimal.)

%	0	1	2	3	4	5	6	7	8	9
0.0	0	0.57	0.81	0.99	1.15−	1.28	1.40	1.52	1.62	1.72
0.1	1.81	1.90	1.99	2.07	2.14	2.22	2.29	2.36	2.43	2.50
0.2	2.56	2.63	2.69	2.75−	2.81	2.87	2.92	2.98	3.03	3.09
0.3	3.14	3.19	3.24	3.29	3.34	3.39	3.44	3.49	3.53	3.58
0.4	3.63	3.67	3.72	3.76	3.80	3.85−	3.89	3.93	3.97	4.01
0.5	4.05+	4.09	4.13	4.17	4.21	4.25+	4.29	4.33	4.37	4.40
0.6	4.44	4.48	4.52	4.55+	4.59	4.62	4.66	4.69	4.73	4.76
0.7	4.80	4.83	4.87	4.90	4.93	4.97	5.00	5.03	5.07	5.10
0.8	5.13	5.16	5.20	5.23	5.26	5.29	5.32	5.35+	5.38	5.41
0.9	5.44	5.47	5.50	5.53	5.56	5.59	5.62	5.65+	5.68	5.71
1	5.74	6.02	6.29	6.55−	6.80	7.04	7.27	7.49	7.71	7.92
2	8.13	8.33	8.53	8.72	8.91	9.10	9.28	9.46	9.63	9.81
3	9.98	10.14	10.31	10.47	10.63	10.78	10.94	11.09	11.24	11.39
4	11.54	11.68	11.83	11.97	12.11	12.25−	12.39	12.52	12.66	12.79
5	12.92	13.05+	13.18	13.31	13.44	13.56	13.69	13.81	13.94	14.06
6	14.18	14.30	14.42	14.54	14.65+	14.77	14.89	15.00	15.12	15.23
7	15.34	15.45+	15.56	15.68	15.79	15.89	16.00	16.11	16.22	16.32
8	16.43	16.54	16.64	16.74	16.85−	16.95+	17.05+	17.16	17.26	17.36
9	17.46	17.56	17.66	17.76	17.85+	17.95+	18.05−	18.15−	18.24	18.34
10	18.44	18.53	18.63	18.72	18.81	18.91	19.00	19.09	19.19	19.28
11	19.37	19.46	19.55+	19.64	19.73	19.82	19.91	20.00	20.09	20.18
12	20.27	20.36	20.44	20.53	20.62	20.70	20.79	20.88	20.96	21.05−
13	21.13	21.22	21.30	21.39	21.47	21.56	21.64	21.72	21.81	21.89
14	21.97	22.06	22.14	22.22	22.30	22.38	22.46	22.55−	22.63	22.71
15	22.79	22.87	22.95−	23.03	23.11	23.19	23.26	23.34	23.42	23.50
16	23.58	23.66	23.73	23.81	23.89	23.97	24.04	24.12	24.20	24.27
17	24.35+	24.43	24.50	24.58	24.65+	24.73	24.80	24.88	24.95+	25.03
18	25.10	25.18	25.25+	25.33	25.40	25.48	25.55−	25.62	25.70	25.77
19	25.84	25.92	25.99	26.06	26.13	26.21	26.28	26.35−	26.42	26.49
20	26.56	26.64	26.71	26.78	26.85+	26.92	26.99	27.06	27.13	27.20
21	27.28	27.35−	27.42	27.49	27.56	27.63	27.69	27.76	27.83	27.90
22	27.97	28.04	28.11	28.18	28.25−	28.32	28.38	28.45+	28.52	28.59
23	28.66	28.73	28.79	28.86	28.93	29.00	29.06	29.13	29.20	29.27
24	29.33	29.40	29.47	29.53	29.60	29.67	29.73	29.80	29.87	29.93
25	30.00	30.07	30.13	30.20	30.26	30.33	30.40	30.46	30.53	30.59
26	30.66	30.72	30.79	30.85+	30.92	30.98	31.05−	31.11	31.18	31.24
27	31.31	31.37	31.44	31.50	31.56	31.63	31.69	31.76	31.82	31.88
28	31.95−	32.01	32.08	32.14	32.20	32.27	32.33	32.39	32.46	32.52
29	32.58	32.65−	32.71	32.77	32.83	32.90	32.96	33.02	33.09	33.15−
30	33.21	33.27	33.34	33.40	33.46	33.52	33.58	33.65−	33.71	33.77
31	33.83	33.89	33.96	34.02	34.08	34.14	34.20	34.27	34.33	34.39
32	34.45−	34.51	34.57	34.63	34.70	34.76	34.82	34.88	34.94	35.00
33	35.06	35.12	35.18	35.24	35.30	35.37	35.43	35.49	35.55−	35.61
34	35.67	35.73	35.79	35.85−	35.91	35.97	36.03	36.09	36.15+	36.21
35	36.27	36.33	36.39	36.45+	36.51	36.57	36.63	36.69	36.75+	36.81
36	36.87	36.93	36.99	37.05−	37.11	37.17	37.23	37.29	37.35−	37.41
37	37.47	37.52	37.58	37.64	37.70	37.76	37.82	37.88	37.94	38.00
38	38.06	38.12	38.17	38.23	38.29	38.35−	38.41	38.47	38.53	38.59
39	38.65−	38.70	38.76	38.82	38.88	38.94	39.00	39.06	39.11	39.17
40	39.23	39.29	39.35−	39.41	39.47	39.52	39.58	39.64	39.70	39.76
41	39.82	39.87	39.93	39.99	40.05−	40.11	40.16	40.22	40.28	40.34
42	40.40	40.46	40.51	40.57	40.63	40.69	40.74	40.80	40.86	40.92
43	40.98	41.03	41.09	41.15−	41.21	41.27	41.32	41.38	41.44	41.50
44	41.55+	41.61	41.67	41.73	41.78	41.84	41.90	41.96	42.02	42.07

continued next page

356 Experimental Research Design and Analysis

%	0	1	2	3	4	5	6	7	8	9
45	42.13	42.19	42.25–	42.30	42.36	42.42	42.48	42.53	42.59	42.65–
46	42.71	42.76	42.82	42.88	42.94	42.99	43.05–	43.11	43.17	43.22
47	43.28	43.34	43.39	43.45+	43.51	43.57	43.62	43.68	43.74	43.80
48	43.85+	43.91	43.97	44.03	44.08	44.14	44.20	44.25+	44.51	44.37
49	44.43	44.46	44.54	44.60	44.66	44.71	44.77	44.83	44.89	44.94
50	45.00	45.06	45.11	45.17	45.23	45.29	45.34	45.40	45.46	45.52
51	45.57	45.63	45.69	45.75–	45.80	45.86	45.92	45.97	46.03	46.09
52	46.15–	46.20	46.26	46.32	46.38	46.43	46.49	46.55–	46.61	46.66
53	46.72	46.78	46.83	46.89	46.95+	47.01	47.06	47.12	47.18	47.24
54	47.29	47.35+	47.41	47.47	47.52	47.58	47.64	47.70	47.75+	47.81
55	47.87	47.93	47.98	48.04	48.10	48.16	48.22	48.27	48.33	48.39
56	48.45–	48.50	48.56	48.62	48.68	48.73	48.79	48.85+	48.91	48.97
57	49.02	49.08	49.14	49.20	49.26	49.31	49.37	49.43	49.49	49.54
58	49.60	49.66	49.72	49.78	49.84	49.89	49.95+	50.01	50.07	50.13
59	50.18	50.24	50.30	50.36	50.42	50.48	50.53	50.59	50.65+	50.71
60	50.77	50.83	50.89	50.94	51.00	51.06	51.12	51.18	51.24	51.30
61	51.35+	51.41	51.47	51.53	51.59	51.65–	51.71	51.77	51.83	51.88
62	51.94	52.00	52.06	52.12	52.18	52.24	52.30	52.36	52.42	52.48
63	52.53	52.59	52.65+	52.71	52.77	52.83	52.89	52.95+	53.01	53.07
64	53.13	53.19	53.25–	53.31	53.37	53.43	53.49	53.55–	53.61	53.67
65	53.73	53.79	53.85–	53.91	53.97	54.03	54.09	54.15+	54.21	54.27
66	54.33	54.39	54.45+	54.51	54.57	54.63	54.70	54.76	54.82	54.88
67	54.94	55.00	55.06	55.12	55.18	55.24	55.30	55.37	55.43	55.49
68	55.55+	55.61	55.67	55.73	55.80	55.86	55.92	55.98	56.04	56.11
69	56.17	56.23	56.29	56.35+	56.42	56.48	56.54	56.60	56.66	56.73
70	56.79	56.85+	56.91	56.98	57.04	57.10	57.17	57.23	57.29	57.35+
71	57.42	57.48	57.54	57.61	57.67	57.73	57.80	57.86	57.92	57.99
72	58.05+	58.12	58.18	58.24	58.31	58.37	58.44	58.50	58.56	58.63
73	58.69	58.76	58.82	58.89	58.95+	59.02	59.08	59.15–	59.21	59.28
74	59.34	59.41	59.47	59.54	59.60	59.67	59.74	59.80	59.87	59.93
75	60.00	60.07	60.13	60.20	60.27	60.33	60.40	60.47	60.53	60.60
76	60.67	60.73	60.80	60.87	60.94	61.00	61.07	61.14	61.21	61.27
77	61.34	61.41	61.48	61.55–	61.62	61.68	61.75+	61.82	61.89	61.96
78	62.03	62.10	62.17	62.24	62.31	62.37	62.44	62.51	62.58	62.65+
79	62.72	62.80	62.87	62.94	63.01	63.08	63.15–	63.22	63.29	63.36
80	63.44	63.51	63.58	63.65+	63.72	63.79	63.87	63.94	64.01	64.08
81	64.16	64.23	64.30	64.38	64.45+	64.52	64.60	64.67	64.75–	64.82
82	64.90	64.97	65.05–	65.12	65.20	65.27	65.35–	65.42	65.50	65.57
83	65.65–	65.73	65.80	65.88	65.96	66.03	66.11	66.19	66.27	66.34
84	66.42	66.50	66.58	66.66	66.74	66.81	66.89	66.97	67.05+	67.13
85	67.21	67.29	67.37	67.45+	67.54	67.62	67.70	67.78	67.86	67.94
86	68.03	68.11	68.19	68.28	68.36	68.44	68.53	68.61	68.70	68.78
87	68.87	68.95+	69.04	69.12	69.21	69.30	69.38	69.47	69.56	69.64
88	69.73	69.82	69.91	70.00	70.09	70.18	70.27	70.36	70.45–	70.54
89	70.63	70.72	70.81	70.91	71.00	71.09	71.19	71.23	71.37	71.47
90	71.56	71.66	71.76	71.85+	71.95+	72.05–	72.15–	72.24	72.34	72.44
91	72.54	72.64	72.74	72.84	72.95–	73.05–	73.15+	73.26	73.36	73.46
92	73.57	73.68	73.78	73.89	74.00	74.11	74.21	74.32	74.44	74.55–
93	74.66	74.77	74.88	75.00	75.11	75.23	75.35–	75.46	75.58	75.70
94	75.82	75.94	76.06	76.19	76.31	76.44	76.56	76.69	76.82	76.95–
95	77.08	77.21	77.34	77.48	77.61	77.75+	77.89	78.03	78.17	78.32
96	78.46	78.61	78.76	78.91	79.06	79.22	79.37	79.53	79.69	79.86
97	80.02	80.19	80.37	80.54	80.72	80.90	81.09	81.28	81.47	81.67
98	81.87	82.08	82.29	82.51	82.73	82.96	83.20	83 45+	83.71	83.98

%	0	1	2	3	4	5	6	7	8	9
99.0	84.26	84.29	84.32	84.35−	84.38	84.41	84.44	84.47	84.50	84.53
99.1	84.56	84.59	84.62	84.65−	84.68	84.71	84.74	84.77	84.80	84.84
99.2	84.87	84.90	84.93	84.97	85.00	85.03	85.07	85.10	85.13	85.17
99.3	85.20	85.24	85.27	85.31	85.34	85.38	85.41	85.45−	85.48	85.52
99.4	85.56	85.60	85.63	85.67	85.71	85.75−	85.79	85.83	85.87	85.91
99.5	85.95−	85.99	86.03	86.07	86.11	86.15−	86.20	86.24	86.28	86.33
99.6	86.37	86.42	86.47	86.51	86.56	86.61	86.66	86.71	86.76	86.81
99.7	86.86	86.91	86.97	87.02	87.08	87.13	87.19	87.25+	87.31	87.37
99.8	87.44	87.50	87.57	87.64	87.71	87.78	87.86	87.93	88.01	88.10
99.9	88.19	88.28	88.38	88.48	88.60	88.72	88.85+	89.01	89.19	89.43
100.0	90.00									

Reproduced from *Principles and Procedures of Statistics* by R. G. D. Steel and J. H. Torrie. 1960.
Printed with the permission of C. I. Bliss. pp. 448—449.

Appendix C

Selected Latin Squares

3 x 3

```
A  B  C
B  C  A
C  A  B
```

4 x 4

(1)

```
A  B  C  D
B  A  D  C
C  D  B  A
D  C  A  B
```

(2)

```
A  B  C  D
B  C  D  A
C  D  A  B
D  A  B  C
```

(3)

```
A  B  C  D
B  D  A  C
C  A  D  B
D  C  B  A
```

(4)

```
A  B  C  D
B  A  D  C
C  D  A  B
D  C  B  A
```

5 x 5

```
A  B  C  D  E
B  A  E  C  D
C  D  A  E  B
D  E  B  A  C
E  C  D  B  A
```

6 x 6

```
A  B  C  D  E  F
B  F  D  C  A  E
C  D  E  F  B  A
D  A  F  E  C  B
E  C  A  B  F  D
F  E  B  A  D  C
```

7 x 7

```
A  B  C  D  E  F  G
B  C  D  E  F  G  A
C  D  E  F  G  A  B
D  E  F  G  A  B  C
E  F  G  A  B  C  D
F  G  A  B  C  D  E
G  A  B  C  D  E  F
```

8 X 8

```
A  B  C  D  E  F  G  H
B  C  D  E  F  G  H  A
C  D  E  F  G  H  A  B
D  E  F  G  H  A  B  C
E  F  G  H  A  B  C  D
F  G  H  A  B  C  D  E
G  H  A  B  C  D  E  F
H  A  B  C  D  E  F  G
```

9 X 9

```
A  B  C  D  E  F  G  H  I
B  C  D  E  F  G  H  I  A
C  D  E  F  G  H  I  A  B
D  E  F  G  H  I  A  B  C
E  F  G  H  I  A  B  C  D
F  G  H  I  A  B  C  D  E
G  H  I  A  B  C  D  E  F
H  I  A  B  C  D  E  F  G
I  A  B  C  D  E  F  G  H
```

10 x 10

```
A  B  C  D  E  F  G  H  I  J
B  C  D  E  F  G  H  I  J  A
C  D  E  F  G  H  I  J  A  B
D  E  F  G  H  I  J  A  B  C
E  F  G  H  I  J  A  B  C  D
F  G  H  I  J  A  B  C  D  E
G  H  I  J  A  B  C  D  E  F
H  I  J  A  B  C  D  E  F  G
I  J  A  B  C  D  E  F  G  H
J  A  B  C  D  E  F  G  H  I
```

11 x 11

```
A  B  C  D  E  F  G  H  I  J  K
B  C  D  E  F  G  H  I  J  K  A
C  D  E  F  G  H  I  J  K  A  B
D  E  F  G  H  I  J  K  A  B  C
E  F  G  H  I  J  K  A  B  C  D
F  G  H  I  J  K  A  B  C  D  E
G  H  I  J  K  A  B  C  D  E  F
H  I  J  K  A  B  C  D  E  F  G
I  J  K  A  B  C  D  E  F  G  H
J  K  A  B  C  D  E  F  G  H  I
K  A  B  C  D  E  F  G  H  I  J
```

12 x 12

A	B	C	D	E	F	G	H	I	J	K	L
B	C	D	E	F	G	H	I	J	K	L	A
C	D	E	F	G	H	I	J	K	L	A	B
D	E	F	G	H	I	J	K	L	A	B	C
E	F	G	H	I	J	K	L	A	B	C	D
F	G	H	I	J	K	L	A	B	C	D	E
G	H	I	J	K	L	A	B	C	D	E	F
H	I	J	K	L	A	B	C	D	E	F	G
I	J	K	L	A	B	C	D	E	F	G	H
J	K	L	A	B	C	D	E	F	G	H	I
K	L	A	B	C	D	E	F	G	H	I	J
L	A	B	C	D	E	F	G	H	I	J	K

Greco-Latin squares

3 x 3

A_1	B_3	C_2
B_2	C_1	A_3
C_3	A_2	B_1

4 X 4

A_1	B_3	C_4	D_2
B_2	A_4	D_3	C_1
C_3	D_1	A_2	B_4
D_4	C_2	B_1	A_3

5 X 5

A_1	B_3	C_5	D_2	E_4
B_2	C_4	D_1	E_3	A_5
C_3	D_5	E_2	A_4	B_1
D_4	E_1	A_3	B_5	C_2
E_5	A_2	B_4	C_1	D_3

7 X 7

A_1	B_5	C_2	D_6	E_3	F_7	G_4
B_2	C_6	D_3	E_7	F_4	G_1	A_5
C_3	D_7	E_4	F_1	G_5	A_2	B_6
D_4	E_1	F_5	G_2	A_6	B_3	C_7
E_5	F_2	G_6	A_3	B_7	C_4	D_1
F_6	G_3	A_7	B_4	C_1	D_5	E_2
G_7	A_4	B_1	C_5	D_2	E_6	F_3

8 X 8

A_1	B_5	C_2	D_3	E_7	F_4	G_8	H_6
B_2	C_8	G_1	F_7	H_3	D_6	C_5	E_4
C_3	G_4	A_7	E_1	D_2	H_5	B_6	F_8
D_4	F_3	E_6	A_5	C_8	B_1	H_7	G_2
E_5	H_1	D_8	C_4	A_6	G_3	F_2	B_7
F_6	D_7	H_4	B_8	G_5	A_2	E_3	C_1
G_7	C_6	B_3	H_2	F_1	E_8	A_4	D_5
H_8	E_2	F_5	G_6	B_4	C_7	D_1	A_3

9 x 9

A_1	B_3	C_2	D_7	E_9	F_8	G_4	H_6	I_5
B_2	C_1	A_3	E_8	F_7	D_9	H_5	I_4	G_6
C_3	A_2	B_1	F_9	D_8	E_7	I_6	G_5	H_4
D_4	E_6	F_5	G_1	H_3	I_2	A_7	B_9	C_8
E_5	F_7	D_6	H_2	I_1	G_3	B_8	C_7	A_9
F_6	D_5	E_4	I_3	G_2	H_1	C_9	A_8	B_7
G_7	H_9	I_8	A_4	B_6	C_5	D_1	E_3	F_2
H_8	I_7	G_9	B_5	C_4	A_6	E_2	F_1	D_3
I_9	G_8	H_7	C_6	A_5	B_4	F_3	D_2	E_1

11 x 11

A_1	B_7	C_2	D_8	E_3	F_9	G_4	H_{10}	I_5	J_{11}	K_6
B_2	C_8	D_3	E_9	F_4	G_{10}	H_5	I_{11}	J_6	K_1	A_7
C_3	D_9	E_4	F_{10}	G_5	H_{11}	I_6	J_1	K_7	A_2	B_8
D_4	E_{10}	F_5	G_{11}	H_6	I_1	J_7	K_2	A_8	B_3	C_9
E_5	F_{11}	G_6	H_1	I_7	J_2	K_8	A_3	B_9	C_4	D_{10}
F_6	G_1	H_7	I_2	J_8	K_3	A_9	B_4	C_{10}	D_5	E_{11}
G_7	H_2	I_8	J_3	K_9	A_4	B_{10}	C_5	D_{11}	E_6	F_1
H_8	I_3	J_9	K_4	A_{10}	B_5	C_{11}	D_6	E_1	F_7	G_2
I_9	J_4	K_{10}	A_5	B_{11}	C_6	D_1	E_7	F_2	G_8	H_3
J_{10}	K_5	A_{11}	B_6	C_1	D_7	E_2	F_8	G_3	H_9	I_4
K_{11}	A_6	B_1	C_7	D_2	E_8	F_3	G_9	H_4	I_{10}	J_5

12 x 12

A_1	B_{12}	C_6	D_7	I_5	J_4	K_{10}	L_{11}	E_9	F_8	G_2	H_3
B_2	A_{11}	D_5	C_8	J_6	I_3	L_9	K_{12}	F_{10}	E_7	H_1	G_4
C_3	D_{10}	A_8	B_5	K_7	L_2	I_{12}	J_9	G_{11}	H_6	E_4	F_1
D_4	C_9	B_7	A_6	L_8	K_1	J_{11}	I_{10}	H_{12}	G_5	F_3	E_2
E_5	F_4	G_{10}	H_{11}	A_9	B_8	C_2	D_3	I_1	J_{12}	K_6	L_7
F_6	E_3	H_9	G_{12}	B_{10}	A_7	D_1	C_4	J_2	I_{11}	L_5	K_8
G_7	H_2	E_{12}	F_9	C_{11}	D_6	A_4	B_1	K_3	L_{10}	I_8	J_5
H_8	G_1	F_{11}	E_{10}	D_{12}	C_5	B_3	A_2	L_4	K_9	J_7	I_6
I_9	J_8	K_2	L_3	E_1	F_{12}	G_6	H_7	A_5	B_4	C_{10}	D_{11}
J_{10}	I_7	L_1	K_4	F_2	E_{11}	H_5	G_8	B_6	A_3	D_9	C_{12}
K_{11}	L_6	I_4	J_1	G_3	H_{10}	E_8	F_5	C_7	D_2	A_{12}	B_9
L_{12}	K_5	J_3	I_2	H_4	G_9	F_7	E_6	D_8	C_1	B_{11}	A_{10}

Reprinted by permission of John Wiley and Sons, Inc. from *Experimental Designs* by William G. Cochran and Gertrude M. Cox, second edition, 1957.

Appendix D

Random digits

85967	73152	14511	85285	36009	95892	36962	67835	63314	50162
07483	51453	11649	86348	76431	81594	95848	36738	25014	15460
96283	01898	61414	83525	04231	13604	75339	11730	85423	60698
49174	12074	98551	37895	93547	24769	09404	76548	05393	96770
97366	39941	21225	93629	19574	71565	33413	56087	40875	13351
90474	41469	16812	81542	81652	45554	27931	93994	22375	00953
28599	64109	09497	76235	41383	31555	12639	00619	22909	29563
25254	16210	89717	65997	82667	74624	36348	44018	64732	93589
28785	02760	24359	99410	77319	73408	58993	61098	04393	48245
84725	86576	86944	93296	10081	82454	76810	52975	10324	15457
41059	66456	47679	66810	15941	84602	14493	65515	19251	41642
67434	41045	82830	47617	36932	46728	71183	36345	41404	81110
72766	68816	37643	19959	57550	49620	98480	25640	67257	18671
92079	46784	66125	94932	64451	29275	57669	66658	30818	58353
29187	40350	62533	73603	34075	16451	42885	03448	37390	96328
74220	17612	65522	80607	19184	64164	66962	82310	18163	63495
03786	02407	06098	92917	40434	60602	82175	04470	78754	90775
75085	55558	15520	27038	25471	76107	90832	10819	56797	33751
09161	33015	19155	11715	00551	24909	31894	37774	37953	78837
75707	48992	64998	87080	39333	00767	45637	12538	67439	94914
21333	48660	31288	00086	79889	75532	28704	62844	92337	99695
65626	50061	42539	14812	48895	11196	34335	60492	70650	51108
84380	07389	87891	76255	89604	41372	10837	66992	93183	56920
46479	32072	80083	63868	70930	89654	05359	47196	12452	38234
59847	97197	55147	76639	76971	55928	36441	95141	42333	67483
31416	11231	27904	57383	31852	69137	96667	14315	01007	31929
82066	83436	67914	21465	99605	83114	97885	74440	99622	87912
01850	42782	39202	18582	46214	99228	79541	78298	75404	63648
32315	89276	89582	87138	16165	15984	21466	63830	30475	74729
59388	42703	55198	80380	67067	97155	34160	85019	03527	78140
58089	27632	50987	91373	07736	20436	96130	73483	85332	24384
61705	57285	30392	23660	75841	21931	04295	00875	09114	32101
18914	98982	60199	99275	41967	35208	30357	76772	92656	62318
11965	94089	34803	48941	69709	16784	44642	89761	66864	62803
85251	48111	80936	81781	93248	67877	16498	31924	51315	79921
66121	96986	84844	93873	46352	92183	51152	85878	30490	15974
53972	96642	24199	58080	35450	03482	66953	49521	63719	57615
14509	16594	78883	43222	23093	58645	60257	89250	63266	90858
37700	07688	65533	72126	23611	93993	01848	03910	38552	17472
85466	59392	72722	15473	73295	49759	56157	60477	83284	56367
52969	55863	42312	67842	05673	91878	82738	36563	79540	61935
42744	68315	17514	02878	97291	74851	42725	57894	81434	62041
26140	13336	67726	61876	29971	99294	96664	52817	90039	53211
95589	56319	14563	24071	06916	59555	18195	32280	79357	04224
39113	13217	59999	49952	83021	47709	53105	19295	88318	41626
41392	17622	18994	98283	07249	52289	24209	91139	30715	06604
54684	53645	79246	70183	87731	19185	08541	33519	07223	97413
89442	61001	36658	57444	95388	36682	38052	46719	09428	94012
36751	16778	54888	15357	68003	43564	90976	58904	40512	07725
98159	02564	21416	74944	53049	88749	02865	25772	89853	88714

Reprinted by permission of Houghton Mifflin Company from *Business Statistics for Management and Economics* by Daniel/Terrell, 6th edition, 1992.

Appendix E

Points for the Distribution of F [5% (light type) and 1% (bold face type)]

f_1, Degrees of freedom (for greater mean square)

f_2	1	2	3	4	5	6	7	8	9	10	11	12	14	16	20	24	30	40	50	75	100	200	500	∞
1	161 **4,052**	200 **4,999**	216 **5,403**	225 **5,625**	230 **5,764**	234 **5,859**	237 **5,928**	239 **5,981**	241 **6,022**	242 **6,056**	243 **6,082**	244 **6,106**	245 **6,142**	246 **6,169**	248 **6,208**	249 **6,234**	250 **6,261**	251 **6,286**	252 **6,302**	253 **6,323**	253 **6,334**	254 **6,352**	254 **6,361**	254 **6,366**
2	18.51 **98.49**	19.00 **99.00**	19.16 **99.17**	19.25 **99.25**	19.30 **99.30**	19.33 **99.33**	19.36 **99.34**	19.37 **99.36**	19.38 **99.38**	19.39 **99.40**	19.40 **99.41**	19.41 **99.42**	19.42 **99.43**	19.43 **99.44**	19.44 **99.45**	19.45 **99.46**	19.46 **99.47**	19.47 **99.48**	19.47 **99.48**	19.48 **99.49**	19.49 **99.49**	19.49 **99.49**	19.50 **99.50**	19.50 **99.50**
3	10.13 **34.12**	9.55 **30.82**	9.28 **29.46**	9.12 **28.71**	9.01 **28.24**	8.94 **27.91**	8.88 **27.67**	8.84 **27.49**	8.81 **27.34**	8.78 **27.23**	8.76 **27.13**	8.74 **27.05**	8.71 **26.92**	8.69 **26.83**	8.66 **26.69**	8.64 **26.60**	8.62 **26.50**	8.60 **26.41**	8.58 **26.35**	8.57 **26.27**	8.56 **26.23**	8.54 **26.18**	8.54 **26.14**	8.53 **26.12**
4	7.71 **21.20**	6.94 **18.00**	6.59 **16.69**	6.39 **15.98**	6.26 **15.52**	6.16 **15.21**	6.09 **14.98**	6.04 **14.80**	6.00 **14.66**	5.96 **14.54**	5.93 **14.45**	5.91 **14.37**	5.87 **14.24**	5.84 **14.15**	5.80 **14.02**	5.77 **13.93**	5.74 **13.83**	5.71 **13.74**	5.70 **13.69**	5.68 **13.61**	5.66 **13.57**	5.65 **13.52**	5.64 **13.48**	5.63 **13.46**
5	6.61 **16.26**	5.79 **13.27**	5.41 **12.06**	5.19 **11.39**	5.05 **10.97**	4.95 **10.67**	4.88 **10.45**	4.82 **10.29**	4.78 **10.15**	4.74 **10.05**	4.70 **9.96**	4.68 **9.89**	4.64 **9.77**	4.60 **9.68**	4.56 **9.55**	4.53 **9.47**	4.50 **9.38**	4.46 **9.29**	4.44 **9.24**	4.42 **9.17**	4.40 **9.13**	4.38 **9.07**	4.37 **9.04**	4.36 **9.02**
6	5.99 **13.74**	5.14 **10.92**	4.76 **9.78**	4.53 **9.15**	4.39 **8.75**	4.28 **8.47**	4.21 **8.26**	4.15 **8.10**	4.10 **7.98**	4.06 **7.87**	4.03 **7.79**	4.00 **7.72**	3.96 **7.60**	3.92 **7.52**	3.87 **7.39**	3.84 **7.31**	3.81 **7.23**	3.77 **7.14**	3.75 **7.09**	3.72 **7.02**	3.71 **6.99**	3.69 **6.94**	3.68 **6.90**	3.67 **6.88**
7	5.59 **12.25**	4.74 **9.55**	4.35 **8.45**	4.12 **7.85**	3.97 **7.46**	3.87 **7.19**	3.79 **7.00**	3.73 **6.84**	3.68 **6.71**	3.63 **6.62**	3.60 **6.54**	3.57 **6.47**	3.52 **6.35**	3.49 **6.27**	3.44 **6.15**	3.41 **6.07**	3.38 **5.98**	3.34 **5.90**	3.32 **5.85**	3.29 **5.78**	3.28 **5.75**	3.25 **5.70**	3.24 **5.67**	3.23 **5.65**
8	5.32 **11.26**	4.46 **8.65**	4.07 **7.59**	3.84 **7.01**	3.69 **6.63**	3.58 **6.37**	3.50 **6.19**	3.44 **6.03**	3.39 **5.91**	3.34 **5.82**	3.31 **5.74**	3.28 **5.67**	3.23 **5.56**	3.20 **5.48**	3.15 **5.36**	3.12 **5.28**	3.08 **5.20**	3.05 **5.11**	3.03 **5.06**	3.00 **5.00**	2.98 **4.96**	2.96 **4.91**	2.94 **4.88**	2.93 **4.86**
9	5.12 **10.56**	4.26 **8.02**	3.86 **6.99**	3.63 **6.42**	3.48 **6.06**	3.37 **5.80**	3.29 **5.62**	3.23 **5.47**	3.18 **5.35**	3.13 **5.26**	3.10 **5.18**	3.07 **5.11**	3.02 **5.00**	2.98 **4.92**	2.93 **4.80**	2.90 **4.73**	2.86 **4.64**	2.82 **4.56**	2.80 **4.51**	2.77 **4.45**	2.76 **4.41**	2.73 **4.36**	2.72 **4.33**	2.71 **4.31**
10	4.96 **10.04**	4.10 **7.56**	3.71 **6.55**	3.48 **5.99**	3.33 **5.64**	3.22 **5.39**	3.14 **5.21**	3.07 **5.06**	3.02 **4.95**	2.97 **4.85**	2.94 **4.78**	2.91 **4.71**	2.86 **4.60**	2.82 **4.52**	2.77 **4.41**	2.74 **4.33**	2.70 **4.25**	2.67 **4.17**	2.64 **4.12**	2.61 **4.05**	2.59 **4.01**	2.56 **3.96**	2.55 **3.93**	2.54 **3.91**
11	4.84 **9.65**	3.98 **7.20**	3.59 **6.22**	3.36 **5.67**	3.20 **5.32**	3.09 **5.07**	3.01 **4.88**	2.95 **4.74**	2.90 **4.63**	2.86 **4.54**	2.82 **4.46**	2.79 **4.40**	2.74 **4.29**	2.70 **4.21**	2.65 **4.10**	2.61 **4.02**	2.57 **3.94**	2.53 **3.86**	2.50 **3.80**	2.47 **3.74**	2.45 **3.70**	2.42 **3.66**	2.41 **3.62**	2.40 **3.60**
12	4.75 **9.33**	3.88 **6.93**	3.49 **5.95**	3.26 **5.41**	3.11 **5.06**	3.00 **4.82**	2.92 **4.65**	2.85 **4.50**	2.80 **4.39**	2.76 **4.30**	2.72 **4.22**	2.69 **4.16**	2.64 **4.05**	2.60 **3.98**	2.54 **3.86**	2.50 **3.78**	2.46 **3.70**	2.42 **3.61**	2.40 **3.56**	2.36 **3.49**	2.35 **3.46**	2.32 **3.41**	2.31 **3.38**	2.30 **3.36**
13	4.67 **9.07**	3.80 **6.70**	3.41 **5.74**	3.18 **5.20**	3.02 **4.86**	2.92 **4.62**	2.84 **4.44**	2.77 **4.30**	2.72 **4.19**	2.67 **4.10**	2.63 **4.02**	2.60 **3.96**	2.55 **3.85**	2.51 **3.78**	2.46 **3.67**	2.42 **3.59**	2.38 **3.51**	2.34 **3.42**	2.32 **3.37**	2.28 **3.30**	2.26 **3.27**	2.24 **3.21**	2.22 **3.18**	2.21 **3.16**

continued next page

f_1, Degrees of freedom (for greater mean square)

(Each cell: upper value = 5% point, lower value in bold = 1% point.)

f_2	1	2	3	4	5	6	7	8	9	10	11	12	14	16	20	24	30	40	50	75	100	200	500	∞
14	4.60/**8.86**	3.74/**6.51**	3.34/**5.56**	3.11/**5.03**	2.96/**4.69**	2.85/**4.46**	2.77/**4.28**	2.70/**4.14**	2.65/**4.03**	2.60/**3.94**	2.56/**3.86**	2.53/**3.80**	2.48/**3.70**	2.44/**3.62**	2.39/**3.51**	2.35/**3.43**	2.31/**3.34**	2.27/**3.26**	2.24/**3.21**	2.21/**3.14**	2.19/**3.11**	2.16/**3.06**	2.14/**3.02**	2.13/**3.00**
15	4.54/**8.68**	3.68/**6.36**	3.29/**5.42**	3.06/**4.89**	2.90/**4.56**	2.79/**4.32**	2.70/**4.14**	2.64/**4.00**	2.59/**3.89**	2.55/**3.80**	2.51/**3.73**	2.48/**3.67**	2.43/**3.56**	2.39/**3.48**	2.33/**3.36**	2.29/**3.29**	2.25/**3.20**	2.21/**3.12**	2.18/**3.07**	2.15/**3.00**	2.12/**2.97**	2.10/**2.92**	2.08/**2.89**	2.07/**2.87**
16	4.49/**8.53**	3.63/**6.23**	3.24/**5.29**	3.01/**4.77**	2.85/**4.44**	2.74/**4.20**	2.66/**4.03**	2.59/**3.89**	2.54/**3.78**	2.49/**3.69**	2.45/**3.61**	2.42/**3.55**	2.37/**3.45**	2.33/**3.37**	2.28/**3.25**	2.24/**3.18**	2.20/**3.10**	2.16/**3.01**	2.13/**2.96**	2.09/**2.93**	2.07/**2.86**	2.04/**2.80**	2.02/**2.77**	2.01/**2.75**
17	4.45/**8.40**	3.59/**6.11**	3.20/**5.18**	2.96/**4.67**	2.81/**4.34**	2.70/**4.10**	2.62/**3.93**	2.55/**3.79**	2.50/**3.68**	2.45/**3.59**	2.41/**3.52**	2.38/**3.45**	2.33/**3.35**	2.29/**3.27**	2.23/**3.16**	2.19/**3.08**	2.15/**3.00**	2.11/**2.92**	2.08/**2.86**	2.04/**2.79**	2.02/**2.76**	1.99/**2.70**	1.97/**2.67**	1.96/**2.65**
18	4.41/**8.28**	3.55/**6.01**	3.16/**5.09**	2.93/**4.58**	2.77/**4.25**	2.66/**4.01**	2.58/**3.85**	2.51/**3.71**	2.46/**3.60**	2.41/**3.51**	2.37/**3.44**	2.34/**3.37**	2.29/**3.27**	2.25/**3.19**	2.19/**3.07**	2.15/**3.00**	2.11/**2.91**	2.07/**2.83**	2.04/**2.78**	2.00/**2.71**	1.98/**2.68**	1.95/**2.62**	1.93/**2.59**	1.92/**2.57**
19	4.38/**8.18**	3.52/**5.93**	3.13/**5.01**	2.90/**4.50**	2.74/**4.17**	2.63/**3.94**	2.55/**3.77**	2.48/**3.63**	2.43/**3.52**	2.38/**3.43**	2.34/**3.36**	2.31/**3.30**	2.26/**3.19**	2.21/**3.12**	2.15/**3.00**	2.11/**2.92**	2.07/**2.84**	2.02/**2.76**	2.00/**2.70**	1.96/**2.63**	1.94/**2.60**	1.91/**2.54**	1.90/**2.51**	1.88/**2.49**
20	4.35/**8.10**	3.49/**5.85**	3.10/**4.94**	2.87/**4.43**	2.71/**4.10**	2.60/**3.87**	2.52/**3.71**	2.45/**3.56**	2.40/**3.45**	2.35/**3.37**	2.31/**3.30**	2.28/**3.23**	2.23/**3.13**	2.18/**3.05**	2.12/**2.94**	2.08/**2.86**	2.04/**2.77**	1.99/**2.69**	1.96/**2.63**	1.92/**2.56**	1.90/**2.53**	1.87/**2.47**	1.85/**2.44**	1.84/**2.42**
21	4.32/**8.02**	3.47/**5.78**	3.07/**4.87**	2.84/**4.37**	2.68/**4.04**	2.57/**3.81**	2.49/**3.65**	2.42/**3.51**	2.37/**3.40**	2.32/**3.31**	2.28/**3.24**	2.25/**3.17**	2.20/**3.07**	2.15/**2.99**	2.09/**2.88**	2.05/**2.80**	2.00/**2.72**	1.96/**2.63**	1.93/**2.58**	1.89/**2.51**	1.87/**2.47**	1.84/**2.42**	1.82/**2.38**	1.81/**2.36**
22	4.30/**7.94**	3.44/**5.72**	3.05/**4.82**	2.82/**4.31**	2.66/**3.99**	2.55/**3.76**	2.47/**3.59**	2.40/**3.45**	2.35/**3.35**	2.30/**3.26**	2.26/**3.18**	2.23/**3.12**	2.18/**3.02**	2.13/**2.94**	2.07/**2.83**	2.03/**2.75**	1.98/**2.67**	1.93/**2.58**	1.91/**2.53**	1.87/**2.46**	1.84/**2.42**	1.81/**2.37**	1.80/**2.33**	1.78/**2.31**
23	4.28/**7.88**	3.42/**5.66**	3.03/**4.76**	2.80/**4.26**	2.64/**3.94**	2.53/**3.71**	2.45/**3.54**	2.38/**3.41**	2.32/**3.30**	2.28/**3.21**	2.24/**3.14**	2.20/**3.07**	2.14/**2.97**	2.10/**2.89**	2.04/**2.78**	2.00/**2.70**	1.96/**2.62**	1.91/**2.53**	1.88/**2.48**	1.84/**2.41**	1.82/**2.37**	1.79/**2.32**	1.77/**2.28**	1.76/**2.26**
24	4.26/**7.82**	3.40/**5.61**	3.01/**4.72**	2.78/**4.22**	2.62/**3.90**	2.51/**3.67**	2.43/**3.50**	2.36/**3.36**	2.30/**3.25**	2.26/**3.17**	2.22/**3.09**	2.18/**3.03**	2.13/**2.93**	2.09/**2.85**	2.02/**2.74**	1.98/**2.66**	1.94/**2.58**	1.89/**2.49**	1.86/**2.44**	1.82/**2.36**	1.80/**2.33**	1.76/**2.27**	1.74/**2.23**	1.73/**2.21**
25	4.24/**7.77**	3.38/**5.57**	2.99/**4.68**	2.76/**4.18**	2.60/**3.86**	2.49/**3.63**	2.41/**3.46**	2.34/**3.32**	2.28/**3.21**	2.24/**3.13**	2.20/**3.05**	2.16/**2.99**	2.11/**2.89**	2.06/**2.81**	2.00/**2.70**	1.96/**2.62**	1.92/**2.54**	1.87/**2.45**	1.84/**2.40**	1.80/**2.32**	1.77/**2.29**	1.74/**2.23**	1.72/**2.19**	1.71/**2.17**
26	4.22/**7.72**	3.37/**5.53**	2.98/**4.64**	2.74/**4.14**	2.59/**3.82**	2.47/**3.59**	2.39/**3.42**	2.32/**3.29**	2.27/**3.17**	2.22/**3.09**	2.18/**3.02**	2.15/**2.96**	2.10/**2.86**	2.05/**2.77**	1.99/**2.66**	1.95/**2.58**	1.90/**2.50**	1.85/**2.41**	1.82/**2.36**	1.78/**2.28**	1.76/**2.25**	1.72/**2.19**	1.70/**2.15**	1.69/**2.13**

Experimental Research Design and Analysis

Appendix E (Continued)

f_1, Degrees of freedom (for greater mean square)

Each cell shows two values (upper / lower).

f_2	1	2	3	4	5	6	7	8	9	10	11	12	14	16	20	24	30	40	50	75	100	200	500	∞
27	4.21/7.68	3.35/5.49	2.96/4.60	2.73/4.11	2.57/3.79	2.46/3.56	2.37/3.39	2.30/3.26	2.25/3.14	2.20/3.06	2.16/2.98	2.13/2.93	2.08/2.83	2.03/2.74	1.97/2.63	1.93/2.55	1.88/2.47	1.84/2.38	1.80/2.33	1.76/2.25	1.74/2.21	1.71/2.16	1.68/2.12	1.67/2.10
28	4.20/7.64	3.34/5.45	2.95/4.57	2.71/4.07	2.56/3.76	2.44/3.53	2.36/3.36	2.29/3.23	2.24/3.11	2.19/3.03	2.15/2.95	2.12/2.90	2.06/2.80	2.02/2.71	1.96/2.60	1.91/2.52	1.87/2.44	1.81/2.35	1.78/2.30	1.75/2.22	1.72/2.18	1.69/2.13	1.67/2.09	1.65/2.06
29	4.18/7.60	3.33/5.42	2.93/4.54	2.70/4.04	2.54/3.73	2.43/3.50	2.35/3.33	2.28/3.20	2.22/3.08	2.18/3.00	2.14/2.92	2.10/2.87	2.05/2.77	2.00/2.68	1.94/2.57	1.90/2.49	1.85/2.41	1.80/2.32	1.77/2.27	1.73/2.19	1.71/2.15	1.68/2.10	1.65/2.06	1.64/2.03
30	4.17/7.56	3.32/5.39	2.92/4.51	2.69/4.02	2.53/3.70	2.42/3.47	2.34/3.30	2.27/3.17	2.21/3.06	2.16/2.98	2.12/2.90	2.09/2.84	2.04/2.74	1.99/2.66	1.93/2.55	1.89/2.47	1.84/2.38	1.79/2.29	1.76/2.24	1.72/2.16	1.69/2.13	1.66/2.07	1.64/2.03	1.62/2.01
32	4.15/7.50	3.30/5.34	2.90/4.46	2.67/3.97	2.51/3.66	2.40/3.42	2.32/3.25	2.25/3.12	2.19/3.01	2.14/2.94	2.10/2.86	2.07/2.80	2.02/2.70	1.97/2.62	1.91/2.51	1.86/2.42	1.82/2.34	1.76/2.25	1.74/2.20	1.69/2.12	1.67/2.08	1.64/2.02	1.61/1.98	1.59/1.96
34	4.13/7.44	3.28/5.29	2.88/4.42	2.65/3.93	2.49/3.61	2.38/3.38	2.30/3.21	2.23/3.08	2.17/2.97	2.12/2.89	2.08/2.82	2.05/2.76	2.00/2.66	1.95/2.58	1.89/2.47	1.84/2.38	1.80/2.30	1.74/2.21	1.71/2.15	1.67/2.08	1.64/2.04	1.61/1.98	1.59/1.94	1.57/1.91
36	4.11/7.39	3.26/5.25	2.86/4.38	2.63/3.89	2.48/3.58	2.36/3.35	2.28/3.18	2.21/3.04	2.15/2.94	2.10/2.86	2.06/2.78	2.03/2.72	1.98/2.62	1.93/2.54	1.87/2.43	1.82/2.35	1.78/2.26	1.72/2.17	1.69/2.12	1.65/2.04	1.62/2.00	1.59/1.94	1.56/1.90	1.55/1.87
38	4.10/7.35	3.25/5.21	2.85/4.34	2.62/3.86	2.46/3.54	2.35/3.32	2.26/3.15	2.19/3.02	2.14/2.91	2.09/2.82	2.05/2.75	2.02/2.69	1.96/2.59	1.92/2.51	1.85/2.40	1.80/2.32	1.76/2.22	1.71/2.14	1.67/2.08	1.63/2.00	1.60/1.97	1.57/1.90	1.54/1.86	1.53/1.84
40	4.08/7.31	3.23/5.18	2.84/4.31	2.61/3.83	2.45/3.51	2.34/3.29	2.25/3.12	2.18/2.99	2.12/2.88	2.07/2.80	2.04/2.73	2.00/2.66	1.95/2.56	1.90/2.49	1.84/2.37	1.79/2.29	1.74/2.20	1.69/2.11	1.66/2.05	1.61/1.97	1.59/1.94	1.55/1.88	1.53/1.84	1.51/1.81
42	4.07/7.27	3.22/5.15	2.83/4.29	2.59/3.80	2.44/3.49	2.32/3.26	2.24/3.10	2.17/2.96	2.11/2.86	2.06/2.77	2.02/2.70	1.99/2.64	1.94/2.54	1.89/2.46	1.82/2.35	1.78/2.26	1.73/2.17	1.68/2.08	1.64/2.02	1.60/1.94	1.57/1.91	1.54/1.85	1.51/1.80	1.49/1.78
44	4.06/7.24	3.21/5.12	2.82/4.26	2.58/3.78	2.43/3.46	2.31/3.24	2.23/3.07	2.16/2.94	2.10/2.84	2.05/2.75	2.01/2.68	1.98/2.62	1.92/2.52	1.88/2.44	1.81/2.32	1.76/2.24	1.72/2.15	1.66/2.06	1.63/2.00	1.58/1.92	1.56/1.88	1.52/1.82	1.50/1.78	1.48/1.75
46	4.05/7.21	3.20/5.10	2.81/4.24	2.57/3.76	2.42/3.44	2.30/3.22	2.22/3.05	2.14/2.92	2.09/2.82	2.04/2.73	2.00/2.66	1.97/2.60	1.91/2.50	1.87/2.42	1.80/2.30	1.75/2.22	1.71/2.13	1.65/2.04	1.62/1.98	1.57/1.90	1.54/1.86	1.51/1.80	1.48/1.76	1.46/1.72
48	4.04/7.19	3.19/5.08	2.80/4.22	2.56/3.74	2.41/3.42	2.30/3.20	2.21/3.04	2.14/2.90	2.08/2.80	2.03/2.71	1.99/2.64	1.96/2.58	1.90/2.48	1.86/2.40	1.79/2.28	1.74/2.20	1.70/2.11	1.64/2.02	1.61/1.96	1.56/1.88	1.53/1.84	1.50/1.78	1.47/1.73	1.45/1.70

continued next page

f_1, Degrees of freedom (for greater mean square)

f_2	1	2	3	4	5	6	7	8	9	10	11	12	14	16	20	24	30	40	50	75	100	200	500	∞
50	4.03 / 7.17	3.18 / 5.06	2.79 / 4.20	2.56 / 3.72	2.40 / 3.41	2.29 / 3.18	2.20 / 3.02	2.13 / 2.88	2.07 / 2.78	2.02 / 2.70	1.98 / 2.62	1.95 / 2.56	1.90 / 2.46	1.85 / 2.39	1.78 / 2.26	1.74 / 2.18	1.69 / 2.10	1.63 / 2.00	1.60 / 1.94	1.55 / 1.86	1.52 / 1.82	1.48 / 1.76	1.46 / 1.71	1.44 / 1.68
55	4.02 / 7.12	3.17 / 5.01	2.78 / 4.16	2.54 / 3.68	2.38 / 3.37	2.27 / 3.15	2.18 / 2.98	2.11 / 2.85	2.05 / 2.75	2.00 / 2.66	1.97 / 2.59	1.93 / 2.53	1.88 / 2.43	1.83 / 2.35	1.76 / 2.23	1.72 / 2.15	1.67 / 2.06	1.61 / 1.96	1.58 / 1.90	1.52 / 1.82	1.50 / 1.78	1.46 / 1.71	1.43 / 1.66	1.41 / 1.64
60	4.00 / 7.08	3.15 / 4.98	2.76 / 4.13	2.52 / 3.65	2.37 / 3.34	2.25 / 3.12	2.17 / 2.95	2.10 / 2.82	2.04 / 2.72	1.99 / 2.63	1.95 / 2.56	1.92 / 2.50	1.86 / 2.40	1.81 / 2.32	1.75 / 2.20	1.70 / 2.12	1.65 / 2.03	1.59 / 1.93	1.56 / 1.87	1.50 / 1.79	1.48 / 1.74	1.44 / 1.68	1.41 / 1.63	1.39 / 1.60
65	3.99 / 7.04	3.14 / 4.95	2.75 / 4.10	2.51 / 3.62	2.36 / 3.31	2.24 / 3.09	2.15 / 2.93	2.08 / 2.79	2.02 / 2.70	1.98 / 2.61	1.94 / 2.54	1.90 / 2.47	1.85 / 2.37	1.80 / 2.30	1.73 / 2.18	1.68 / 2.09	1.63 / 2.00	1.57 / 1.90	1.54 / 1.84	1.49 / 1.76	1.46 / 1.71	1.42 / 1.64	1.39 / 1.60	1.37 / 1.56
70	3.98 / 7.01	3.13 / 4.92	2.74 / 4.08	2.50 / 3.60	2.35 / 3.29	2.23 / 3.07	2.14 / 2.91	2.07 / 2.77	2.01 / 2.67	1.97 / 2.59	1.93 / 2.51	1.89 / 2.45	1.84 / 2.35	1.79 / 2.28	1.72 / 2.15	1.67 / 2.07	1.62 / 1.98	1.56 / 1.88	1.53 / 1.82	1.47 / 1.74	1.45 / 1.69	1.40 / 1.62	1.37 / 1.56	1.35 / 1.53
80	3.96 / 6.96	3.11 / 4.88	2.72 / 4.04	2.48 / 3.56	2.33 / 3.25	2.21 / 3.04	2.12 / 2.87	2.05 / 2.74	1.99 / 2.64	1.95 / 2.55	1.91 / 2.48	1.88 / 2.41	1.82 / 2.32	1.77 / 2.24	1.70 / 2.11	1.65 / 2.03	1.60 / 1.94	1.54 / 1.84	1.51 / 1.78	1.45 / 1.70	1.42 / 1.65	1.38 / 1.57	1.35 / 1.52	1.32 / 1.49
100	3.94 / 6.90	3.09 / 4.82	2.70 / 3.98	2.46 / 3.51	2.30 / 3.20	2.19 / 2.99	2.10 / 2.82	2.03 / 2.69	1.97 / 2.59	1.92 / 2.51	1.88 / 2.43	1.85 / 2.36	1.79 / 2.26	1.75 / 2.19	1.68 / 2.06	1.63 / 1.98	1.57 / 1.89	1.51 / 1.79	1.48 / 1.73	1.42 / 1.64	1.39 / 1.59	1.34 / 1.51	1.30 / 1.46	1.28 / 1.43
125	3.92 / 6.84	3.07 / 4.78	2.68 / 3.94	2.44 / 3.47	2.29 / 3.17	2.17 / 2.95	2.08 / 2.79	2.01 / 2.65	1.95 / 2.56	1.90 / 2.47	1.86 / 2.40	1.83 / 2.33	1.77 / 2.23	1.72 / 2.15	1.65 / 2.03	1.60 / 1.94	1.55 / 1.85	1.49 / 1.75	1.45 / 1.68	1.39 / 1.59	1.36 / 1.54	1.31 / 1.46	1.27 / 1.40	1.25 / 1.37
150	3.91 / 6.81	3.06 / 4.75	2.67 / 3.91	2.43 / 3.44	2.27 / 3.14	2.16 / 2.92	2.07 / 2.76	2.00 / 2.62	1.94 / 2.53	1.89 / 2.44	1.85 / 2.37	1.82 / 2.30	1.76 / 2.20	1.71 / 2.12	1.64 / 2.00	1.59 / 1.91	1.54 / 1.83	1.47 / 1.72	1.44 / 1.66	1.37 / 1.56	1.34 / 1.51	1.29 / 1.43	1.25 / 1.37	1.22 / 1.33
200	3.89 / 6.76	3.04 / 4.71	2.65 / 3.88	2.41 / 3.41	2.26 / 3.11	2.14 / 2.90	2.05 / 2.73	1.98 / 2.60	1.92 / 2.50	1.87 / 2.41	1.83 / 2.34	1.80 / 2.28	1.74 / 2.17	1.69 / 2.09	1.62 / 1.97	1.57 / 1.88	1.52 / 1.79	1.45 / 1.69	1.42 / 1.62	1.35 / 1.53	1.32 / 1.48	1.26 / 1.39	1.22 / 1.33	1.19 / 1.28
400	3.86 / 6.70	3.02 / 4.66	2.62 / 3.83	2.39 / 3.36	2.23 / 3.06	2.12 / 2.85	2.03 / 2.69	1.96 / 2.55	1.90 / 2.46	1.85 / 2.37	1.81 / 2.29	1.78 / 2.23	1.72 / 2.12	1.67 / 2.04	1.60 / 1.92	1.54 / 1.84	1.49 / 1.74	1.42 / 1.64	1.38 / 1.57	1.32 / 1.47	1.28 / 1.42	1.22 / 1.32	1.16 / 1.24	1.13 / 1.19
1000	3.85 / 6.66	3.00 / 4.62	2.61 / 3.80	2.38 / 3.34	2.22 / 3.04	2.10 / 2.82	2.02 / 2.66	1.95 / 2.53	1.89 / 2.43	1.84 / 2.34	1.80 / 2.26	1.76 / 2.20	1.70 / 2.09	1.65 / 2.01	1.58 / 1.89	1.53 / 1.81	1.47 / 1.71	1.41 / 1.61	1.36 / 1.54	1.30 / 1.44	1.26 / 1.38	1.19 / 1.28	1.13 / 1.19	1.08 / 1.11
∞	3.84 / 6.64	2.99 / 4.60	2.60 / 3.78	2.37 / 3.32	2.21 / 3.02	2.09 / 2.80	2.01 / 2.64	1.94 / 2.51	1.88 / 2.41	1.83 / 2.32	1.79 / 2.24	1.75 / 2.18	1.69 / 2.07	1.64 / 1.99	1.57 / 1.87	1.52 / 1.79	1.46 / 1.69	1.40 / 1.59	1.35 / 1.52	1.28 / 1.41	1.24 / 1.36	1.17 / 1.25	1.11 / 1.15	1.00 / 1.00

Reprinted by permission from STATISTICAL METHODS by George W. Snedecor and William G. Cochran, sixth edition (c) 1967 by Iowa State University Press. Ames, Iowa

Appendix F

BASIC PLANS FOR BALANCED AND PARTIALLY BALANCED LATTICE DESIGNS

PLANS

Plan 1   3 × 3 balanced lattice

$$t = 9, k = 3, r = 4, b = 12, \lambda = 1 *$$

Block	Rep. I				Rep. II				Rep. III				Rep IV		
(1)	1	2	3	(4)	1	4	7	(7)	1	5	9	(10)	1	8	6
(2)	4	5	6	(5)	2	5	8	(8)	7	2	6	(11)	4	2	9
(3)	7	8	9	(6)	3	6	9	(9)	4	8	3	(12)	7	5	3

Plan 2   4 × 4 balanced lattice

$$t = 16, k = 4, r = 5, b = 20, \lambda = 1$$

Block	Rep. I					Rep. II					Rep. III			
(1)	1	2	3	4	(5)	1	5	9	13	(9)	1	6	11	16
(2)	5	6	7	8	(6)	2	6	10	14	(10)	5	2	15	12
(3)	9	10	11	12	(7)	3	7	11	15	(11)	9	14	3	8
(4)	13	14	15	16	(8)	4	8	12	16	(12)	13	10	7	4

Block	Rep. IV					Rep. V			
(13)	1	14	7	12	(17)	1	10	15	8
(14)	13	2	11	8	(18)	9	2	7	16
(15)	5	10	3	16	(19)	13	6	3	12
(16)	9	6	15	4	(20)	5	14	11	4

Plan 3   5 × 5 balanced lattice

$$t = 25, k = 5, r = 6, b = 30, \lambda = 1$$

Block	Rep. I						Rep. II						Rep. III				
(1)	1	2	3	4	5	(6)	1	6	11	16	21	(11)	1	7	13	19	25
(2)	6	7	8	9	10	(7)	2	7	12	17	22	(12)	21	2	8	14	20
(3)	11	12	13	14	15	(8)	3	8	13	18	23	(13)	16	22	3	9	15
(4)	16	17	18	19	20	(9)	4	9	14	19	24	(14)	11	17	23	4	10
(5)	21	22	23	24	25	(10)	5	10	15	20	25	(15)	6	12	18	24	5

Block	Rep. IV						Rep. V						Rep. VI				
(16)	1	12	23	9	20	(21)	1	17	8	24	15	(26)	1	22	18	14	10
(17)	16	2	13	24	10	(22)	11	2	18	9	25	(27)	6	2	23	19	15
(18)	6	17	3	14	25	(23)	21	12	3	19	10	(28)	11	7	3	24	20
(19)	21	7	18	4	15	(24)	6	22	13	4	20	(29)	16	12	8	4	25
(20)	11	22	8	19	5	(25)	16	7	23	14	5	(30)	21	17	13	9	5

*The symbol λ denotes the number of times that two treatments appear in the same block.

PLANS

Plan 4   7 × 7 balanced lattice

$$t = 49, k = 7, r = 8, b = 56, \lambda = 1$$

Block		Rep. I								Rep. II					
(1)	1	2	3	4	5	6	7	(8)	1	8	15	22	29	36	43
(2)	8	9	10	11	12	13	14	(9)	2	9	16	23	30	37	44
(3)	15	16	17	18	19	20	21	(10)	3	10	17	24	31	38	45
(4)	22	23	24	25	26	27	28	(11)	4	11	18	25	32	39	46
(5)	29	30	31	32	33	34	35	(12)	5	12	19	26	33	40	47
(6)	36	37	38	39	40	41	42	(13)	6	13	20	27	34	41	48
(7)	43	44	45	46	47	48	49	(14)	7	14	21	28	35	42	49

		Rep. III								Rep. IV					
(15)	1	9	17	25	33	41	49	(22)	1	37	24	11	47	34	21
(16)	43	2	10	18	26	34	42	(23)	15	2	38	25	12	48	35
(17)	36	44	3	11	19	27	35	(24)	29	16	3	39	26	13	49
(18)	29	37	45	4	12	20	28	(25)	43	30	17	4	40	27	14
(19)	22	30	38	46	5	13	21	(26)	8	44	31	18	5	41	28
(20)	15	23	31	39	47	6	14	(27)	22	9	45	32	19	6	42
(21)	8	16	24	32	40	48	7	(28)	36	23	10	46	33	20	7

		Rep. V								Rep. VI					
(29)	1	30	10	39	19	48	28	(36)	1	23	45	18	40	13	35
(30)	22	2	31	11	40	20	49	(37)	29	2	24	46	19	41	14
(31)	43	23	3	32	12	41	21	(38)	8	30	3	25	47	20	42
(32)	15	44	24	4	33	13	42	(39)	36	9	31	4	26	48	21
(33)	36	16	45	25	5	34	14	(40)	15	37	10	32	5	27	49
(34)	8	37	17	46	26	6	35	(41)	43	16	38	11	33	6	28
(35)	29	9	38	18	47	27	7	(42)	22	44	17	39	12	34	7

		Rep. VII								Rep. VIII					
(43)	1	16	31	46	12	27	42	(50)	1	44	38	32	26	20	14
(44)	36	2	17	32	47	13	28	(51)	8	2	45	39	33	27	21
(45)	22	37	3	18	33	48	14	(52)	15	9	3	46	40	34	28
(46)	8	23	38	4	19	34	49	(53)	22	16	10	4	47	41	35
(47)	43	9	24	39	5	20	35	(54)	29	23	17	11	5	48	42
(48)	29	44	10	25	40	6	21	(55)	36	30	24	18	12	6	49
(49)	15	30	45	11	26	41	7	(56)	43	37	31	25	19	13	7

LATTICE DESIGNS

Plan 5   8 × 8 balanced lattice

$$t = 64, k = 8, r = 9, b = 72, \lambda = 1$$

Block				Rep. I				
(1)	1	2	3	4	5	6	7	8
(2)	9	10	11	12	13	14	15	16
(3)	17	18	19	20	21	22	23	24
(4)	25	26	27	28	29	30	31	32
(5)	33	34	35	36	37	38	39	40
(6)	41	42	43	44	45	46	47	48
(7)	49	50	51	52	53	54	55	56
(8)	57	58	59	60	61	62	63	64

				Rep. II				
(9)	1	9	17	25	33	41	49	57
(10)	2	10	18	26	34	42	50	58
(11)	3	11	19	27	35	43	51	59
(12)	4	12	20	28	36	44	52	60
(13)	5	13	21	29	37	45	53	61
(14)	6	14	22	30	38	46	54	62
(15)	7	15	23	31	39	47	55	63
(16)	8	16	24	32	40	48	56	64

				Rep. III				
(17)	1	10	19	28	37	46	55	64
(18)	9	2	51	44	61	30	23	40
(19)	17	50	3	36	29	62	15	48
(20)	25	42	35	4	21	14	63	56
(21)	33	58	27	20	5	54	47	16
(22)	41	26	59	12	53	6	39	24
(23)	49	18	11	60	45	38	7	32
(24)	57	34	43	52	13	22	31	8

				Rep. IV				
(25)	1	18	27	44	13	62	39	56
(26)	17	2	35	60	53	46	31	16
(27)	25	34	3	12	45	54	23	64
(28)	41	58	11	4	29	22	55	40
(29)	9	50	43	28	5	38	63	24
(30)	57	42	51	20	37	6	15	32
(31)	33	26	19	52	61	14	7	48
(32)	49	10	59	36	21	30	47	8

				Rep. V				
(33)	1	26	43	60	21	54	15	40
(34)	25	2	11	52	37	62	47	24
(35)	41	10	3	20	61	38	31	56
(36)	57	50	19	4	45	30	39	16
(37)	17	34	59	44	5	14	55	32
(38)	49	58	35	28	13	6	23	48
(39)	9	42	27	36	53	22	7	64
(40)	33	18	51	12	29	46	63	8

				Rep. VI				
(41)	1	34	11	20	53	30	63	48
(42)	33	2	59	28	45	22	15	56
(43)	9	58	3	52	21	46	39	32
(44)	17	26	51	4	13	38	47	64
(45)	49	42	19	12	5	62	31	40
(46)	25	18	43	36	61	6	55	16
(47)	57	10	35	44	29	54	7	24
(48)	41	50	27	60	37	14	23	8

PLANS

Plan 5 *(Continued)* 8 × 8 balanced lattice

Rep. VII

(49)	1	42	59	52	29	38	23	16
(50)	41	2	19	36	13	54	63	32
(51)	57	18	3	28	53	14	47	40
(52)	49	34	27	4	61	46	15	24
(53)	25	10	51	60	5	22	39	48
(54)	33	50	11	44	21	6	31	64
(55)	17	58	43	12	37	30	7	56
(56)	9	26	35	20	45	62	55	8

Rep. VIII

(57)	1	50	35	12	61	22	47	32
(58)	49	2	43	20	29	14	39	64
(59)	33	42	3	60	13	30	55	24
(60)	9	18	59	4	37	54	31	48
(61)	57	26	11	36	5	46	23	56
(62)	17	10	27	52	45	6	63	40
(63)	41	34	51	28	21	62	7	16
(64)	25	58	19	44	53	38	15	8

Rep. IX

(65)	1	58	51	36	45	14	31	24
(66)	57	2	27	12	21	38	55	48
(67)	49	26	3	44	37	22	63	16
(68)	33	10	43	4	53	62	23	32
(69)	41	18	35	52	5	30	15	64
(70)	9	34	19	60	29	6	47	56
(71)	25	50	59	20	13	46	7	40
(72)	17	42	11	28	61	54	39	8

Plan 6 9 × 9 balanced lattice

$t = 81, k = 9, r = 10, b = 90, \lambda = 1$

Rep. I

Block

(1)	1	2	3	4	5	6	7	8	9
(2)	10	11	12	13	14	15	16	17	18
(3)	19	20	21	22	23	24	25	26	27
(4)	28	29	30	31	32	33	34	35	36
(5)	37	38	39	40	41	42	43	44	45
(6)	46	47	48	49	50	51	52	53	54
(7)	55	56	57	58	59	60	61	62	63
(8)	64	65	66	67	68	69	70	71	72
(9)	73	74	75	76	77	78	79	80	81

Rep. II

(10)	1	10	19	28	37	46	55	64	73
(11)	2	11	20	29	38	47	56	65	74
(12)	3	12	21	30	39	48	57	66	75
(13)	4	13	22	31	40	49	58	67	76
(14)	5	14	23	32	41	50	59	68	77
(15)	6	15	24	33	42	51	60	69	78
(16)	7	16	25	34	43	52	61	70	79
(17)	8	17	26	35	44	53	62	71	80
(18)	9	18	27	36	45	54	63	72	81

Rep. III

(19)	1	20	12	58	77	69	34	53	45
(20)	10	2	21	67	59	78	43	35	54
(21)	19	11	3	76	68	60	52	44	36
(22)	28	47	39	4	23	15	61	80	72
(23)	37	29	48	13	5	24	70	62	81
(24)	46	38	30	22	14	6	79	71	63
(25)	55	74	66	31	50	42	7	26	18
(26)	64	56	75	40	32	51	16	8	27
(27)	73	65	57	49	41	33	25	17	9

Rep. IV

(28)	1	11	21	31	41	51	61	71	81
(29)	19	2	12	49	32	42	79	62	72
(30)	10	20	3	40	50	33	70	80	63
(31)	55	65	75	4	14	24	34	44	54
(32)	73	56	66	22	5	15	52	35	45
(33)	64	74	57	13	23	6	43	53	36
(34)	28	38	48	58	68	78	7	17	27
(35)	46	29	39	76	59	69	25	8	18
(36)	37	47	30	67	77	60	16	26	9

LATTICE DESIGNS

Plan   6 (Continued)   9 × 9 balanced lattice

	Rep. V								
(37)	1	29	57	22	50	78	16	44	72
(38)	55	2	30	76	23	51	70	17	45
(39)	28	56	3	49	77	24	43	71	18
(40)	10	38	66	4	32	60	25	53	81
(41)	64	11	39	58	5	33	79	26	54
(42)	37	65	12	31	59	6	52	80	27
(43)	19	47	75	13	41	69	7	35	63
(44)	73	20	48	67	14	42	61	8	36
(45)	46	74	21	40	68	15	34	62	9

	Rep. VI								
(46)	1	56	30	13	68	42	25	80	54
(47)	28	2	57	40	14	69	52	26	81
(48)	55	29	3	67	41	15	79	53	27
(49)	19	74	48	4	59	33	16	71	45
(50)	46	20	75	31	5	60	43	17	72
(51)	73	47	21	58	32	6	70	44	18
(52)	10	65	39	22	77	51	7	62	36
(53)	37	11	66	49	23	78	34	8	63
(54)	64	38	12	76	50	24	61	35	9

	Rep. VII								
(55)	1	47	66	76	14	33	43	62	27
(56)	64	2	48	31	77	15	25	44	63
(57)	46	65	3	13	32	78	61	26	45
(58)	37	56	21	4	50	69	79	17	36
(59)	19	38	57	67	5	51	34	80	18
(60)	55	20	39	49	68	6	16	35	81
(61)	73	11	30	40	59	24	7	53	72
(62)	28	74	12	22	41	60	70	8	54
(63)	10	29	75	58	23	42	52	71	9

	Rep. VIII								
(64)	1	74	39	67	32	24	52	17	63
(65)	37	2	75	22	68	33	61	53	18
(66)	73	38	3	31	23	69	16	62	54
(67)	46	11	57	4	77	42	70	35	27
(68)	55	47	12	40	5	78	25	71	36
(69)	10	56	48	76	41	6	34	26	72
(70)	64	29	21	49	14	60	7	80	45
(71)	19	65	30	58	50	15	43	8	81
(72)	28	20	66	13	59	51	79	44	9

	Rep. IX								
(73)	1	65	48	40	23	60	79	35	18
(74)	46	2	66	58	41	24	16	80	36
(75)	64	47	3	22	59	42	34	17	81
(76)	73	29	12	4	68	51	43	26	63
(77)	10	74	30	49	5	69	61	44	27
(78)	28	11	75	67	50	6	25	62	45
(79)	37	20	57	76	32	15	7	71	54
(80)	55	38	21	13	77	33	52	8	72
(81)	19	56	39	31	14	78	70	53	9

	Rep. X								
(82)	1	38	75	49	59	15	70	26	36
(83)	73	2	39	13	50	60	34	71	27
(84)	37	74	3	58	14	51	25	35	72
(85)	64	20	30	4	41	78	52	62	18
(86)	28	65	21	76	5	42	16	53	63
(87)	19	29	66	40	77	6	61	17	54
(88)	46	56	12	67	23	33	7	44	81
(89)	10	47	57	31	68	24	79	8	45
(90)	55	11	48	22	32	69	43	80	9

PLANS

Plan 7   6 × 6 triple lattice

Block	Rep. I							Rep. II					
(1)	1	2	3	4	5	6	(7)	1	7	13	19	25	31
(2)	7	8	9	10	11	12	(8)	2	8	14	20	26	32
(3)	13	14	15	16	17	18	(9)	3	9	15	21	27	33
(4)	19	20	21	22	23	24	(10)	4	10	16	22	28	34
(5)	25	26	27	28	29	30	(11)	5	11	17	23	29	35
(6)	31	32	33	34	35	36	(12)	6	12	18	24	30	36

Block	Rep. III					
(13)	1	8	15	22	29	36
(14)	31	2	9	16	23	30
(15)	25	32	3	10	17	24
(16)	19	26	33	4	11	18
(17)	13	20	27	34	5	12
(18)	7	14	21	28	35	6

Plan 8   10 × 10 triple lattice

Block	Rep. I										Block	Rep. II									
(1)	1	2	3	4	5	6	7	8	9	10	(11)	1	11	21	31	41	51	61	71	81	91
(2)	11	12	13	14	15	16	17	18	19	20	(12)	2	12	22	32	42	52	62	72	82	92
(3)	21	22	23	24	25	26	27	28	29	30	(13)	3	13	23	33	43	53	63	73	83	93
(4)	31	32	33	34	35	36	37	38	39	40	(14)	4	14	24	34	44	54	64	74	84	94
(5)	41	42	43	44	45	46	47	48	49	50	(15)	5	15	25	35	45	55	65	75	85	95
(6)	51	52	53	54	55	56	57	58	59	60	(16)	6	16	26	36	46	56	66	76	86	96
(7)	61	62	63	64	65	66	67	68	69	70	(17)	7	17	27	37	47	57	67	77	87	97
(8)	71	72	73	74	75	76	77	78	79	80	(18)	8	18	28	38	48	58	68	78	88	98
(9)	81	82	83	84	85	86	87	88	89	90	(19)	9	19	29	39	49	59	69	79	89	99
(10)	91	92	93	94	95	96	97	98	99	100	(20)	10	20	30	40	50	60	70	80	90	100

Block	Rep. III									
(21)	1	12	23	34	45	56	67	78	89	100
(22)	91	2	13	24	35	46	57	68	79	90
(23)	81	92	3	14	25	36	47	58	69	80
(24)	71	82	93	4	15	26	37	48	59	70
(25)	61	72	83	94	5	16	27	38	49	60
(26)	51	62	73	84	95	6	17	28	39	50
(27)	41	52	63	74	85	96	7	18	29	40
(28)	31	42	53	64	75	86	97	8	19	30
(29)	21	32	43	54	65	76	87	98	9	20
(30)	11	22	33	44	55	66	77	88	99	10

LATTICE DESIGNS

Plan 9   12 × 12 quadruple lattice

Block   Rep. I

(1)	1	2	3	4	5	6	7	8	9	10	11	12
(2)	13	14	15	16	17	18	19	20	21	22	23	24
(3)	25	26	27	28	29	30	31	32	33	34	35	36
(4)	37	38	39	40	41	42	43	44	45	46	47	48
(5)	49	50	51	52	53	54	55	56	57	58	59	60
(6)	61	62	63	64	65	66	67	68	69	70	71	72
(7)	73	74	75	76	77	78	79	80	81	82	83	84
(8)	85	86	87	88	89	90	91	92	93	94	95	96
(9)	97	98	99	100	101	102	103	104	105	106	107	108
(10)	109	110	111	112	113	114	115	116	117	118	119	120
(11)	121	122	123	124	125	126	127	128	129	130	131	132
(12)	133	134	135	136	137	138	139	140	141	142	143	144

Rep. II

(13)	1	13	25	37	49	61	73	85	97	109	121	133
(14)	2	14	26	38	50	62	74	86	98	110	122	134
(15)	3	15	27	39	51	63	75	87	99	111	123	135
(16)	4	16	28	40	52	64	76	88	100	112	124	136
(17)	5	17	29	41	53	65	77	89	101	113	125	137
(18)	6	18	30	42	54	66	78	90	102	114	126	138
(19)	7	19	31	43	55	67	79	91	103	115	127	139
(20)	8	20	32	44	56	68	80	92	104	116	128	140
(21)	9	21	33	45	57	69	81	93	105	117	129	141
(22)	10	22	34	46	58	70	82	94	106	118	130	142
(23)	11	23	35	47	59	71	83	95	107	119	131	143
(24)	12	24	36	48	60	72	84	96	108	120	132	144

Rep. III

(25)	1	14	27	40	57	70	83	96	101	114	127	140
(26)	2	13	28	39	58	69	84	95	102	113	128	139
(27)	3	16	25	38	59	72	81	94	103	116	125	138
(28)	4	15	26	37	60	71	82	93	104	115	126	137
(29)	5	18	31	44	49	62	75	88	105	118	131	144
(30)	6	17	32	43	50	61	76	87	106	117	132	143
(31)	7	20	29	42	51	64	73	86	107	120	129	142
(32)	8	19	30	41	52	63	74	85	108	119	130	141
(33)	9	22	35	48	53	66	79	92	97	110	123	136
(34)	10	21	36	47	54	65	80	91	98	109	124	135
(35)	11	24	33	46	55	68	77	90	99	112	121	134
(36)	12	23	34	45	56	67	78	89	100	111	122	133

PLANS

Plan 9 *(Continued)*   12 × 12 quadruple lattice

Rep. IV

(37)	1	24	30	43	53	64	82	95	105	116	122	135
(38)	2	23	29	44	54	63	81	96	106	115	121	136
(39)	3	22	32	41	55	62	84	93	107	114	124	133
(40)	4	21	31	42	56	61	83	94	108	113	123	134
(41)	5	16	34	47	57	68	74	87	97	120	126	139
(42)	6	15	33	48	58	67	73	88	98	119	125	140
(43)	7	14	36	45	59	66	76	85	99	118	128	137
(44)	8	13	35	46	60	65	75	86	100	117	127	138
(45)	9	20	26	39	49	72	78	91	101	112	130	143
(46)	10	19	25	40	50	71	77	92	102	111	129	144
(47)	11	18	28	37	51	70	80	89	103	110	132	141
(48)	12	17	27	38	52	69	79	90	104	109	131	142

Plan 10   3 × 4 rectangular lattice

Block	Rep. X		
X1	1	2	3
X2	4	5	6
X3	7	8	9
X4	10	11	12

	Rep. Y		
Y1	4	7	10
Y2	1	8	11
Y3	2	5	12
Y4	3	6	9

	Rep. Z		
Z1	6	8	12
Z2	2	9	10
Z3	3	4	11
Z4	1	5	7

Plan 11   4 × 5 rectangular lattice

Block	Rep. X			
X1	1	2	3	4
X2	5	6	7	8
X3	9	10	11	12
X4	13	14	15	16
X5	17	18	19	20

	Rep. Y			
Y1	5	9	13	17
Y2	1	10	14	18
Y3	2	6	15	19
Y4	3	7	11	20
Y5	4	8	12	16

	Rep. Z			
Z1	8	11	15	18
Z2	2	9	16	20
Z3	4	7	14	17
Z4	1	5	12	19
Z5	3	6	10	13

Plan 12   5 × 6 rectangular lattice

Block	Rep. X				
X1	1	2	3	4	5
X2	6	7	8	9	10
X3	11	12	13	14	15
X4	16	17	18	19	20
X5	21	22	23	24	25
X6	26	27	28	29	30

	Rep. Y				
Y1	6	11	16	21	26
Y2	1	12	17	22	27
Y3	2	7	18	23	28
Y4	3	8	13	24	29
Y5	4	9	14	19	30
Y6	5	10	15	20	25

	Rep. Z				
Z1	7	13	19	25	27
Z2	5	14	16	23	29
Z3	1	8	20	21	30
Z4	2	9	15	22	26
Z5	3	10	11	17	28
Z6	4	6	12	18	24

LATTICE DESIGNS

Plan   *13*   6 × 7 rectangular lattice

Block	Rep. X							Block	Rep. Y					
X1	1	2	3	4	5	6		Y1	7	13	19	25	31	37
X2	7	8	9	10	11	12		Y2	1	14	20	26	32	38
X3	13	14	15	16	17	18		Y3	2	8	21	27	33	39
X4	19	20	21	22	23	24		Y4	3	9	15	28	34	40
X5	25	26	27	28	29	30		Y5	4	10	16	22	35	41
X6	31	32	33	34	35	36		Y6	5	11	17	23	29	42
X7	37	38	39	40	41	42		Y7	6	12	18	24	30	36

Block	Rep. Z					
Z1	12	17	22	28	33	38
Z2	2	13	24	29	35	40
Z3	4	9	20	25	36	42
Z4	6	11	16	27	32	37
Z5	1	7	18	23	34	39
Z6	3	8	14	19	30	41
Z7	5	10	15	21	26	31

Plan   *14*   7 × 8 rectangular lattice

Block	Rep. X								Block	Rep. Y						
X1	1	2	3	4	5	6	7		Y1	8	15	22	29	36	43	50
X2	8	9	10	11	12	13	14		Y2	1	16	23	30	37	44	51
X3	15	16	17	18	19	20	21		Y3	2	9	24	31	38	45	52
X4	22	23	24	25	26	27	28		Y4	3	10	17	32	39	46	53
X5	29	30	31	32	33	34	35		Y5	4	11	18	25	40	47	54
X6	36	37	38	39	40	41	42		Y6	5	12	19	26	33	48	55
X7	43	44	45	46	47	48	49		Y7	6	13	20	27	34	41	56
X8	50	51	52	53	54	55	56		Y8	7	14	21	28	35	42	49

Block	Rep. Z						
Z1	9	17	25	33	41	49	51
Z2	7	18	26	34	38	43	53
Z3	1	10	27	35	36	47	55
Z4	2	11	19	29	42	44	56
Z5	3	12	20	28	37	45	50
Z6	4	13	21	22	30	46	52
Z7	5	14	15	23	31	39	54
Z8	6	8	16	24	32	40	48

PLANS

Plan   15   8 × 9 rectangular lattice

Block				Rep. X				
X1	1	2	3	4	5	6	7	8
X2	9	10	11	12	13	14	15	16
X3	17	18	19	20	21	22	23	24
X4	25	26	27	28	29	30	31	32
X5	33	34	35	36	37	38	39	40
X6	41	42	43	44	45	46	47	48
X7	49	50	51	52	53	54	55	56
X8	57	58	59	60	61	62	63	64
X9	65	66	67	68	69	70	71	72

				Rep. Y				
Y1	9	17	25	33	41	49	57	65
Y2	1	18	26	34	42	50	58	66
Y3	2	10	27	35	43	51	59	67
Y4	3	11	19	36	44	52	60	68
Y5	4	12	20	28	45	53	61	69
Y6	5	13	21	29	37	54	62	70
Y7	6	14	22	30	38	46	63	71
Y8	7	15	23	31	39	47	55	72
Y9	8	16	24	32	40	48	56	64

				Rep. Z				
Z1	16	23	30	37	45	52	59	66
Z2	2	17	32	39	46	54	61	68
Z3	4	11	26	33	48	55	63	70
Z4	6	13	20	35	42	49	64	72
Z5	8	15	22	29	44	51	58	65
Z6	1	9	24	31	38	53	60	67
Z7	3	10	18	25	40	47	62	69
Z8	5	12	19	27	34	41	56	71
Z9	7	14	21	28	36	43	50	57

LATTICE DESIGNS

Plan   16   9 × 10 rectangular lattice

Block				Rep. X								Rep. Y								
X1	1	2	3	4	5	6	7	8	9	Y1	10	19	28	37	46	55	64	73	82
X2	10	11	12	13	14	15	16	17	18	Y2	1	20	29	38	47	56	65	74	83
X3	19	20	21	22	23	24	25	26	27	Y3	2	11	30	39	48	57	66	75	84
X4	28	29	30	31	32	33	34	35	36	Y4	3	12	21	40	49	58	67	76	85
X5	37	38	39	40	41	42	43	44	45	Y5	4	13	22	31	50	59	68	77	86
X6	46	47	48	49	50	51	52	53	54	Y6	5	14	23	32	41	60	69	78	87
X7	55	56	57	58	59	60	61	62	63	Y7	6	15	24	33	42	51	70	79	88
X8	64	65	66	67	68	69	70	71	72	Y8	7	16	25	34	43	52	61	80	89
X9	73	74	75	76	77	78	79	80	81	Y9	8	17	26	35	44	53	62	71	90
X10	82	83	84	85	86	87	88	89	90	Y10	9	18	27	36	45	54	63	72	81

Rep. Z

Z1	11	21	31	41	51	61	71	81	83
Z2	9	22	32	42	52	62	64	76	84
Z3	1	12	33	43	53	63	68	73	87
Z4	2	13	23	44	54	55	65	79	89
Z5	3	14	24	34	46	56	72	75	90
Z6	4	15	25	35	45	57	67	74	82
Z7	5	16	26	36	37	47	66	77	85
Z8	6	17	27	28	38	48	58	78	86
Z9	7	18	19	29	39	49	59	69	88
Z10	8	10	20	30	40	50	60	70	80

Appendix G

Fractional Factorial Design Plans

Plan 1. 2^4 factorial in 8 units (1/2 replicate)

Defining Contrast: $ABCD$

Estimable 2-factor interactions: $AB = CD, AC = BD, AD = BC$.

---

(1)
ab
cd
ace
bce
ade
bde
abcd

Effects	d.f.
Main	4
2-factor	3
Total	7

Plan 2. 2^5 factorial in 8 units (1/4 replicate)

Defining contrasts: $ABE, CDE, ABCD$

Main effects have 2-factors as aliases. The only estimable 2-factors are $AC = BD$ and $AD = BC$.

---

(1)
ab
cd
ace
bce
ade
bde
abcd

Effects	d.f.
Main	5
2-factor	2
Total	7

Plan 3. 2^5 factorial in 16 units (1/2 replicate)

Defining contrast: $ABCDE$

Blocks of 4 units

Estimable 2-factors: All except CD, CE, and DE (confounded with blocks).

Blocks	(1)	(2)	(3)	(4)	Effects	d.f.
	(1)	ac	ae	ad	Block	3
	ab	bc	be	bd	Main	5
	acde	de	cd	ce	2-factor	7
	bcde	abde	abcd	abce	Total	15

Blocks of 8 units

Estimable 2-factors: All except DE.

Combine blocks 1 and 2; and blocks 3 and 4. DE is confounded.

Effects	d.f.
Block	1
Main	5
2-factor	9
Total	15

Blocks of 16 units

Estimable 2-factors: All.

Combine blocks 1 and 4.

Effects	d.f.
Main	5
2-factor	10
Total	15

Plan 4. 2^6 factorial in 8 units (1/8 replicate)

Defining contrast: $ACE, ADF, BCF, BDE, ABCD, ABEF, CDEF$

Main effects have 2-factors as aliases. The only estimable 2-factor is the set $AB = CD = EF$.

(1)		Effects	d.f.
acf		Main	6
ade		2-factor ($AB = CD = EF$)	1
bce		Total	7
bdf			
abcd			
abef			
cdef			

Plan 5. 2^6 factorial in 16 units (1/4 replicate)

Defining contrast: $ACE, ADF, BCF, BDE, ABCD, ABEF, CDEF$

Blocks of 4 units

Estimable 2-factors: The alias sets $AC = BE$, and $AD = BF, AE = BC$, $AF = BD, CD = EF, CF = DE$.

Blocks	(1)	(2)	(3)	(4)
	(1)	acd	ab	acf
	abce	aef	ce	ade
	abdf	bcf	df	bcd
	cdef	bde	abcdef	bef

Effects	d.f.
Block	3
Main	6
2-factor	6
Total	15

AB, ACF, BCF confounded.

Blocks of 8 units

Estimable 2-factors: Same as in blocks of 4 units, plus the set $AB = CE = DF$.

Combine blocks 1 and 2; and blocks 3 and 4. *ACF* is confounded.

Effects	d.f.
Block	1
Main	6
2-factor	7
3-factor	1
Total	15

Blocks of 16 units

Estimable 2-factors: Same as in blocks of 8 units

Combine blocks 1 - 4.

Effects	d.f.
Main	6
2-factor	7
3-factor	2
Total	15

Plan 6. 2^6 factorial in 32 units (1/2 replicate)

Defining contrast: *ABCDEF*

Blocks of 4 units

Estimable 2-factors: All except *AE*, *BF*, and *CD* (confounded with blocks).

Blocks	(1)	(2)	(3)	(4)	(5)	(6)	(7)	(8)
	(1)	ab	ac	bc	ae	af	ad	bd
	abef	ef	de	df	bf	be	ce	cf
	acde	acdf	abdf	acef	cd	abcd	abcf	abce
	bcdf	bcde	bcef	abde	abcdef	cdef	bdef	adef

AE, BF, CD, ABC, ABD, ACF, ADF confounded.

Effects	d.f.
Block	7
Main	6
2-factor	12
Higher order	6
Total	31

Blocks of 8 units

Estimable 2-factors: All except *CD*.

Combine blocks 1 and 2; and blocks 3 and 4; blocks 5 and 6; and blocks 7 and 8. *CD, ABC, ABD* confounded.

Effects	d.f.
Block	3
Main	6
2-factor	14
3-factor	8
Total	31

Blocks of 16 units

Estimable 2-factors: All.

Estimable 3-factors: *ABC = DEF* is lost by confounding. The others are in alias pairs, e.g., *ABD =CEF*.

Combine blocks 1 - 4; and blocks of 5-8. *ABC* confounded.

Effects	d.f.
Block	1
Main	6
2-factor	15
3-factor	9
Total	31

Blocks of 32 units

Estimable 2-factors: All.

Estimable 3-factors: These are arranged in 10 alias pairs.

Combine blocks 1 - 8.

Effects	d.f.
Main	6
2-factor	15
3-factor	10
Total	31

Plan 7. 2^7 factorial in 8 units (1/16 replicate)

Defining contrast: *ABG, ACE, ADF, BCF, BDE, CDG, EFG, ABCD, ABEF, ACFG, ADEG, BCEG, BDFG, CDEF, ABCDEFG.*

Main effects have 2-factors as aliases. No 2-factors are estimable.

(1)		Effects	d.f.
abcd		Main	7
abef		Total	7
acfg			
adeg			
bceg			
bdfg			
cdef			

Plan 8. 2^7 factorial in 16 units (1/8 replicate)

Defining contrast: *ABCD, ABEF, ACEG, ADFG, BCFG, BDEG, CDEF.*

Blocks of 4 units

Estimable 2-factors: Only the alias sets $AE = BF = CG$; $AF = BE = DG$; $AG = CE = DF$; $BG = DE = CF$.

Blocks	(1)	(2)	(3)	(4)	Effects	d.f.
	(1)	*abg*	*acf*	*ade*	Block	3
	efg	*cdg*	*bdf*	*bce*	Main	7
	abcd	*abef*	*aceg*	*adfg*	2-factor	4
	abcdefg	*edef*	*bdeg*	*bcfg*	Higher order	1
					Total	15

$AB = CD = EF,$
$AC = BD = EG,$
$AD = BC = FG$ confounded.

Plan 9. 2^7 factorial in 16 units (1/8 replicate)

Defining contrast: *ABCD, ABEF, ACEG, ADFG, BCFG, BDEG, CDEF.*

Blocks of 8 units

Estimable 2-factors: Same as blocks of 4 units, plus the alias sets $AB = CD = EF$, $AC = BD = EG$; $AD = BC = FG$.

Blocks	(1)	(2)	Effects	d.f.
	(1)	*abg*	Block	1
	abcd	*acf*	Main	7
	abef	*ade*	2-factor	7
	aceg	*bce*	Total	15
	adfg	*bdf*		
	bcfg	*cdg*		
	bdeg	*efg*		
	cdef	*abcdefg*		

ABG confounded.

Blocks of 16 units

Estimable 2-factors: Same as in blocks of 8 units.

Combine blocks 1 and 2 of the plan for blocks of 8 units.

Effects	d.f.
Main	7
2-factor	7
Higher order	1
Total	15

Plan 10. 2^7 factorial in 32 units (1/4 replicate)

Defining contrast: *ABCDE, ABCFG, DEFG*.

Blocks of 4 units

Estimable 2-factors: *AB, AC, BC,* and *DF* $=EG$ are lost by confounding. All other 2-factors are estimable, except that $DE = FG$ and $DG = EF$, so that members of these alias pairs cannot be separated.

Blocks (1)	(2)	(3)	(4)	(5)	(6)	(7)	(8)
(1)	*de*	*ab*	*cdg*	*ac*	*bdg*	*bc*	*adg*
defg	*fg*	*cdf*	*cef*	*bdf*	*bef*	*adf*	*aef*
abcdf	*abcdg*	*ceg*	*abde*	*beg*	*acde*	*aeg*	*bcfg*
abceg	*abcef*	*abdefg*	*abfg*	*abcdefg*	*acfg*	*bcdefg*	*bcde*

AB, AC, BC, and *DF* $=EG$, *ADG, BDG, CDG* confounded.

Effects	d.f.
Block	7
Main	7
2-factor	14
Higher order	3
Total	31

Plan 11. 2^7 factorial in 32 units (1/4 replicate)

Defining contrast: *ABCDE, ABCFG, DEFG.*

Blocks of 8 units

Estimable 2-factors: All except *DF* = *EG* (confounded with blocks). However, *DE* = *FG* and DG = *EF* are alias pairs which cannot be separated.

Blocks	(1)	(2)	(3)	(4)		Effects	d.f.
	(1)	*bdg*	*ab*	*de*		Block	3
	bc	*bef*	*ac*	*fg*		Main	7
	adf	*cef*	*bdf*	*adg*		2-factor	17
	aeg	*abfg*	*beg*	*aef*		Higher order	4
	defg	*acfg*	*cdf*	*bcde*		Total	31
	abcdf	*abde*	*ceg*	*bcfg*			
	abceg	*acde*	*acdefg*	*abcdg*			
	bcdefg	*cdg*	*abdefg*	*abcef*			

DF = *EG, ADE, AEF* confounded.

Blocks of 16 units

Estimable 2-factors: All except that *DE* = *FG, DG* = *EF*, and *DF* = *EG* are alias pairs.

Combine blocks 1 and 2; and blocks 3 and 4. *AEF* confounded.

Effects	d.f.
Block	1
Main	7
2-factor	18
Higher order	5
Total	31

Blocks of 32 units

Estimable 2-factors: Same as in blocks of 16 units.

Combine blocks 1 - 4.

Effects	d.f.
Main	7
2-factor	18
3-factor	6
Total	31

Plan 12. 2^7 factorial in 64 units (1/2 replicate)

Defining contrast: *ABCDEFG*

Blocks of 4 units

Estimable 2-factors: All except *AB, AC, BC, EF, EG,* and *FG* (confounded with blocks).

Blocks (1)	(2)	(3)	(4)	(5)	(6)	(7)	(8)
(1)	ab	ac	bc	ae	be	ce	abce
abcd	cd	bd	ad	bcde	acde	abde	de
defg	abdefg	acdefg	bcdefg	adfg	bdfg	cdfg	abcdfg
abcefg	cefg	befg	aefg	bcfg	acfg	abfg	fg

(9)	(10)	(11)	(12)	(13)	(14)	(15)	(16)
af	bf	cf	abcf	ef	abef	acef	bcef
bcdf	acdf	abdf	df	abcdef	cdef	bdef	adef
adeg	bdeg	cdeg	abcdeg	dg	abdg	acdg	bcdg
bceg	aceg	abeg	eg	abcg	cg	bg	ag

AB, AC, BC, EF, EG, FG, ADE, ADF, ADG, BDE, BDF, BDG, CDE, CDF, CDG confounded.

Effects	d.f.
Block	15
Main	7
2-factor	15
Higher order	26
Total	63

Plan 13. 2^7 factorial in 64 units (1/2 replicate)

Defining contrast: *ABCDEFG*

Blocks of 8 units

Estimable 2-factors: All.

Estimable 3-factors: All except *ABC, ADE, AFG, BDF, EG, BEG, CDG, CEF* confounded.

Blocks	(1)	(2)	(3)	(4)	(5)	(6)	(7)	(8)
	(1)	bc	ac	ab	ag	af	ae	ad
	abdg	de	df	dg	bd	be	bf	bg
	abef	fg	eg	ef	ce	cd	cg	cf
	acdf	abdf	abde	acde	abcf	abcg	abcd	abce
	aceg	abeg	abfg	acfg	adef	adeg	adfg	aefg
	bcde	acdg	bcdg	bcdf	befg	bdfg	bdeg	bdef
	bcfg	acef	bcef	bceg	cdfg	cefg	cdef	cdeg
	defg	bcdefg	acdefg	abdefg	abcdeg	abcdef	abcefg	abcdfg

Effects	d.f.
Block	7
Main	7
2-factor	21
Higher order	2̲8̲
Total	63

Blocks of 16 units

Estimable 2-factors: All.

Estimable 3-factors: All except *ABC, ADE, AFG* (confounded).

Combine blocks 1 and 2; and blocks 3 and 4; blocks 5 and 6; and blocks 7 and 8.

Effects	d.f.
Block	3
Main	7
2-factor	21
Higher order	3̲2̲
Total	63

Blocks of 32 units

Estimable 2-factors: All.

Estimable 3-factors: All except *ABC* (confounded).

Combine blocks 1 - 4; and blocks 5-8.

Effects	d.f.
Block	1
Main	7
2-factor	21
3-factor	34
Total	63

Blocks of 64 units

Estimable 2-factors: All.

Estimable 3-factors: All.

Combine blocks 1 - 8.

Effects	d.f.
Main	7
2-factor	21
3-factor	35
Total	63

Plan 14. 2^8 factorial in 16 units (1/16 replicate)

Defining contrast: *ABCD, ABEF, ABGH, ACEH, ACFG, ADEG, ADFH, BCEG, BCFH, BDEH, BDFG, CDEF, CDGH, EFGH, ABCDEFGH.*

Blocks of 4 units

Estimable 2-factors: Only the 4 alias sets AE =BF =CH =DG ;AF =BE =CG =DH; AG =BH =CF =DE; AH =BG =CE =DF. Except in special circumstances, only main effects are estimable.

Blocks	(1)	(2)	(3)	(4)
	(1)	abef	adeg	aceh
	abcd	abgh	adfh	acfg
	efgh	cdef	bceg	bdeh
	abcdefgh	cdgh	bcfh	bdfg

Effects	d.f.
Block	3
Main	8
2-factor	4
Total	15

AB, AC, AD confounded.

Blocks of 8 units

Estimable 2-factors: As in blocks of 4 units, plus the alias sets $AC = BD$ $= EH = FG$; and $AD = EC = EG = FH$.

Combine blocks 1 and 2; and blocks 3 and 4. AB confounded.

Effects	d.f.
Block	1
Main	8
2-factor	6
Total	15

Blocks of 16 units

Estimable 2-factors: As in blocks of 8 units, plus the alias sets $AB = CD$ $= EF = GH$.

Combine blocks 1 and 4.

Effects	d.f.
Main	8
2-factor	7
Total	15

Plan 15. 2^8 factorial in 32 units (1/8 replicate)

Defining contrast: *BCDH, BDFG, CFGH, ABCEF, ABEGH, ACDEG, ADEFH*

Blocks of 4 units

Estimable 2-factors: All interactions of A and E. All other 2-factors are lost by confounding.

Blocks (1)	(2)	(3)	(4)	(5)	(6)	(7)	(8)
(1)	abd	dgh	cdf	afg	ach	bcg	bfh
æ	bde	abcf	abgh	efg	ceh	adfh	acdg
abcdfgh	cfgh	bcef	begh	bcdh	bdfg	defh	cdeg
bcdefgh	acefgh	adegh	acdef	abcdeh	abdefg	abceg	abefh

$BD, BF, BH, CF, DF, DH, FH$ and their aliases confounded.

Effects	d.f.
Block ·	7
Main	8
2-factor	13
Higher order	3
Total	31

Blocks of 8 units

Estimable 2-factors: All interactions of A and E, and the alias pairs $BC = DH, BF = DG$, and $BG = DF, BH = CD$.

Combine blocks 1 and 2; and blocks 3 and 4; blocks 5 and 6; and blocks 7 and 8. BD, CF, FH confounded.

Effects	d.f.
Block	3
Main	8
2-factor	17
Higher order	3
Total	31

Blocks of 16 units

Estimable 2-factors: As in blocks of 8 units, plus the alias sets $CG = FH, BD = CH = FG$.

Combine blocks 1 and 4; and blocks 5 - 8. CF confounded.

Effects	d.f.
Block	1
Main	8
2-factor	19
Higher order	3
Total	31

Blocks of 32 units

Estimable 2-factors: All interactions of A and E, plus the alias sets $BC_a = DH, BF = DG, BG = DF, BH = CD, CG = FH, CF = GH, BD = CH = FG$.

Combine blocks 1 and 8.

Effects	d.f.
Main	8
2-factor	20
Higher order	3
Total	31

Plan 16. 2^8 factorial in 64 units (1/4 replicate)

Defining contrast: *ABCEG, ABDFH, CDEFGH,*

Blocks of 4 units

Estimable 2-factors: All except *AF, AH, BC, BG, CG, DE*, and *FH* (confounded with blocks).

Blocks (1)	(2)	(3)	(4)	(5)	(6)	(7)	(8)
(1)	adg	ach	beh	eg	fh	bef	acf
adefh	abce	bfg	abdf	bcd	ade	abdh	bgh
bcdeg	efgh	cdef	cdgh	adfgh	abcg	cdfg	cdeh
abcfgh	bcdfh	abdegh	acefg	abcefh	bcdefgh	acegh	abdefg

(9)	(10)	(11)	(12)	(13)	(14)	(15)	(16)
ab	ce	afg	df	acd	cg	dh	agh
cfgh	bdg	bch	aeh	abeg	bde	aef	bcf
acdeg	acdfh	degh	bcefg	cefh	abfh	bcegh	defg
bdefh	abefgh	abcdef	abcdgh	bdfgh	acdefgh	abcdfg	abcdeh

AF, AH, BC, BG, CG, DE, FH, ACD, ADG, BEF, BEH, CDF, CEF, DFG, EFG confounded.

Effects	d.f.
Block	15
Main	8
2-factor	21
Higher order	<u>19</u>
Total	63

Blocks of 8 units

Estimable 2-factors: All except *BC* and *FH* (confounded with blocks).

Combine blocks 1 and 2; and blocks 3 and 4; blocks 5 and 6; blocks 7 and 8; blocks 9 and 10; blocks 11 and 12; blocks 13 and 14; and blocks 15 and 16. *BC, FH, ACD, BEF, BEH, CEF, DFG* confounded.

Effects	d.f.
Block	7
Main	8
2-factor	26
Higher order	<u>22</u>
Total	63

Blocks of 16 units

Estimable 2-factors: All.

Combine blocks 1- 4; and blocks 5 - 8; blocks 9 - 12; and blocks 13 - 16. *ACD, BEF, DFG* confounded.

Effects	d.f.
Block	3
Main ·	8
2-factor	28
Higher order	24
Total	63

Blocks of 32 units

Estimable 2-factors: All.

Combine blocks 1- 8; and blocks 9 - 16. *ACD* confounded.

Effects	d.f.
Block	1
Main	8
2-factor	28
Higher order	26
Total	63

Blocks of 64 units

Estimable 2-factors: All.

Combine blocks 1- 16.

Effects	d.f.
Main	8
2-factor	28
Higher order	27
Total	63

Plan 17. 2^8 factorial in 128 units (1/2 replicate)

Defining contrast: *ABCDEFGH*

Blocks of 16 units

Estimable 2-factors: All.

Estimable 3-factors: All.

Blocks (1)	(2)	(3)	(4)	(5)	(6)	(7)	(8)
(1)	acfh	fh	ac	af	ch	ah	cf
abcd	bdfh	abcdfh	bd	bcdf	abdh	bcdh	abdf
adeg	abefgh	abefgh	abeg	defg	begh	degh	befg
bceg	cdefgh	bcefgh	cdeg	abcefg	acdegh	abcegh	acdefg
adfh	ab	ad	abfh	dh	bf	df	bh
bcfh	cd	bc	cdfh	abch	acdf	abcf	acdh
efgh	aceg	eg	acefgh	aegh	cefg	aefg	cegh
abcdefgh	bdeg	abcdeg	bdefgh	bcdegh	abdefg	bcdefg	abdegh
abgh	ef	abfg	eh	bfgh	ae	bg	aefh
aceh	adfg	acef	adgh	cefh	dg	ce	dfgh
bdeh	bcfg	bdef	bcgh	abdefh	abcg	abde	abcfgh
cdgh	abcdef	cdfg	abcdeh	acdfgh	bcde	acdg	bcdefh
abef	gh	abeh	fg	be	afgh	befh	ag
acfg	adeh	acgh	adef	cg	defh	cfgh	de
bdfg	bceh	bdgh	bcef	abdg	abcefh	abdfgh	abce
cdef	abcdgh	cdeh	abcdfg	acde	bcdfgh	acdefh	bcdg

ABCD . ABEF . ACFG, ADEG, BCEG, BDFG, CDEF confounded.

Effects	d.f.
Block	7
Main	8
2-factor	28
3-factor	56
Higher order	28
Total	127

Blocks of 32 units

Estimable 2-factors: All.

Estimable 3-factors: All.

Combine blocks 1 and 2; and blocks 3 and 4; blocks 5 and 6; blocks 7 and 8. *ABCD, ABEF, CDEF* confounded.

Effects	d.f.
Block	3
Main	8
2-factor	28
3-factor	56
Higher order	32
Total	127

Blocks of 64 units

Estimable 2-factors: All.

Estimable 3-factors: All.

Combine blocks 1- 4; and blocks 5 - 8. *ABCD* confounded.

Effects	d.f.
Block	1
Main	8
2-factor	28
3-factor	56
Higher order	34
Total	127

Blocks of 128 units

Estimable 2-factors: All.

Estimable 3-factors: All.

Combine blocks 1- 8.

Effects	d.f.
Main	8
2-factor	28
3-factor	56
Higher order	35
Total	127

Blocks of 8 units

Estimable 2-factors: All except *EG* and *FH* (confounded). Start with the plan for blocks of 32 units. The first 4 rows of blocks 1 and 2 form the first block of 8 units: i.e., this block contains 1, *abcd, adeg, bceg, acfh, bdfh, abefgh, cdefgh*. Similarly, rows 5 - 8 of blocks 1 and 2 give the second block, rows 9-12 the third and rows 13-16 the fourth. The remaining 12 blocks are formed likewise from blocks 3 and 4; blocks 5 and 6; and blocks 7 and 8.

Effects	d.f.
Block	15
Main	8
2-factor	26
Higher order	78
Total	127

Plan 18. 3^4 factorial in 27 units (1/3 replicate)

Defining contrast: 2 d.f. from *ABCD*, equivalent to putting $D = ABC$ (*Y*)

Blocks of 9 units

Estimable 2-factors: 16 of the 24 d.f. are clear. *CD* (*I*) is lost by confounding. Also *AB* (*J*) = *CD* (*J*); *AC* (*I*) =*BD* (*I*); *AD* (*I*) = *BC* (*I*).

Blocks	(1)	(2)	(3)
	0000	0021	0012
	0122	0110	0101
	0211	0202	0220
	1022	1010	1001
	1111	1102	1120
	1200	1221	1212
	2011	2002	2020
	2100	2121	2112
	2222	2210	2201

Effects	d.f.
Block	2
Main	8
2-factor	16
Total	26

If all interactions of D are negligible, the analysis may be written:

Effects	d.f.
Block	2
Main	8
AB, AC, BC	12
Error (from interactions of D)	4
Total	26

Plan 19. 3^5 factorial in 81 units (1/3 replicate)

Defining contrast: 2 d.f. from $ABCDE$

Blocks of 9 units

Estimable 2-factors: All except AE (J), which is confounded with blocks.

Blocks	(1)	(2)	(3)	(4)	(5)	(6)	(7)	(8)	(9)
ab	cde	cde	cde	cde	cde	cde	cde	cde	cde
00	000	201	102	120	021	222	111	012	210
10	122	020	221	212	110	011	200	101	002
20	211	112	010	001	202	100	022	220	121
01	110	011	212	200	101	002	221	122	020
11	202	100	001	022	220	121	010	211	112
21	021	222	120	111	012	210	102	000	201
02	220	121	022	010	211	112	001	202	100
12	012	210	111	102	000	201	120	021	222
22	101	002	200	221	122	020	212	110	011

Effects	d.f.
Block	8
Main	10
2-factor	38
Higher order	24
Total	80

Blocks of 27 units

Estimable 2-factors: All.

Combine blocks 1- 3; blocks 4-6; blocks 7-9.

Effects	d.f.
Block	2
Main	10
2-factor	40
Higher order	28
Total	80

Blocks of 81 units

Estimable 2-factors: All.

Combine blocks 1- 9.

Effects	d.f.
Main	10
2-factor	40
Higher order	30
Total	80

Plan 20. 4×2^4 factorial in 32 units (1/2 replicate)

Defining contrast: $A'''BCDE$

Blocks of 8 units

Estimable 2-factors: All except DE (confounded with blocks).

Blocks	(1)	(2)	(3)	(4)
ab	cde	cde	cde	cde
00	100	010	111	001
01	011	101	000	110
10	011	101	000	110
11	100	010	111	011
20	111	001	100	010
21	000	110	011	101
30	000	110	011	101
31	111	001	100	010

Effects	d.f.
Block	3
Main	7
2-factor	17
Higher order	4
Total	31

Blocks of 16 units

Estimable 2-factors: All.

Combine blocks 1, 2; and blocks 3, 4

Effects	d.f.
Block	1
Main	7
2-factor	18
Higher order	5
Total	31

Blocks of 32 units

Estimable 2-factors: All.

Combine blocks 1, 2, 3, and 4.

Effects	d.f.
Main	7
2-factor	18
Higher order	6
Total	31

Appendix H

Significant Studentized Ranges for 5% and 1% Level New Multiple Range Test

p = number of means for range being tested

Error df	Protection level	2	3	4	5	6	7	8	9	10	12	14	16	18	20
1	.05	18.0	18.0	18.0	18.0	18.0	18.0	18.0	18.0	18.0	18.0	18.0	18.0	18.0	18.0
	.01	90.0	90.0	90.0	90.0	90.0	90.0	90.0	90.0	90.0	90.0	90.0	90.0	90.0	90.0
2	.05	6.09	6.09	6.09	6.09	6.09	6.09	6.09	6.09	6.09	6.09	6.09	6.09	6.09	6.09
	.01	14.0	14.0	14.0	14.0	14.0	14.0	14.0	14.0	14.0	14.0	14.0	14.0	14.0	14.0
3	.05	4.50	4.50	4.50	4.50	4.50	4.50	4.50	4.50	4.50	4.50	4.50	4.50	4.50	4.50
	.01	8.26	8.5	8.6	8.7	8.8	8.9	8.9	9.0	9.0	9.0	9.1	9.2	9.3	9.3
4	.05	3.93	4.01	4.02	4.02	4.02	4.02	4.02	4.02	4.02	4.02	4.02	4.02	4.02	4.02
	.01	6.51	6.8	6.9	7.0	7.1	7.1	7.2	7.2	7.3	7.3	7.4	7.4	7.5	7.5
5	.05	3.64	3.74	3.79	3.83	3.83	3.83	3.83	3.83	3.83	3.83	3.83	3.83	3.83	3.83
	.01	5.70	5.96	6.11	6.18	6.26	6.33	6.40	6.44	6.5	6.6	6.6	6.7	6.7	6.8
6	.05	3.46	3.58	3.64	3.68	3.68	3.68	3.68	3.68	3.68	3.68	3.68	3.68	3.68	3.68
	.01	5.24	5.51	5.65	5.73	5.81	5.88	5.95	6.00	6.0	6.1	6.2	6.2	6.3	6.3
7	.05	3.35	3.47	3.54	3.58	3.60	3.61	3.61	3.61	3.61	3.61	3.61	3.61	3.61	3.61
	.01	4.95	5.22	5.37	5.45	5.53	5.61	5.69	5.73	5.8	5.8	5.9	5.9	6.0	6.0
8	.05	3.26	3.39	3.47	3.52	3.55	3.56	3.56	3.56	3.56	3.56	3.56	3.56	3.56	3.56
	.01	4.74	5.00	5.14	5.23	5.32	5.40	5.47	5.51	5.5	5.6	5.7	5.7	5.8	5.8
9	.05	3.20	3.34	3.41	3.47	3.50	3.52	3.52	3.52	3.52	3.52	3.52	3.52	3.52	3.52
	.01	4.60	4.86	4.99	5.08	5.17	5.25	5.32	5.36	5.4	5.5	5.5	5.6	5.7	5.7
10	.05	3.15	3.30	3.37	3.43	3.46	3.47	3.47	3.47	3.47	3.47	3.47	3.47	3.47	3.48
	.01	4.48	4.73	4.88	4.96	5.06	5.13	5.20	5.24	5.28	5.36	5.42	5.48	5.54	5.55
11	.05	3.11	3.27	3.35	3.39	3.43	3.44	3.45	3.46	3.46	3.46	3.46	3.46	3.47	3.48
	.01	4.39	4.63	4.77	4.86	4.94	5.01	5.06	5.12	5.15	5.24	5.28	5.34	5.38	5.39
12	.05	3.08	3.23	3.33	3.36	3.40	3.42	3.44	3.44	3.46	3.46	3.46	3.46	3.47	3.48
	.01	4.32	4.55	4.68	4.76	4.81	4.92	4.96	5.02	5.07	5.13	5.17	5.22	5.24	5.26
13	.05	3.06	3.21	3.30	3.35	3.38	3.41	3.42	3.44	3.45	3.45	3.46	3.46	3.47	3.47
	.01	4.26	4.48	4.62	4.69	4.74	4.84	4.88	4.94	4.98	5.04	5.08	5.13	5.14	5.15
14	.05	3.03	3.18	3.27	3.33	3.37	3.39	3.41	3.42	3.44	3.45	3.46	3.46	3.47	3.47
	.01	4.21	4.42	4.55	4.63	4.70	4.78	4.83	4.87	4.91	4.96	5.00	5.04	5.06	5.07
15	.05	3.01	3.16	3.25	3.31	3.36	3.38	3.40	3.42	3.43	3.44	3.45	3.46	3.47	3.47
	.01	4.17	4.37	4.50	4.58	4.64	4.72	4.77	4.81	4.84	4.90	4.94	4.97	4.99	5.00

continued next page

Error df	Protection level	\multicolumn{14}{c}{p = number of means for range being tested}													
		2	3	4	5	6	7	8	9	10	12	14	16	18	20
16	.05	3.00	3.15	3.23	3.30	3.34	3.37	3.39	3.41	3.43	3.44	3.45	3.46	3.47	3.47
	.01	4.13	4.34	4.45	4.54	4.60	4.67	4.72	4.76	4.79	4.84	4.88	4.91	4.93	4.94
17	.05	2.98	3.13	3.22	3.28	3.33	3.36	3.38	3.40	3.42	3.44	3.46	3.46	3.47	3.47
	.01	4.10	4.30	4.41	4.50	4.56	4.63	4.68	4.72	4.76	4.80	4.83	4.86	4.88	4.89
18	.05	2.97	3.12	3.21	3.27	3.32	3.35	3.37	3.39	3.41	3.43	3.45	3.46	3.47	3.47
	.01	4.07	4.27	4.38	4.46	4.53	4.59	4.64	4.68	4.71	4.76	4.79	4.82	4.84	4.85
19	.05	2.96	3.11	3.19	3.26	3.31	3.35	3.37	3.39	3.41	3.43	3.44	3.46	3.47	3.47
	.01	4.05	4.24	4.35	4.43	4.50	4.56	4.61	4.64	4.67	4.72	4.76	4.79	4.81	4.82
20	.05	2.95	3.10	3.18	3.25	3.30	3.34	3.36	3.38	3.40	3.43	3.44	3.46	3.47	3.47
	.01	4.02	4.22	4.33	4.40	4.47	4.53	4.58	4.61	4.65	4.69	4.73	4.76	4.79	4.79
22	.05	2.93	3.08	3.17	3.24	3.29	3.32	3.35	3.37	3.39	3.42	3.44	3.45	3.46	3.47
	.01	3.99	4.17	4.28	4.36	4.42	4.48	4.53	4.57	4.60	4.65	4.68	4.71	4.74	4.75
24	.05	2.92	3.07	3.15	3.22	3.28	3.31	3.34	3.37	3.38	3.41	3.44	3.45	3.46	3.47
	.01	3.96	4.14	4.24	4.33	4.39	4.44	4.49	4.53	4.57	4.62	4.64	4.67	4.70	4.72
26	.05	2.91	3.06	3.14	3.21	3.27	3.30	3.34	3.36	3.38	3.41	3.43	3.45	3.46	3.47
	.01	3.93	4.11	4.21	4.30	4.36	4.41	4.46	4.50	4.53	4.58	4.62	4.65	4.67	4.69
28	.05	2.90	3.04	3.13	3.20	3.26	3.30	3.33	3.35	3.37	3.40	3.43	3.45	3.46	3.47
	.01	3.91	4.08	4.18	4.28	4.34	4.39	4.43	4.47	4.51	4.56	4.60	4.62	4.65	4.67
30	.05	2.89	3.04	3.12	3.20	3.25	3.29	3.32	3.35	3.37	3.40	3.43	3.44	3.46	3.47
	.01	3.89	4.06	4.16	4.22	4.32	4.36	4.41	4.45	4.48	4.54	4.58	4.61	4.63	4.65
40	.05	2.86	3.01	3.10	3.17	3.22	3.27	3.30	3.33	3.35	3.39	3.42	3.44	3.46	3.47
	.01	3.82	3.99	4.10	4.17	4.21	4.30	4.34	4.37	4.41	4.46	4.51	4.54	4.57	4.59
60	.05	2.83	2.98	3.08	3.14	3.20	3.24	3.28	3.31	3.33	3.37	3.40	3.43	3.45	3.47
	.01	3.76	3.92	4.03	4.12	4.17	4.23	4.27	4.31	4.34	4.39	4.44	4.47	4.50	4.53
100	.05	2.80	2.95	3.05	3.12	3.18	3.22	3.26	3.29	3.32	3.36	3.40	3.42	3.45	3.47
	.01	3.71	3.86	3.98	4.06	4.11	4.17	4.21	4.25	4.29	4.35	4.38	4.42	4.45	4.48
∞	.05	2.77	2.92	3.02	3.09	3.15	3.19	3.23	3.26	3.29	3.34	3.38	3.41	3.44	3.47
	.01	3.64	3.80	3.90	3.98	4.04	4.09	4.14	4.17	4.20	4.26	4.31	4.34	4.38	4.41

Source: Reproduced from *Principles and Procedures of Statistics* by R. G. D. Steel and J. H. Torrie, 1960. McGraw Hill Book Co. Inc. New York. Printed with the permission of the publisher. Biometric Society. North Carolina

Appendix I

Student *t* Distribution

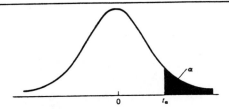

The following table provides the values of t_α that correspond to a given upper-tail area α and a specified number of degrees of freedom.

Degrees of Freedom	Upper-Tail Area α									
	.4	.25	.1	.05	.025	.01	.005	.0025	.001	.0005
1	0.325	1.000	3.078	6.314	12.706	31.821	63.657	127.32	318.31	636.62
2	.289	.816	1.886	2.920	4.303	6.965	9.925	14.089	22.327	31.598
3	.277	.765	1.638	2.353	3.182	4.541	5.841	7.453	10.214	12.924
4	.271	.741	1.533	2.132	2.776	3.747	4.604	5.598	7.173	8.610
5	0.267	0.727	1.476	2.015	2.571	3.365	4.032	4.773	5.893	6.869
6	.265	.718	1.440	1.943	2.447	3.143	3.707	4.317	5.208	5.959
7	.263	.711	1.415	1.895	2.365	2.998	3.499	4.029	4.785	5.408
8	.262	.706	1.397	1.860	2.306	2.896	3.355	3.833	4.501	5.041
9	.261	.703	1.383	1.833	2.262	2.821	3.250	3.690	4.297	4.781
10	0.260	0.700	1.372	1.812	2.228	2.764	3.169	3.581	4.144	4.587
11	.260	.697	1.363	1.796	2.201	2.718	3.106	3.497	4.025	4.437
12	.259	.695	1.356	1.782	2.179	2.681	3.055	3.428	3.930	4.318
13	.259	.694	1.350	1.771	2.160	2.650	3.012	3.372	3.852	4.221
14	.258	.692	1.345	1.761	2.145	2.624	2.977	3.326	3.787	4.140
15	0.258	0.691	1.341	1.753	2.131	2.602	2.947	3.286	3.733	4.073
16	.258	.690	1.337	1.746	2.120	2.583	2.921	3.252	3.686	4.015
17	2.57	.689	1.333	1.740	2.110	2.567	2.898	3.222	3.646	3.965
18	.257	.688	1.330	1.734	2.101	2.552	2.878	3.197	3.610	3.922
19	.257	.688	1.328	1.729	2.093	2.539	2.861	3.174	3.579	3.883
20	0.257	0.687	1.325	1.725	2.086	2.528	2.845	3.153	3.552	3.850
21	.257	.686	1.323	1.721	2.080	2.518	2.831	3.135	3.527	3.819
22	.256	.686	1.321	1.717	2.074	2.508	2.819	3.119	3.505	3.792
23	.256	.685	1.319	1.714	2.069	2.500	2.807	3.104	3.485	3.767
24	.256	.685	1.318	1.711	2.064	2.492	2.797	3.091	3.467	3.745
25	0.256	0.684	1.316	1.708	2.060	2.485	2.787	3.078	3.450	3.725
26	.256	.684	1.315	1.706	2.056	2.479	2.779	3.067	3.435	3.707
27	.256	.684	1.314	1.703	2.052	2.473	2.771	3.057	3.421	3.690
28	.256	.683	1.313	1.701	2.048	2.467	2.763	3.047	3.408	3.674
29	.256	.683	1.311	1.699	2.045	2.462	2.756	3.038	3.396	3.659
30	0.256	0.683	1.310	1.697	2.042	2.457	2.750	3.030	3.385	3.646
40	.255	.681	1.303	1.684	2.021	2.423	2.704	2.971	3.307	3.551
60	.254	.679	1.296	1.671	2.000	2.390	2.660	2.915	3.232	3.460
120	.254	.677	1.289	1.658	1.980	2.358	2.617	2.860	3.160	3.373
∞	.253	.674	1.282	1.645	1.960	2.326	2.576	2.807	3.090	3.291

Source: E. S. Pearson and H. O. Hartley, *Biometrika Tables for Statisticians*, Vol. I. London: Cambridge University Press, 1966. Partly derived from Table III of Fisher and Yates, *Statistical Tables for Biological, Agricultural and Medical Research*, published by Longman Group Ltd., London (previously published by Oliver & Boyd, Edinburgh, 1963). Reproduced with permission of the authors and publishers.

Appendix J

Coefficients and the Sum of Squares of Sets of Orthogonal Polynomials When There Are Equal Interval Treatments

Degree of Polynomials	Comparison	Number of Levels							Sum of Squares of the Coefficients
		T_1	T_2	T_3	T_4	T_5	T_6	T_7	(Σc^2)
1	Linear	-1	+1						2
2	Linear	-1	0	+1					2
	Quadratic	+1	-2	+1					6
3	Linear	-3	-1	+1	+3				20
	Quadratic	+1	-1	-1	+1				4
	Cubic	-1	+3	-3	+1				20
4	Linear	-2	-1	0	+1	+2			10
	Quadratic	+2	-1	-2	-1	+2			14
	Cubic	-1	+2	0	-2	+1			10
	Quartic	+1	-4	+6	-4	+1			70
5	Linear	-5	-3	-1	+1	+3	+5		70
	Quadratic	+5	-1	-4	-4	-1	+5		84
	Cubic	-5	+7	+4	-4	-7	+5		180
	Quartic	+1	-3	+2	+2	-3	+1		28
	Quintic	-1	+5	-10	+10	-5	+1		252
6	Linear	-3	-2	-1	0	+1	+2	+3	28
	Quadratic	+5	0	-3	-4	-3	0	+5	84
	Cubic	-1	+1	+1	0	-1	-1	+1	6
	Quartic	+3	-7	+1	+6	+1	-7	+3	154
	Quintic	-1	+4	-5	0	+5	-4	+1	84
	Sextic	+1	-6	+15	-20	+15	-6	+1	924

Reprinted by permission from *Statistical Methods* by George W. Snedecor and William G. Cochran, 6th Edition, © 1971 by Iowa State University Press, Ames, Iowa.

INDEX

401